CRISPR-BASED GENE EDITING

A Solution for Global Food Security

Modern Trends in Molecular Biology

CRISPR-BASED GENE EDITING

A Solution for Global Food Security

Shakeel Ahmad, PhD
Asif Ali Khan, PhD
Muhammad Arshad Javed, PhD
Editors

AAP | APPLE ACADEMIC PRESS

First edition published 2025

Apple Academic Press Inc.
1265 Goldenrod Circle, NE,
Palm Bay, FL 32905 USA

760 Laurentian Drive, Unit 19,
Burlington, ON L7N 0A4, Canada

CRC Press
2385 NW Executive Center Drive,
Suite 320, Boca Raton FL 33431

4 Park Square, Milton Park,
Abingdon, Oxon, OX14 4RN UK

Library and Archives Canada Cataloguing in Publication

Title: CRISPR-based gene editing : a solution for global food security / Shakeel Ahmad, PhD, Asif Ali Khan, PhD, Muhammad Arshad Javed, PhD, editors.
Names: Ahmad, Shakeel (Agriculturist), editor. | Khan, Asif Ali (Lecturer in agriculture), editor. | Javed, Muhammad Arshad, editor.
Series: Modern trends in molecular biology.
Description: First edition. | Series statement: Modern trends in molecular biology | Includes bibliographical references and index.
Identifiers: Canadiana (print) 20250133075 | Canadiana (ebook) 20250133091 | ISBN 9781774915981 (hardcover) | ISBN 9781774915998 (softcover) | ISBN 9781032712147 (ebook)
Subjects: LCSH: Crops—Genetic engineering. | LCSH: Food crops—Genetics. | LCSH: CRISPR activation. | LCSH: CRISPR (Genetics) | LCSH: Crop improvement.
Classification: LCC SB123.57 .C75 2025 | DDC 631.5/233—dc23

Library of Congress Cataloging-in-Publication Data

CIP data on file with US Library of Congress

ISBN: 978-1-77491-598-1 (hbk)
ISBN: 978-1-77491-599-8 (pbk)
ISBN: 978-1-03271-214-7 (ebk)

ABOUT THE BOOK SERIES: MODERN TRENDS IN MOLECULAR BIOLOGY

Series Editor:
Devarajan Thangadurai, PhD
Associate Professor, Karnatak University, Dharwad, South India
Email: drthangaduraid@gmail.com / drthanga.kud@gmail.com

Advancements in molecular biology have stimulated a great impact on all life sciences investigation to improve the trends in global biodiversity condition. This dynamic force involves the development of highly sophisticated experimental methods in molecular biology with broad interdisciplinary applications, the quickly expanding information flow of scientific discoveries and technical novelties across scientific community, and the development of precise computational tools with constantly updated biological databases for, respectively, analyzing gene expression level, cytogenetic profile, storing data for genotypes, and other molecular features. This has transformed the approach of scientific research allowing revolutionized discoveries in the area of molecular biology. Advances in high-throughput tools, recombinant technology, and NGS made it feasible to unveil genome complexity and to explain the detailed mechanisms for transmission of genetic information. Moreover disease analysis, treatment, and prevention; manipulation of animals and plants for improved phenotypic traits; and synthesis of new peptides and proteins are some of the applications of molecular biology that are frequently being adopted.

There are several approaches from which knowledge in the field of molecular biology is increasingly growing, and it could be challenging for researchers coming into life sciences field. Accordingly there is a need for a platform for preparing molecular biologists to keep themselves abreast of modern trends in molecular biology. Thus, this book series, Modern Trends in Molecular Biology, includes the most of the dynamic pieces of research work being done across the globe on numerous themes associated with molecular biology. This book series attempts to enlighten readers regarding novel concepts and findings in the molecular biology field. The scope of this book series includes recent findings and advancements in the areas of transcriptomics, proteomics, genomics, micro RNA, molecular markers, DNA, RNA and protein isolation, CRISPR, gene editing molecular methods, gene cloning, construction of genome and cDNA library, DNA methylation role for gene expression analysis, AFLP, RAPD, RFLP, VNTRs, genome sequencing tools, phylogenetic analyses, protein modeling, restriction enzymes, metagenomics, metatranscriptomics, metaproteomics, DNA repair, replication and recombination, DNA forensics, pharmacogenomics, function genomics, structural genomics, and computational analyses for proteins and genes.

This book series, Modern Trends in Molecular Biology, aims to relay state-of-the-art knowledge in field of molecular biology. This book series will provide important resources for college and university students, lecturers and professors, and researchers and scientists in almost all life sciences related disciplines. The book series will offer useful a practical and theoretical basis with which to understand the knowledge gaps in relation to a wide range of scientific investigations exploiting molecular biology methods. Overall, presented in concise, organized, and simple format, this series allows smart access to highly considerate topics.

This book series will offer peer-reviewed resources on a variety of topics covering modern aspects of molecular biology, with the goal of helping to keep researchers, scientists, and students abreast of recent advances in the field. Topics will include:

- Molecular Biology: Fundamentals and Applications Molecular Parasitology
- Molecular Entomology
- Molecular Virology
- Molecular Phycology
- Molecular Ecology
- Molecular Nematology
- Molecular Physiology
- Molecular Epidemiology
- Molecular Breeding
- Molecular Agriculture
- Molecular Aquaculture
- Molecular Pathology
- Molecular Medicine
- Molecular Genetics
- Molecular Markers

- Molecular Diagnostics
- Molecular Evolution
- Molecular Immunology
- Molecular Microbiology
- Molecular Pharmacology
- Molecular Biodiversity
- Molecular Enzymology
- Molecular Phylogenetics

BOOKS IN THE SERIES

Hematopoiesis: Biochemical, Cellular, Molecular, and Genomic Perspectives
Editor: Ravi Kumar Gutti, PhD

Cervical Cancer: Signaling Pathways, Molecular Mechanisms, Natural Products, and Therapeutic Approaches
Pratibha Pandey, PhD, Fahad Khan, PhD, Saher Islam, PhD, and Devarajan Thangadurai, PhD

CRISPR-Based Gene Editing: A Solution for Global Food Security
Editors: Shakeel Ahmad, PhD, Asif Ali Khan, PhD, and Muhammad Arshad Javed, PhD

ABOUT THE EDITORS

Shakeel Ahmad, PhD
Seed Center and Plant Genetic Resources Bank, Ministry of Environment, Water and Agriculture, Riyadh, Saudi Arabia.

Shakeel Ahmad, PhD, is a Senior Scientific Officer and Consultant in Biotechnology and Plant Genetic Resources at the Seed Center and Plant Genetic Resources Bank, Ministry of Environment, Water, and Agriculture, Riyadh, Saudi Arabia. Dr. Ahmad earned his Ph.D. in Crop Breeding and Genetics from the Graduate School of the Chinese Academy of Agricultural Sciences (GSCAAS), Beijing, in 2021. His doctoral research focused on cloning and characterizing the *PGL10 (Pale Green Leaf 10)* gene, which regulates chlorophyll and chloroplast development in *indica* rice, employing advanced molecular tools and functional genomics.

In addition to *PGL10*, Dr. Ahmad has cloned and characterized key genes such as *DTN1 (Decreased Tiller Number 1), YGL3 (Yellow-green Leaf 3), qMq-1 and qMq-2* (QTLs for *Milling Quality 1* and *2*), and NRL3 (Narrow and Rolled Leaf 3). He has developed novel, blast- and bacterial leaf blight (BLB)-resistant lines of Chinese rice cultivars *(Yuzhengxiang, Zhongjian-100, and Ninggeng-3)* using the CRISPR/Cas9 multiplex gene-editing system. He also co-created aromatic three-line hybrid rice and improved rice lines with enhanced grain appearance quality using the CRISPR/Cas9-based gene editing system, as well as co-developed the superior *japonica* rice variety *YJ144* with enhanced blast resistance and increased head rice yield (HRY) using extensive molecular breeding and genetic engineering. His research has been widely published in renowned peer-reviewed journals, including Physiologia Plantarum, New Phytologist, Plant Biotechnology Journal, Rice, The Crop Journal, and IJMS. His contributions to rice research earned him the prestigious CNRRI (China National Rice Research Institute) Student's S&T Award in 2021 from the CNRRI.

Following his Ph.D., Dr. Ahmad completed a postdoctoral fellowship in Plant Genome Editing at the U.S.-Pakistan Center for Advanced Studies in Agriculture and Food Security, UAF. He honed his expertise in plant tissue culture, gene editing, bioinformatics, project management, and interpersonal communication there. Currently, his laboratory is curating the Saudi National

Gene Bank, using cutting-edge technologies such as Genotyping-by-Sequencing and Whole Genome Re-Sequencing. His lab also uses advanced biotechnology and plant breeding tools to breed trait-specific, climate-resilient crop varieties. In his role, he liaises with national and international partners, fostering scientific collaboration and ensuring smooth coordination of research projects.

Dr. Ahmad is the recipient of multiple prestigious scholarships, including the King Saud University Scholarship (2015), GSCAAS Scholarship (2017), and the Beijing Government Scholarship (2020). He has published over a dozen book chapters with leading publishers such as Springer, Elsevier, and CRC Press (Taylor & Francis Group). In addition, he serves on the editorial boards of *Frontiers in Genetics, Frontiers in Genome Editing, Agriculture,* and *Rice,* and reviews for several high-impact scientific journals and funding agencies, including the Biotechnology and Biological Sciences Research Council of UK Research and Innovation.

Dr. Ahmad is an active member and ambassador of the American Society of Plant Biologists (ASPB), a member of the Australasian Grain Science Association (AGSA), the American Phytopathological Society (APS), and the Scientific Committee of the Seed Center and Plant Genetic Resources Bank. His research interests focus on gene identification and editing, stress physiology, molecular breeding, and conservation and utilization of plant genetic resources to develop climate-resilient crops and to contribute to global food security and nutrition.

Asif Ali Khan, PhD
Chairperson, National Seed Development and Regulatory Authority, Islamabad, Pakistan

Asif Ali Khan, PhD, is the Chairperson of the National Seed Development and Regulatory Authority, Islamabad, Pakistan. He is a Professor Emeritus, UNESCO Chair, and former faculty member of MNS University of Agriculture, Multan, Pakistan. He is an outstanding geneticist and plant breeder and has supervised 17 PhD students and authored 13 books, chapters, and manuals, as well as over 150 research articles in high-quality journals. He developed several varieties of crops and vegetables for general cultivation, including Okra 3A, Rachna Maize, QS Super Sorghum, Chenab Pasta Wheat, and Spinach. He has contributed to breeding FH 87, a high oil content cotton variety, and many promising cotton lines like B622, B630,

PB38, PB896, and MNH 1045. Dr. Ali has also contributed to improving protocols for efficient screening of crop plants for salinity tolerance, protocols for non-radioactive microsatellite enrichment, rapid methods of DNA extraction, and a low-cost indigenous CEL 1 production protocol. He has published over 150 peer-reviewed scientific papers. He was appointed as Vice Chancellor of MNS University of Agriculture, Multan, in 2016 for two terms. In addition, he successfully launched various R&D, outreach, and capacity-building projects with international and local organizations. His accomplishments have been acknowledged by the Higher Education Commission with a Best Teacher Award and by the Government of Pakistan by conferring the Civil Award (Tamgha e Imtiaz) in 2020. Dr. Ali earned his PhD degree in Breeding and Genomics from the University of Liverpool, UK, and an MBA in Information Management Systems. He also served as a visiting professor at the University of California, Davis, USA.

Muhammad Arshad Javed, PhD
Professor and Head, Department of Plant Breeding and Genetics, University of the Punjab, Lahore, Pakistan

Muhammad Arshad Javed, PhD, is currently serving as Professor and Head of the Department of Plant Breeding and Genetics at the University of the Punjab, Lahore, Pakistan. He earned his MS and PhD degrees from Kobe University, Japan. He has over 29 years of teaching, research, and development (R&D) experience. His expertise in plant sciences covers molecular genetics, tissue culture, haploid plant breeding, and agronomy. He served at Universiti Teknologi Malaysia (UTM), Malaysia (2011–2019). In addition, Prof. Javed has been awarded an Excellent Performance Award (Citra Krishma), including a cash prize, for his outstanding performance in teaching, research, and student development. He also received a Konvensyen Felo Award from IPF Malaysia in recognition of his services as a student advisor. He served in the Department of Agriculture, Government of the Punjab, and developed a fine rice variety "Shaheen Basmati," which was approved by the Government of Pakistan for general cultivation in normal and saline soils. He has over 10 research grants from national and international agencies. He has published over 65 research articles in international peer-reviewed journals and presented over 60 conference papers at various national and international conferences.

CONTENTS

CONTRIBUTORS

Khadija Aaliya
Center of Excellence in Molecular Biology, University of the Punjab, Lahore, Pakistan

Kamel A. Abd-Elsalam
Plant Pathology Research Institute, Agricultural Research Center (ARC), Giza, Egypt

Muhammad Afzal
College of Food and Agriculture Sciences, Plant Production Department, King Saud University, Riyadh, Saudi Arabia

Furqan Ahmad
Institute of Plant Breeding and Biotechnology, MNS University of Agriculture, Multan, Pakistan

Iqrar Ahmad
Centre for Advanced Studies in Agriculture and Food Security, University of Agriculture, Faisalabad, Pakistan

Shakeel Ahmad
Seed Centre and Plant Genetic Resources Bank, Ministry of Environment, Water and Agriculture, Riyadh, Saudi Arabia

Farah K. Ahmed
Biotechnology English Program, Faculty of Agriculture, Cairo University, Giza, Egypt

Umar Akram
Biotechnology Research Institute, Chinese Academy of Agricultural Sciences, Beijing, P.R. China; Institute of Plant Breeding and Biotechnology, MNS University of Agriculture, Multan, Pakistan

Víctor Manuel Cetina Alcalá
Colegio de Postgraduados, Campus Montecillo, Texcoco, Mexico

Salem Safer Alghamdi
College of Food and Agriculture Sciences, Plant Production Department, King Saud University, Riyadh, Saudi Arabia

Zulfiqar Ali
Institute of Plant Breeding and Biotechnology, MNS University of Agriculture, Multan, Pakistan

Ghalib T G Alsubaie
Seed Centre and Plant Genetic Resources Bank, Ministry of Environment, Water and Agriculture, Riyadh, Saudi Arabia

Syed Muhammad Hassan Askri
Zhejiang Key Laboratory of Crop Germplasm Resource, Department of Agronomy, College of Agriculture and Biotechnology, Zhejiang University, Hangzhou, 310058, China

Fidaa Asslam
National Center of Excellence in Molecular Biology (CEMB), Punjab University, Lahore, Pakistan

Rana Muhammad Atif
Department of Plant Sciences, University of California Davis, California, United States; Department of Plant Breeding and Genetics, University of Agriculture, Faisalabad 38000, Pakistan; Precision Agriculture and Analytics Lab, Centre for Advanced Studies in Agriculture and Food Security, National Centre in Big Data and Cloud Computing, University of Agriculture, Faisalabad, Pakistan

Muhammad Ashar Ayub
Institute of Agro-Industry and Environment, The Faculty of Agriculture and Environment, The Islamia University of Bahawalpur, 63100, Punjab, Pakistan

Iftikhar Hussain Badar
Department of Meat Science and Technology, Faculty of Animal Production and Technology, University of Veterinary and Animal Sciences, Lahore, Pakistan

Hebert Jair Barrales-Cureño
Instituto Tecnológico de Estudios Superiores de Zamora, Ingeniería en Innovación Agrícola Sustentable, Zamora, Michoacan, Mexico

Sutar Suhas Bharat
Institute of Crop Sciences, Chinese Academy of Agricultural Sciences, Beijing, P.R. China

Luan Luong Chu
National Key Laboratory of Enzyme and Protein Technology, University of Science, Vietnam National University, Hanoi (VNU), 334 Nguyen Trai, Thanh Xuan, Hanoi 10000, Vietnam

Ehab Elharty
College of Food and Agriculture Sciences, Plant Production Department, King Saud University, Riyadh, Saudi Arabia

Sajid Faiz
Department of Plant Breeding and Genetics, The University of Haripur, Khyber Pakhtunkhwa, Pakistan

Awais Farooq
Department of Plant Breeding and Genetics, University of Agriculture, Faisalabad, Pakistan

Mubeen Fatima
Department of Forensic Science, Lahore University of Biological & Applied Sciences (UBAS), Lahore, Punjab, Pakistan; Center for Applied Molecular Biology and Forensic Sciences, University of Punjab, Lahore, Pakistan

Kabeer Haneef
School of Life Science, Tsinghua University, Beijing, China

Braulio Edgar Herrera-Cabrera
Colegio de Postgraduados, Campus Puebla, Boulevard Forjadores, Puebla, Mexico

Nguyen Quang Huy
Faculty of Biology, University of Science, Vietnam National University, Hanoi (VNU), 334 Nguyen Trai, Thanh Xuan, Hanoi 10000, Vietnam

Sehrish Iftikhar
Microfluidics Innovation Center, Elvesys, Paris, France

Muawuz Ijaz
Department of Animal Sciences, College of Veterinary and Animal Sciences, Jhang, University of Veterinary and Animal Sciences, Lahore, Pakistan

Ayesha Imran
National Center of Excellence in Molecular Biology (CEMB), Punjab University, Lahore, Pakistan

Muhammad Zaffar Iqbal
Ayub Agricultural Research Institute, Faisalabad, Punjab, Pakistan

Shakra Jamil
Agricultural Biotechnology Research Institute, AARI, Faisalabad, Pakistan

Rabia Kalsoom
School of Biological Science and Technology, Beijing Forestry University, Beijing, P.R. China

Asif Ali Khan
Institute of Plant Breeding and Biotechnology, MNS University of Agriculture, Multan, Pakistan

Muhammad Altaf Khan
College of Food and Agriculture Sciences, Plant Production Department, King Saud University, Riyadh, Saudi Arabia

Zulqurnain Khan
Institute of Plant Breeding and Biotechnology, MNS University of Agriculture, Multan, Pakistan

Luis Germán López-Valdez
Laboratorio de Productos Naturales, Universidad Autónoma Chapingo, Texcoco, State of Mexico, Mexico

Mubarik Mahmood
Department of Animal Sciences, College of Veterinary and Animal Sciences, Jhang, University of Veterinary and Animal Sciences, Lahore, Pakistan

Waqas Malik
Genomics Lab, Department of Plant Breeding and Genetics, Bahauddin Zakariya University, Multan, Pakistan

Hussein H. Migdadi
College of Food and Agriculture Sciences, Plant Production Department, King Saud University, Riyadh, Saudi Arabia; National Agricultural Research Center, Baqa, Jordan

Jorge Montiel-Montoya
Centro Interdisciplinario de Investigación Para el Desarrollo Integral Regional, CIIDIR, Instituto Politécnico Nacional Unidad Sinaloa, Guasave, Sinaloa, Mexico

Hira Nawaz
Plant Pathology, Faculty of Agriculture, University of Agriculture, Faisalabad, Pakistan

Amina Nisar
Department of Plant Breeding and Genetics, University of Agriculture, Faisalabad, Punjab, Pakistan

Víctor Manuel Ocaño-Higuera
Universidad de Sonora, Hermosillo, Sonora, Mexico

Deepu Pandita
Government Department of School Education, Jammu, Jammu and Kashmir, India

Shoaib Ur Rehman
Agricultural Biotechnology Research Institute, AARI, Faisalabad, Pakistan

Adel A. Rezk
Department of Biotechnology, College of Agricultural and Food Sciences, King Faisal University, Al-Ahsa, Kingdom of Saudi Arabia

Sahar Sadaqat
Center of Excellence in Molecular Biology, University of the Punjab, Lahore, Pakistan

Farah Saeed
Institute of Agricultural Sciences (IAGS), University of the Punjab, Quaid-e-Azam Campus, Lahore, Pakistan

Rabia Saleem
Department of Biological Sciences, Forman Christian College and University, Lahore, Pakistan

Rafael Salgado-Garciglia
Department of Biological Sciences, Forman Christian College and University, Lahore-54600, Pakistan

Ibrahim Bala Salisu
Department of Animal Science, Faculty of Agriculture, Federal University, Dutse, Nigeria;
Center of Excellence in Molecular Biology, University of the Punjab, Lahore, Pakistan

Leticia Mónica Sánchez-Herrera
Instituto de Investigaciones Químico-Biológicas, Universidad Michoacana de San Nicolás de Hidalgo, Morelia, Michoacán, Mexico

Muhamma Naeem Sattar
Central Laboratories, King Faisal University, Al-Ahsa, Kingdom of Saudi Arabia

Muhammad Shaban
Genomics Lab, Department of Plant Breeding and Genetics, Bahauddin Zakariya University, Multan, Pakistan

Waqar Shafaqat
Department of Forestry, College of Forest Resources, Mississippi State University, Mississippi State, USA

Sameen Shahid
Center of Applied Molecular Biology (CAMB), Punjab University, Lahore, Pakistan

Rahil Shahzad
Agricultural Biotechnology Research Institute, AARI, Faisalabad, Pakistan

Wajid Umar
Leibniz Institute of Agricultural Engineering and Bioeconomy, 14469 Potsdam, Germany

Amina Yaqoob
Center of Excellence in Molecular Biology, University of the Punjab, Lahore, Pakistan

Muhammad Kashif Yar
Department of Animal Sciences, College of Veterinary and Animal Sciences, Jhang, University of Veterinary and Animal Sciences, Lahore, Pakistan; Department of Meat Science and Technology, Faculty of Animal Production and Technology, University of Veterinary and Animal Sciences, Lahore, Pakistan

Sidqua Zafar
Center for Applied Molecular Biology and Forensic Sciences, University of Punjab, Lahore, Pakistan

ABBREVIATIONS

4-HT	4-hydroxytamoxifen
AAAS	American Association for the Advancement of Science
AAV	adeno-associated virus
ABA	abscisic acid
ABE	adenine base editors
AC	amylose content
ACMV	African cassava mosaic virus
AD	activation domain
AHAS	acetohydroxyacid synthase
AITR	ABA-induced transcription repressors
ALC	Alcatraz
ALS	actolactate synthase
AMR	anti-microbial resistance
AREB1	ABA-responsive element-binding protein 1
ARG	auxin-regulated gene
ARGOS	auxin regulated gene involved in organ size
AZFN	artificial zinc finger nuclease
BBTV	banana bunchy top virus
BCTV	beet curly top virus
BeYDV	bean yellow dwarf virus
BLG	B-lactoglobulin
BR	brassinosteroid
BRI	brassinosteroid insensitive
BSCTV	beet severe curly top virus
BSV	banana streak virus
BZR1	brassinazole resistant 1
C2H2	Cys2-His2
CaLCuV	cabbage leaf curl virus
CaMV	cauliflower mosaic virus
Cas9	CRISPR-associated protein 9
CATs	catalases
CBE	cytosine base editors
CBF1	C-repeat binding factor 1
CBSV	cassava brown streak virus
CCNs	cereal cyst nematodes
ChiLCV	chili leaf curl virus
CJEU	Court of Justice of the European Union

CLCuKoV	cotton leaf curl Kokhran virus
CLCuMuB	cotton leaf curl Multan betasatellite
CMV	cucumber mosaic virus
CNTs	carbon nanotubes
CP	coat protein
CRISPR	clustered regularly interspaced short palindromic repeats
CRISPRa	CRISPR activation
CRISPRi	CRISPR interference
crRNA	CRISPR RNA
CTL	complex trait locus
CVYV	cucumber vein yellowing virus
DBD	DNA-binding domain
DEP1	dense and erect panicle 1
DH	double haploid
DMR6	downy mildew resistance gene 6
DSB	double strand break
dsDNA	double-stranded DNA
DST	drought and salt tolerance
DTM	desired target mutator
EBEs	effecter binding sites
EMNS	engineered mega nucleases
ETI	effectors-triggered immunity
ETS	effectors-triggered susceptibility
F3H	flavanone-3-hydroxylase
FAD2	fatty acid desaturase 2 enzyme
FAO	Food and Agriculture Organization
FDA	Food and Drug Administration
FMDV	foot-and-mouth disease virus
FNS-II	flavone synthase II
GAI	gibberellic acid insensitive
GE	genome editing
GFP	green fluorescent protein
GM	genetically modified
GMOs	genetically modified organisms
GOI	gene of interest
gRNA	guided RNA
GRs	glutathione reductases
GSTs	glutathione S-transferases
GWS	genome-wide selection
HDR	homology directed repair
HIV	human immunodeficiency virus
HNH	histidine-asparagine-histidine
HR	homologous recombination

HR	hypersensitive response
IFS	isoflavone synthase
IMGE	inducer-mediated genome editing
IND	indehiscent
IPK1	inositol-1,3,5,6-pentakisphosphate kinase 1
IR	intergenic region
JAZ1	jasmonate-zim-domain protein 1
KI	knock-in
LIG1	liguleless 1
LPAT	lysophosphatidic acid acyltransferase
MAB	marker assisted breeding
MAPKs	mitogen-activated protein kinase
MeMV	merremia mosaic virus
MG	mega nuclease
MGE	multiplex genome editing
MNs	meganucleases
MP	movement protein
MSTN	myostatin
MSV	maize streak virus
NBTs	new breeding techniques
NGOs	non-governmental organizations
NHEJ	non-homologous end joining
NIH	Nation Health Institute
NLS	nuclear localization signals
NPBT	new plant breeding technologies
ORF	open reading frame
OST2	open stomata 2
OXPT	oxathiapiprolin
PAM	protospacer adjacent motif
PAO	polyamine oxidase
PBS	primer binding site
PCD	programmed cell death
PDS	phytoene desaturase
PEBV	pea early browning virus
PEG	polyethylene glycol
PHYB	phytochrome B
PLCPs	papain-like cysteine proteases
PODs	peroxidases
PPNs	plant-parasitic nematodes
PRRS	porcine reproductive and respiratory syndrome
PRSV	papaya ring spot virus
PRSV-W	papaya ring spot mosaic virus-W
PTI	PAMP-triggered immunity

PUFA	polyunsaturated fatty acid
PVX	potato virus X
PVY	potato virus Y
RCas9	RNA processing Cas9
RFN	dimerization-dependent activity
RGR	ribozyme-gRNA-ribozyme gene
RMCE	recombinase mediated cassette exchange
RNP	ribonucleoprotein
ROS	reactive oxygen species
RS	resistant starch
RSMV	rice stripe mosaic virus
RT Domain	reverse transcriptase domain
RTSV	rice tungro spherical virus
RVDs	repeat variable residues
SD	short-day
SD1	semi-dwarf 1
SDG2	sustainable development goal 2
SDN	site-directed nuclease
SFR	seed fluorescence reporter
sgRNA	single-guide RNA
shRNAs	small hairpin RNAs
SMV	soybean mosaic virus
SP5G	self-pruning 5G
SRBSDV	Southern rice black-streaked dwarf virus
SRK	site receptor kinase
ssDNA	single-stranded DNA
SSI	site specific integration
SSN	site-specific nucleases
SSNs	sequence-specific nuclease
ssODN	single-stranded DNA oligonucleotide
STEMES	saturated targeted endogenous mutagenesis editors
TALEN	transcription activator-like effector nuclease
TFs	transcription factors
TIFs	translation initiation factors
TMV	tobacco mosaic virus
TOM1	tobamovirus multiplication 1
tracrRNA	trans-activating crRNAs
TRE	tetracycline response element
trugRNA	transformation/truncated gRNA
TRV	tobacco rattle virus
TuMV	turnip mosaic virus
TYLCV	tomato yellow leaf curl virus
UCP1	uncoupling protein 1

UGI	uracil N-glycosylase
USDA	US Department of Agriculture
WAKs	wall-associated kinases
WDV	wheat dwarf virus
ZFAs	zinc finger arrays
ZFN	zinc finger nuclease
ZFP	zinc finger protein
ZYMV	zucchini yellow mosaic virus

FOREWORD

The world has witnessed substantial progress in crop improvements since life started on earth to meet the challenge of global food security. It is estimated that the world has to feed about 10 billion humans population by the end of 2050 with lesser arable land and natural resources than we have today. To feed the ever-growing human population, it is the need of the hour to put our agricultural efforts on the right and fast track to develop such climate resilient and high yielding crop varieties that could help us meet the global food demand.

Consequential efforts have been made to meet the challenges of food security in deteriorating resources and environment, but it seems that we would not be able to meet the global food demand by 2030 or even by 2050 until we do not embrace the power of new plant breeding technologies (NPBTs). In fact, the efforts start from conventional breeding and then go a long way to crop improvements through mutation breeding, transgenic breeding, and marker assisted breeding.

However, nowadays, the application of state-of-the-art NPBTs including gene editing, speed breeding, etc., has revolutionized not only the field of agriculture but also other fields of life. Particularly, Clustered Regularly Interspaced Palindromic Repeats (CRISPR)-based gene editing technology has amazingly gained popularity all over the world in a short period of time and has really put the crop breeding programs on the fast lane. This technology has proven characteristics, like it is more robust, efficient, and less-labor intensive as compared to previous ones, due to which it is being applied in research and crop breeding programs more widely in recent times.

CRISPR-based gene editing system, discovered from a bacterium, has successfully been explored in plants and trees for their improvements at different occasions. Hence, several researchers have reported that this system has significantly helped in improving several important traits of different crops and/or trees, which ultimately enhanced their productivity.

In 2015, United Nations (UN) set 17 sustainable development goals (SDGs), also known as Global Goals, for developing a better world by 2030 and "Zero Hunger" is second among these goals. It is reported by the UN that the world cannot achieve second sustainable development goal by 2030 if the developments and improvements will go with the same pace. Thus, it is believed that NPBTs like gene editing may help researchers develop the desired crop varieties more efficiently and rapidly and can play a critical role in eliminating hunger from the world. Interestingly, this book is a collection of different chapters in which different applications of CRISPR-based gene editing systems have been reported and explained.

The editors together with the authors, have tried to connect the dots between CRISPR systems and global food security, and proposed this new technology as a solution to meet the global food demand by 2030 and/or by 2050. The authors have also discussed about its pros and cons, regulatory aspects, and limitations in different chapters which add value to this book and would be of great interest to students, academia, policy makers, and researchers working on gene editing technologies.

Besides all the facts, proof-of-concepts, and applications of this NPBT in the development of desired transgene-free agricultural crop lines, the products of this technology are facing problems regarding their regulation and acceptability. No doubt several developed countries have embraced the power of this technology and declared the products being generated through this technology as non-GMO and have given them exemption from GMO legislations, but still there are many countries that need to re-think and revise their policies regarding CRISPR-based gene editing systems.

Finally, I would like to congratulate the editors who really worked hard on this collection of CRISPR-based chapters and tried to showcase the potential of CRISPR system to end hunger globally. Of course, a special thanks to all the authors who are an integral part of this book and without whom the editors would not have been able to complete their project.

—**Salem S. Alghamdi**
Legume Research Leader,
College of Food and Agriculture Sciences,
King Saud University,
Saudi Arabia

PREFACE

There are two groups of people on earth that are affected by food shortage: (i) people suffering from chronic hunger; and (ii) people suffering from acute hunger. According to the report titled "The State of Food Security and Nutrition in the World (SOFI) 2024," jointly published in July 2024 by United Nations agencies including the Food and Agriculture Organization (FAO), International Fund for Agricultural Development (IFAD), UNICEF, the UN World Food Programme (WFP), and the World Health Organization (WHO), about 733 million people faced hunger in 2023, representing 1 in 11 people globally and 1 in 5 in Africa. Moreover, the Global Report on Food Crises (GRFC) 2024, released in April 2024, highlights the immense difficulty of achieving the target of eliminating hunger by 2030.

In 2023, approximately 282 million people, representing 21.5% of the population studied across 59 countries or territories, experienced severe or acute hunger, necessitating immediate food and livelihood support. People of this group are in danger, fighting for their survival, and cannot afford any food shortage or disruptions in food supply; therefore, they need more attention. Besides this, it is expected that the total human population would also rise up to 9 billion by 2050. Hence, to feed the ever-growing hungry people and to end food insecurity has become one of the biggest challenges for humanity.

Indeed, plant researchers are struggling hard to meet the global food demand; nevertheless, the changing climate and natural disasters (i.e., COVID-19) are challenging crop productivity all over the world. Moreover, several other factors have also been reported in literature involved in yield losses in different agricultural crops, globally. Thus, it is important to quickly develop such crop varieties, using new plant breeding technologies (NPBTs) that could withstand those factors and produce high yield with good quality.

Currently, plant scientists are using gene editing tools, particularly clustered regularly interspaced short palindromic repeats (CRISPR)-based systems, to improve plants' characteristics to enhance their yield, product quality, and resistance against different environmental factors with greater pace and efficiency. CRISPR has revolutionized agriculture and is being used in several important food crops. Therefore, it is expected that the CRISPR crops could help fulfill the world food demand and they could emerge as a key factor in addressing the global food security challenges.

Given the above, an effort has been made to bring forth a comprehensive volume in the book *CRISPR-based Gene Editing: A Solution for Global Food*

Security, highlighting the new frontiers and applications of the cutting-edge CRISPR-based gene editing systems in the area of crop improvement. The book begins with an introduction to the technique, its discovery, evolution, developments, classification, and diversity. Then, it overviews the new frontiers in CRISPR toolbox. Next, it gives an updated review of the applications of CRISPR-Cas systems in plants' improvement in connection with biotic stress tolerance, abiotic stress tolerance, nutritional, and yield improvement, etc.

The book subsequently reviews the potential and applications of CRISPR-Cas system in the development of desirable livestock and large animals to achieve global food security. In addition, the regulatory aspects and the risks associated with this technology are discussed as well. Briefly, the book encapsulates the current status and potential of the CRISPR-based gene editing tools for improving the food crops and animals to meet the global food demand in future. Thus, as far as its scope is concerned, it will not only attract readership from the field of plant biotechnology but also the researchers and students associated with food science, plant breeding and genetics, agronomy, veterinary science, etc., will certainly benefit from the book.

We are extremely thankful to all the authors who have shown their trust in us and contributed their work for this book. Indeed, this project would not have been possible without their patience, cooperation, and determined efforts. We are grateful to Dr. Tariq Aftab, an Assistant Professor at Department of Botany, Aligarh Muslim University, India, for all his sincere support in proposal writing, guidance, and motivation. We acknowledge our reviewers for their precious time and support, and for all their sincere efforts to improve the quality of the chapters and ultimately the book. Our heartfelt appreciation for the publisher and its whole editorial staff for extending their cooperation for the timely production of this book even during such a difficult time as the COVID-19 pandemic.

—Shakeel Ahmad, PhD
Senior Scientific Officer cum Consultant
Biotechnology and Plant Genetic Resources
Seed Center and Plant Genetic Resources Bank,
Ministry of Environment, Water, and Agriculture, Saudi Arabia

Asif Ali Khan, PhD
Chairperson National Seed Development and Regularity Authority,
Islamabad, Pakistan

Muhammad Arshad Javed, PhD
Professor and Chairman,
Department of Plant Breeding and Genetics,
University of the Punjab, Lahore, Pakistan

CHAPTER 1

3D'S OF GENOME EDITING SYSTEMS: DISCOVERY, DEVELOPMENTS, AND DIVERSITY

ZULQURNAIN KHAN[1] and SHAKEEL AHMAD[2]

[1]*Institute of Plant Breeding and Biotechnology, MNS University of Agriculture, Multan, Pakistan*

[2]*Seed Centre and Plant Genetic Resources Bank, Ministry of Environment, Water and Agriculture, Riyadh, Saudi Arabia*

ABSTRACT

Genome editing (GE) is the method of changing DNA sequence precisely and efficiently to modify genetic code within a cell. Generally, creating a double-stranded cleavage in DNA at unique sites with DNA endonucleases is the basic principle of GE. Clustered regularly interspaced short palindromic repeats (CRISPR), transcription activator-like effector nucleases (TALENs), zinc finger nucleases (ZFNs), and meganucleases (MNs) are four GE technologies based on the above-mentioned principle that have been successfully used for gene deletion, addition, correction, and replacement. CRISPR/Cas is advanced equipment in the GE toolkit with high diversity and classification. It has many advantages over the past and contemporary tools including simplicity, easy designing and cloning, cost-effectiveness, and diversity, and unique features such as multiplexing, targeting multiple genes/loci simultaneously. Following the discovery of new variants and significant changes to the current CRISPR scheme, many new methods have been introduced to the CRISPR toolbox for a variety of GE applications in various species. Base editors and primer editors are the most recent and

CRISPR-Based Gene Editing: A Solution for Global Food Security. Shakeel Ahmad, Asif Ali Khan, and Muhammad Arshad Javed (Eds.)

well-known GE methods, with reports of effective GE in a variety of species. A brief introduction, history, evolution, classification, and applications of GE tools will be addressed in this introductory chapter.

1.1 INTRODUCTION

Genome editing (GE), or the modification of desired sequence of DNA in a genome, has achieved attention in recent years. Cells' phenotypic features can be changed by GE [1]. Researchers also concluded that minute differences in the sequence of DNA can result in disease since detection of the double helical structure of DNA in the mid-20th century and a clear appreciation of the transmission of genetic material to children from parents. The word "genome" refers toward an organism's whole DNA sequence, which is influenced by both various factors such as errors in DNA replication throughout cell division process and sun Ultraviolet rays. Mutations in DNA can be passed from generation to generation across ages and can be found in nearly any cell in the body. Polymorphisms are DNA variations that do not lead to in the formation of an irregular protein. While these DNA polymorphisms can trigger phenotypic variations like color of hair color and form, eye color, and type of blood, they have no impact on a person's health. Certain DNA variants, on the other hand, are found to elevate the risk of certain diseases and even result in severe health problems. Researchers developed approaches to eliminate or reverse genetic defects after finding disease-causing "DNA errors."

The idea of GE, or modification of trait was developed in the end of 1980s based on this theory, resulting in the invention of various GE technologies. Capecchi et al. used homologous recombination (HR)-specific genes to produce knock-in (KI) and knock-out cells in mouse embryonic stem cells for the first time. The procedure was innovative, but expensive because it involved rigorous sorting and testing to find the 1 in 1,000,000 cell that expressed the mutated gene. Despite the low level of recombination due to the low rate of HR in mammalian cells, this solid evidence analysis demonstrated that accurate gene targeting was possible. In the 1990s, a sequence-specific nuclease (SSNs) was used for the initial time to accomplish targeted DNA changes. Cre-lox is a mechanism that uses a site-specific DNA endonuclease called Cre, which recognizes the 34-nt locus loxP, to create Double Stranded DNA Breaks known as DSBs and also helps in performing recombination at these specific sites. This technique enabled the gene knockout at any targeted place and proven to be the most powerful GE method present at that moment for creating mouse models of transgenic in nature.

Although, the Cre-lox method was easier to use than HR for measuring gene expression, it was ineffective due to the significant genetic gap between the loxP sites. The search for a robust way to generate a DSB at any required location for selective editing started after the invention of this popular GE methodology. ZFNs, that have a nuclease domain and a DNA binding domain, were utilized to improve *Drosophila melanogaster* genome in 2002 and also to modify a human genome modification in 2005. Notwithstanding its labor-intensive existence and time, this approach is increasingly being used for a wide range of purposes. Yet another GE tool TALENs, was invented in 2010 and has since been effectively further used change the genome of over 40 different species. However, both TALENs and ZFNs have drawbacks, leading researchers to seek out alternative gene targeting strategies. CRISPR-Cas9, the most effective and now commonly applied GE technique, was developed in 2012.

A requirement for GE is the presence of DBS which is site-specific. As a consequence, by employing the DNA repair machinery present in organisms, methods that produce DSBs in DNA could be utilized to specifically modify the genome. Researchers started to identify nucleases (enzymes that cleave nucleic acids) that can produce DNA DSBs, along with methods to guide these nucleases to specific regions of the genome, after the progress of Cre-lox DNA endonuclease-based gene-editing technologies in 1992. Over the next two decades, four GE approaches appeared. Designer/programmable nucleases are being used to produce DSBs, and the cellular DNA repair process is being used to accurately alter genomic sequences [2]. The target (nucleotide sequences, epigenetic markers), precision for the target, and regulation of the target's function are the three most significant features of a successful GE technique. The four main groups of adjustable DNA-binding proteins described below are presently available for GE, and the variables mentioned before could be used to assess their popularity, applicability, efficacy, and drawbacks.

1.2 GENOME EDITING (GE) TOOLS

Researchers can now alter almost any genomic sequence with targeted nucleases, making it possible to establish isogenic cell lines and animal models for the research of individual disease and creating new avenues for human gene therapy. Three fundamental innovations are CRISPR, TALENs, and ZFNs. Many genome engineering achievements facilitated by these instruments are illustrated, as are the engineering breakthroughs that helped their development. Artificial transcription factors (TFs) are also explored, explaining how this technology could be used in synthetic biology and gene therapy to complement targeted nucleases. In recent times, researchers have been able to integrate sequence-specific changes rapidly and cost-effectively into the genomes of a broad variety of cell types and organisms thanks to the development of extremely versatile genome-editing technologies – (i) clustered regularly interspaced short palindromic repeats (CRISPR)-CRISPR-associated protein 9 (Cas9); (ii) transcription activator-like effector nucleases (TALENs); (iii) zinc-finger nucleases (ZFNs); and (iv) homing endonucleases or mega nucleases (MGs) are the most popular technologies now used to promote GE.

1.3 MEGANUCLEASES (MNS)

Meganucleases (MNs) are formed through microbial mobile genetic components and are also recognized as homing nucleases [3]. The re-engineering of the DNA-binding affinity of naturally found endonucleases is the basis for meganuclease technology. The LAGLIDADG family, that contains the well-known and widely utilized I-CreI and I-SceI enzymes, is the biggest class among homing endonucleases. These homing endonucleases may be re-engineered for targeting novel sequences using a mix of rational selection and design. While several studies indicate potential for use of the MNs in GE, the DNA-binding and cleavage domains of homing endonucleases are hard to separate, and the comparative complexity of creating proteins with unique specificities has historically limited their usage.

To overcome this constraint, chimeric proteins including combinations of MNs, Zing fingers, and TALEs have been created to produce new monomeric enzymes that benefit from the binding affinity of Zing fingers and TALEs as well as the cleavage affinity of MNs. One significant benefit of meganuclease technology is that DSB creation by these enzymes produces a 3′ overhang, that may be more recombinogenic for HDR than that of the 5′ overhang produced by FokI cleavage. Furthermore, because MNs are indeed the smallest family of designed nucleases, they may be accessible to all traditional gene delivery techniques. In fact, numerous meganuclease monomers might be easily packed into one viral vector to generate numerous DSBs at the same time.

The LHE family of endonucleases can attach to DNA sequences as large as 24-nt and are available in both monomeric and homodimeric forms. MNs utilize protein-DNA interactions to identify various DNA sequences [3, 4]. Meganuclease have yet to achieve widespread recognition as a GE technique. Another downside to this method is the high chance of off-target coupling [5]. MNs, on the other side, have a marginal edge in terms of weight (approximately 40 kDa). They're ideal for viral vectors with a small number of coding sequences.

1.4 ZINC-FINGER NUCLEASES (ZFNS)

To be helpful for genetic modification, an endonuclease should have a unique combination of properties: precise identification as well as flexibility for retargeting to customer sequences. The ZFN architecture satisfies these requirements by connecting the DNA-binding domain (DBD) of a diverse class of eukaryotic TFs known as zinc finger proteins (ZFPs) with the nuclease domain of the FokI restriction enzyme (Figure 1.1). The DNA binding precision and cleavage activity that is strong but restricted in the lack of a particular binding event while keeping functional modularity, due to their structure. As a result, it makes retargeting and platform enhancement efforts easier [6].

ZFNs work as dimers, with every monomer recognizing a unique "half site" sequence of DNA (usually 09 to 18 base pairs via the zinc-finger DBD. The FokI cleavage domain breaks DNA inside a 5 to 7 bp spacer region that separates the two-flanking zinc-finger binding sites, allowing the ZFN proteins to dimerize. Each ZFN is normally made up of 03 or 04 zinc-finger domains, with every domain containing 30 amino acids. Off-target mutations are a serious problem with the use of ZFNs. As a consequence, a variety of ways have been developed to improve their specificity. Like, the developments of obligatory heterodimeric ZFN designs that depend on charge to charge repel to prohibit undesired homodimerization of the FokI cleavage domain [7]. Furthermore, ZFN inside cells as protein is for enhancing ZFN selectivity. ZFN In the lack of a DSB, ZFNickases could also aid in gene repair [8].

Unlike CRISPR and TALENs, the difficulties of creating zinc-finger arrays have prevented their broad use in non-specialized laboratories. It is particularly difficult to develop zinc-finger domains capable of recognizing all DNA triplets. Nonetheless, recent clinical studies based on ZFN-mediated deletion of the HIV coreceptor CCR5 for AIDS therapy to facilitate personalized and targeted GE.

In crop breeding, the wild corn gene *ZmIPK1* was interrupted by inserting *PAT* gene cassettes, for herbicide resistance and a change in the inositol phosphate composition of growing maize seeds. ZFN-mediated selective transgene insertion, as an established method, was also employed for trait stacking in maize for sustainable agriculture. Scientists then determine safe locations for gene insertion in rice for future genetic engineering and trait stacking. Nonetheless, the engineering of ZFNs remains a complex and technically hard procedure, with frequently low effectiveness [9].

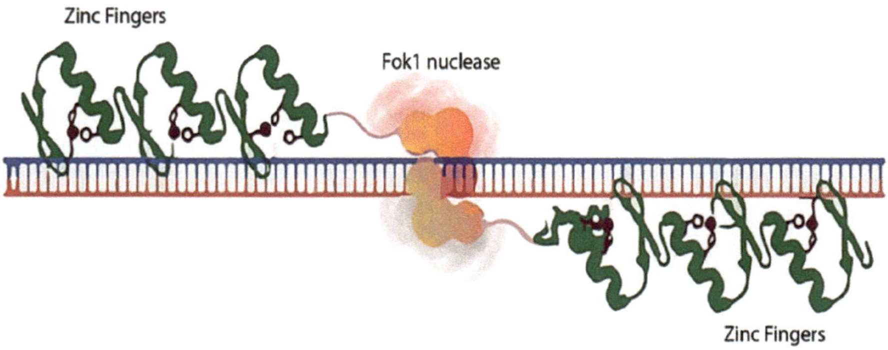

FIGURE 1.1 Artificial zinc finger nuclease (AZFN). Zinc fingers are composed of multiple monomers of protein; each monomer is specific to three DNA bases. Dimerization of two fused FokI nuclease domains results in creation of DSB at the desired DNA sequence.

1.5 TRANSCRIPTION ACTIVATOR-LIKE EFFECTOR NUCLEASES (TALENS)

Transcription activator-like effector nucleases (TALENs) are restriction enzymes, identical to ZFNs (Figure 1.2). The bacteria of *Xanthomonas* genus were discovered to produce effector proteins (TALEs) into the cytoplasm of plants that influence plant cell processes and enhance sensitivity to the disease. TALE proteins are made up of a core domain that binds to DNA, a nuclear localization signal, and a component that triggers targeted gene transcription [10]. The capacity of TALE proteins to attach to DNA was first discovered in 2007, and only a year after, two research groups cracked the code for TALE protein recognition of target DNA. The DBD was shown to be made up of monomers, each of which binds a single nucleotide in the targeted nucleotide sequence. Monomers are tandem repetitions of 34 amino acid residues, 2 of whom are extremely variable and positioned at positions 12 and 13, and they are crucial for the identification of a particular nucleotide. This code is defective; various RVDs can attach to multiple nucleotides with varying degrees of efficiency.

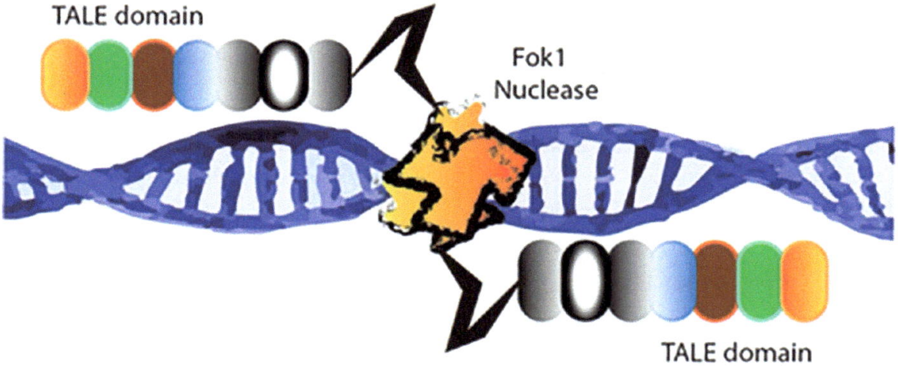

FIGURE 1.2 TALENs on DNA target site. TALE RVDs with DNA binding specificity. Each TALE RVD is specific to a DNA base. Fused FokI endonuclease which creates a DSB at target site on dimerization. Spacer sequence between two TALEs, left and right, is critical for dimerization and cleavage on targeted DNA site.

The target DNA molecule typically includes the very same nucleotide, named thymidine, just before 5′-end of a sequence attached by a TALE monomer. It impacts the binding effectiveness. The final tandem repeat which attaches a nucleotide at recognition site is just 20 amino acid residues long and is hence referred to as a half-repeat. First ever research on the building of chimeric TALEN nucleases were initiated after cracking the code of DNA recognizing by TALE proteins, which drew the interest of researchers around the globe due to its simplicity. To do this, the sequence encoding TALE's DBD was introduced into a plasmid vector formerly being used to create ZFN [11].

TALENs function in pairs, with their binding sites designed to be on opposing DNA strands and isolated by a short piece known as a spacer sequence. When synthetic nucleases enter the nucleus, they attach to target sites. Utilizing synthetic TALEN nucleases, a double-strand break may be created in almost any genomic sequence with identified recognition sites on DBDs [12]. The sole restriction on TALEN nuclease site selection is the requirement for thymine before the 5′-end of the target sequence. In most situations, though, site selection can be accomplished by altering the length of the spacer sequence. The W232 residue in the DBD's N terminal was shown to engage with thymine at 5′ end, reducing the effectiveness of TALEN adherence to the target site. This constraint, though, can be addressed by selecting mutant forms of the TALEN N-terminal domain which can attach to adenine, guanine, or cytosine [13].

1.6 CLUSTERED REGULARLY INTERSPACED SHORT PALINDROMIC REPEATS (CRISPR)/CRISPR ASSOCIATED PROTEIN 9 (CAS9) SYSTEM

The CRISPR/Cas method has lately been presented as a more effective and transparent alternative to ZFNs and TALENs for triggering targeted genetic changes. CRISPR, a microbial adapted immune system, may use a short guidance RNA to attack any genome site [14] (Figure 1.3). A sequence of short repeated sequences (repeats) is interspersed with distinct, short non-repetitive sequences (spacers) originating from MGEs such as bacteriophages and transposons, as well as an AT-rich leader sequences and several Chromatin Cas genes that express Cas proteins [15]. CRISPR-Cas (clustered sometimes interspaced short palindromic repeats) structures are now a popular issue in biology. When studying the gene accountable for alkaline phosphatase isozyme conversion, one of researchers, Y. Ishino, discovered the first CRISPRs in *Escherichia coli* 30 years ago. Due to a complete lack of DNA sequence information at the point, this was hard to predict the biological mechanism of these rare repetitive sequences, especially for mobile genetic elements. Until the mid-2000s, the importance of this one-of-a-kind series remained unclear. CRISPRs were first detected in archaea, originally in *Haloferax mediterranei*, in 1993, and have now been found in a growing number of bacterial and archaeal genomes as life science has progressed into the genomic era. It was important to understand the significance of these sequences in light of the fact that they were maintained in two of life's three domains. When concrete strength between CRISPR spacer regions and sequences of bacteriophages, archaeal viruses, and plasmids was reported in the early 2000s, CRISPR's role as an immune system was unveiled [16].

No doubt, CRISPR has much improved the overall efficiency of genetic manipulation experiments but still off targets remained a great hurdle for achieving the ideal level of results. Different variants of Cas proteins like Cas12 and Cas13 have been utilized to overcome this problem. A detailed comparison of these variants with other GE tools have been provided in the table.

1.7 DISCOVERY AND DEVELOPMENT OF GENOME EDITING (GE) TOOLS

1.7.1 MEGANUCLEASES (MNS)

Meganucleases (MNs) were first discovered as sequence-specific endonucleases that recognized huge targets (>12 b), and the name meganuclease was assigned to them to describe them [17]. In terms of the homing mechanism, this general definition also included artificial endonucleases like zinc finger nucleases (ZFNs), but regular MNs, also known as Homing Endonucleases, are becoming largely synonymous with mega nucleases (HEs). Over the last 20 years, this protein family has grown to include many 100 recognized members present in eukaryotes, bacteria, and archaea, with the significant exception of metazoans [18]. Most of them, including class I introns and exons are encoded by mobile genetic components, but some of them have also been shown to trigger homing events. As a consequence, homing has become a popular biological mechanism in modern times.

1.7.2 ZFNS

The first targeted nuclease to achieve widespread use was ZFNs, which are fusions of a custom-designed Cys2-His2 (C2H2) zinc-finger protein and the cleavage domain of the FokI restriction endonuclease [19]. ZFNs work in pairs, with each monomer recognizing a unique "half-site" sequence (typically nine to 18 bps of DNA). ZFN is composed of [20] three to four zinc-finger domains, each with a 30-amino-acid-residue motif [21]. The DNA recognition residues are found in the helical domain and interact with three bps of DNA on median, with some domain overlap [22].

A significant number of zinc-finger domains that identify distinct DNA triplets have been discovered using techniques such as phage display [23]. Using a canonical linker peptide, these domains can be combined in tandem to build polydactyl zinc-finger proteins that can attack a wide variety of DNA sequences [20, 24]. There were also discoveries of selection-based approaches for constructing zinc-finger proteins [25], such as oligomerized pool engineering (OPEN), which recognizes context-dependent correlations between adjacent zinc-finger domains [8]. In fact, specialized collections of validated two-finger, zinc-finger modules were used to construct zinc-finger arrays, including those which provide for context-dependent results [20, 26].

While using ZFNs for GE, off-target mutation (as in all targeted nucleases) is a big concern [27]. As a result, many steps to improving their reliability have been taken. The development of obligate heterodimeric ZFN architectures that depend on charge-charge repulsion to prevent unnecessary homodimerization of the FokI cleavage domain, decreasing the potential of ZFNs to dimerize at off-target sites, is among the most exciting of these [28]. Besides that, protein engineering methods were also used to improve the FokI cleavage domain's potential to cleave [29].

1.7.2.1 DELIVERY OF ZFNS

Delivering ZFNs as a protein into cells is one especially promising method for enhancing ZFN specificity. Due to the inherent cell-penetrating action of zinc-finger domains, ZFN proteins are generally cell-permeable [30], and when injected directly into cells as purified protein instead of when generated inside cells from nucleic acids, gene editing with less off-target effects could be possible [31]. Since then, scientists also found modified ZFN proteins with enhanced cell-penetrating activity [30]. In the lack of a DSB, ZFNickases may help with gene correction [32]. These enzymes, that are mainly composed of one catalytically inactivated ZFN monomer and a second native ZFN monomer, induce HDR by nicking or cleaving one strand of DNA. They are focused on a hypothesis first shown by Stoddard and associates utilizing homing endonucleases [33].

The challenge of assembling zinc-finger arrays, unlike TALENs and CRISPR-Cas9, has hampered their widespread use in non-specialized laboratories. It is especially difficult to develop zinc-finger domains that can recognize all DNA triplets, especially those of the 5′-CNN-3′ and 5′-TNN-3′ types. As a result, ZFNs do not have the targeting versatility that more advanced genome-editing systems do. Nonetheless, clinical trials for HIV/AIDS therapy focused on ZFN-mediated knockout of the human immunodeficiency virus (HIV)-1 coreceptor CCR5 are underway [34]. ZFNs have the ability to mediate specific and effective gene therapy, as shown by a proposed clinical trial focused on site-specific incorporation of the factor IX gene into the albumin gene [35].

1.7.3 TALENS

TALE proteins are bacterial effectors. TALE proteins identify DNA using a code that was discovered in 2009 [36]. This breakthrough opened the opportunities for sustainable development of custom TALENs capable of altering almost any gene in a matter of months. TALENs, like ZFNs, have a TALE DBD fused to a carboxy-terminal FokI cleavage domain, allowing them structurally and functionally modular [37]. TALEN dimerization, like ZFN dimerization, is mediated by the FokI cleavage domain, which cuts inside a 12- to 19-bp spacer sequence that divides each TALE binding site [37(b)]. TALEs are normally designed to accept 12- to 20-bps of DNA, with more bases resulting in more accurate GE [38].

The TALE-binding domain is composed of a sequence of repeat domains, each weighing 34 residues. The amino acid residues at positions 12 and 13 that come into contact with DNA are regarded as repeat variable residues (RVDs) [36]. Unlike zinc fingers, which recognize DNA triplets, each TALE repeat recognizes only a single bp, with little or no overlap between domains in target sites [39]. The RVDs NI for adenine, HD for cytosine, NG for thymine, and NN or HN for guanine or adenine are by far the most widely used when constructing synthetic TALE arrays [36, 40].

TALE DBDs can be built in a number of ways, the most basic however is the Golden Gate assembly [41]. Just a few high-throughput TALE assembly approaches, such as FLASH assembly [42], iterative capped assembly [43] and ligation independent cloning, have been developed [44]. Recent advancements in TALEN assembly, but in the other hand, have centered on the advancement of methods to enhance their accuracy, such as specificity profiling to distinguish nonconventional RVDs that improve TALEN operation [45], directed evolution to optimize TALE specificity [46], and even fusing TALENs. When it relates to GE, TALENs has two different benefits over ZFNs. TALE arrays, for example, do not involve selection or guided evolution, which reduces the time and experience required to construct a functioning nuclease. Second, TALENs have indeed been found to be more specific and less toxic than some other ZFNs [47], likely due to a higher energetic penalty for associating with base mismatches or a higher preference for target DNA [48]. TALENs are difficult to deliver to cells using lentivirus [12(b)] or a single adeno-associated virus (AAV) particle due to its larger size and highly repetitive form.

To overcome these shortcomings [49], TALENs can be introduced into cells as mRNA [50] or protein [50(b)], and alternative codon use, and amino acid degeneracy could be used to produce RVD arrays which are less vulnerable to recombination [51]. Adenoviral vectors have also shown to be particularly good at transmitting TALEN to difficult-to-transfect cell types [12(b), 52].

1.7.4 CRISPR/CAS SYSTEMS

Mojica discovered that the CRISPR loci belonged to the adaptive immune system, which defends microbes from infections from pathogenic organisms, after performing a systematic bioinformatics study of these repeats in 2003 [15(d), 52]. Horvath et al. discovered a connection between CRISPR activity and the adaptive immune response of microorganisms in 2005. To explore the process of phage resistance in this bacterium, they used a phage-sensitive Streptococcus thermophilus strain.

They found that immune strains accumulated phage-derived sequences at their CRISPR loci, which resulted in higher infection tolerance [15(a)]. They also looked at the functions of the Cas7 and Cas9 proteins. Cas7 was found to be needed for the generation of new spacers and repeats, and not for evoking an immune response. Contrary to popular belief, this is not the case. Cas9 has been discovered to have nuclease function and to play a role in adaptive immunity (Cas9 is referred to as Cas5 in CRISPR-related publications before 2012). Van der Oost et al. investigated the *E. coli* CRISPR system in 2008 by integrating it into another *E. coli* strain that lacked an endogenous CRISPR system. They found the most challenging CRISPR method (Class 1, Type I), which uses Cas3 rather than Cas9. CRISPR arrays are also transcribed into small crRNAs with individual spacers to direct Cas nuclease action, as per the researchers.

In the same year, Cas from the Staphylococcus epidermidis type III-A CRISPR method was reported to have DNA nuclease activity but not RNA cleavage activity [53]. Pyrococcus furious' type III-B CRISPR device, but at the other hand, revealed crRNA-directed RNA cleavage activity as well as DNA nuclease activity [54]. It has been shown that the PAM sequence is critical for the acquisition of type II Cas9 spacers and nuclease operation [55]. They also showed that by prohibiting PAM sequences from interfering with direct repeat sequences, type I and II CRISPR systems are able to avoid self-targeting. In type III CRISPR systems, although, interference is caused by a mismatch between the 5′-end of crRNA and the target DNA [56]. According to a 2010 study of the native type II CRISPR system's functional mechanism, only Cas9 has DNA cleavage action [57]. Non-coding tracrRNAs have been found to hybridize with CRISPR crRNA, allowing Cas9 to be targeted via RNA. This discovery aided in the creation of an RNA-programmable DNA endonuclease for GE. Besides that, the development of mature crRNA from the CRISPR array primary transcript requires the dual RNA hybrid structure of Cas9 and endogenous RNAIII [58].

CRISPR-based GE in yeast, fruit flies, nematodes, rodents, zebrafish, and monkeys has since been written. Furthermore, new CRISPR systems have been found, their biology has been thoroughly explored, and current GE engineering has also been developed [59].

It's worth recalling that the composition and distribution of CRISPR varies considerably between closely associated strains. CRISPR, for instance, is included in the genus *Mycobacterium Tuberculosis* but not in the genus Mycobacterium Leprae. Certain bacteria, such as *E. coli Archaeoglobus fulgidus*, *Methanothermobacter thermoautotrophicus*, and *Mycobacterium avium* have the same CRISPR series [60]. The amount of CRISPR arrays in one genome ranges from 1 to 18, and the number of repeat units in one CRISPR array ranges from 2 to 374 [56]. As of May 2017, CRISPRs were present in 202 (87%) of the 232 archaeal species and 3,059 (45%) of the 6,782 bacterial species, based on the current CRISPR database (http://crispr.i2bc.paris-saclay.fr). According to a sample of 1,724 genome designs, CRISPR-Cas structures are much less common in microbial environmental communities (10.4% in bacteria and 10.1% in archaea). The absence of CRISPR-Cas structures along big bacterial lines with no cultivated members resulted in a large gap in prevalence estimated from full genomes of cultivated microbes as well as that of uncultivated microbes [61].

Centered on the encoding effector proteins, the CRISPR-Cas system is currently classified into two classes [62]. Figure 1.3 shows a general description of categories and shapes, along with instances.

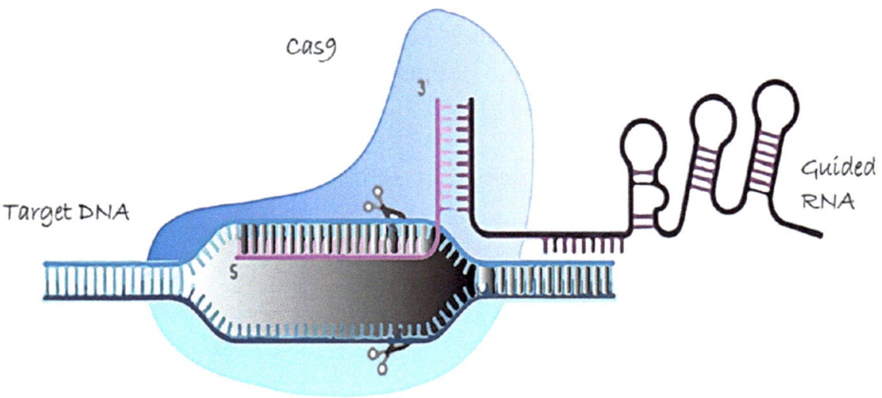

FIGURE 1.3 CRISPR/Cas9 protein on DNA target site. Single guided RNA makes a hybrid with 20 DNA bases of the target site on the bases of availability of PAM. Two endonuclease domains, HNH and RuvC, are responsible for the creation of DSB on the target site.

1.7.4.1 CLASSES OF CRISPR-CAS

Class 1 CRISPR-Cas structures contain four to seven Cas proteins and are built on multi-subunit effector complexes with unequal stoichiometry. This process is common in archaea and bacteria, and it involves all hyperthermophiles, accounting for approximately 90% of all CRISPR-Cas loci found. Class 2 comprises the remaining 10%, employs a single multidomain effector enzyme, which is found only in bacteria [63].

1.7.4.2 SUBDIVISION OF CRISPR CLASSES

CRISPR classes are further subdivided into three types: I, III, and IV in Class 1 and II, V, and VI in Class 2. The inclusion of a particular signature protein, Cas3, makes it easy to distinguish between the types: Cas3 for type I, Cas9 for type II, and Cas10 for type III. The CRISPR-associated antiviral defense complex (Cascade), that comprises of Type I and Type III complexes based on Csm/Cmr complexes and multimeric effector complexes, is evolutionarily and architecturally linked [64]. Unlike most of the other CRISPR-Cas systems, the Type IV system is technically uncharacterized and excludes the Cas1 and Cas2 nuclease adaptation modules [65]. Subtype III-B systems, for example, use spacers supported by type I systems, highlighting the modularity of CRISPR-Cas systems [66]. Despite the fact that many genomes encoding type IV systems lack identifiable CRISPR loci, it's not unlikely that type IV systems are similar to subtype III-B systems, and that crRNAs from various CRISPR arrays will be used until they're accessible [67]. Finally, each subclass is subdivided into different subtypes based on characteristic gene arrangements and additional signature genes (I-A to F and U and III-A to D in class 1; II-A to C, V-A to E and U, and VI-A to C in Class 2) [68].

1.7.4.3 CRISPR TOOLKIT

The Cas9 nuclease is directed to the right target site by a gRNA (guided RNA). RNA-guided Cas9 technology is widely used in cell cultures, animals, and plants. On the plant hand, Cas9-guided RNA-guided genome engineering was used in a number of techniques, including single and multiple gene knockouts and selective donor sequence insertions [69] or even combining transcriptional activation and repressor domains in an inactive Cas9 to target transcriptional regulation. The ability of gRNA-Cas9 to simultaneously target several sequences is a special feature.

1.7.4.4 CLASS 2 CRISPR-CAS

Class 2 CRISPR-Cas solution is a more appealing alternative for GE. Type II Cas9, Type (V) Cas12a (formerly Cpf1), Type (V) Cas12b (C2c1), and Type (VI) Cas13a (C2c2), Cas13b are all Class 2 effectors (C2c3). RuvC and HNH, two unrelated domains of crRNA-dependent endonucleases, are responsible for cleaving targeted and non-targeted DNA strands, respectively, in the crRNA-target DNA complex. The two unrelated domains of crRNA-dependent endonucleases, RuvC, and HNH, are responsible for the cleavage of targeted RNA, and Cas9 is most well and widely used multidomain effector protein.

1.7.4.5 TYPE II

From enough CRISPR to encrypt a trans-activating crRNA, a Type II CRISPR-Cas locus developed (tracrRNA). It's needed for pre-crRNA processing and also the targeted identification of Type II systems. The Cas9-crRNA complex, as seen in Figure 1.3, is a molecular pathway for cleaving targeted DNA. At the atomic level, the crystal structure of the DNA-Cas9-crRNA complex [70] was determined. Cpf1 is a gene located adjacent to Cas1, Cas2, and a CRISPR sequence in the genomes of several bacteria and archaea [62].

1.7.4.6 TYPES III AND VI

Only the Type III-A and Type III-B multi-component structures target the RNA [71]. Type VI effectors, such as Cas13a and Cas13b, target RNA directly and mediate RNA interference. Cas13a and Cas13b contain low-eucaryotic (HEPN) domains instead of RuvC-like nuclear domains, Type II, and Type V-factors [72].

1.7.4.7 TYPE V

Cas12a (Cpf1) is a type V effector template that contains two nuclease-like domains, but one of them, HNH, is missing. However, structural study of the Cas12a-crRNA-target DNA complex showed a unique fold in the second nuclease domain that is functionally similar to the HNH domain of Cas9 [70]. Cas12a is a single RNA-driven nuclease that does not need Cas9 action in the recognition of neighboring protospacer motif (PAM) and its cleavage configuration that specifies the targeted strand, which is required for tracrRNA [73]. The identification of Class 2, which contains the two distantly related Cas9 and Cas12a effector proteins, means that other variants of such structures may occur. Cas12b (Type V) and Cas13a and Cas13b (Type VI), which are not related to Cas9 and Cas12a, have recently been subjected to a direct bioinformatic search [74]. For the guided operation of Type V effectors like Cas9, tracrRNA is needed.

1.7.4.8 CRISPR-CAS TOOLBOX

The CRISPR-Cas toolbox was used widely over the last seven years and CRISPR/Cas has emerged as a cutting-edge tool for genetic modification of any genome region using single-guide RNA (sgRNA) to target DNA [75]. The CRISPR RNA (crRNA)-single protein system relies on a natural microbial adaptive immune system that uses RNA-guided nuclease to identify and cleave foreign DNA components. A chimeric sgRNA and a CRISPR-associated protein (Cas9) unwind and cleave the target DNA specifically, with complementarity with the sgRNA defining the cleavage site [76]. The only limitation to targeting a DNA sequence in this operation is the presence of the adjacent protospacer motif (PAM) region. For site-specific genome engineering, the CRISPR approach has proven to be incredibly useful. The deactivated Cas9 protein, dCas9 for programmable RNA-dependent DNA-binding protein, was recently developed in human and bacterial cells [77].

Since the CRISPR/Cas9 system is so fast, accurate, and efficient, there are a few off-target cases. Therefore, it seems that this approach is the most effective for altering plant genomes. The CRISPR/Cas9 architecture is supposed to have a big impact on plant biology science, both applied and fundamental. It would undoubtedly have a major impact on crop breeding. Since it prevents time-consuming backcrossing in conventional breeding schemes and enables for continuous changes in best cultivars, GE is the most efficient genome engineering strategy.

Gene knockout via NHEJ is by far the most systematic extension of CRISPR/Cas9. Negative regulators of disease tolerance and grain growth can be manipulated to boost crop production as well as provide resistance to pathogens of interest. The most critical step in CRISPR/Cas9-mediated gene editing is the formation of a special sgRNA expression cassette, which serves as a reference system for the Cas9/sgRNA complex, which consists of 98 nucleotides with a targeted sequence based on 20 nucleotides [70]. Plants transcribe sgRNA with RNA polymerase III, and its expression is primarily regulated by U3 or U6 promoters. Since the sgRNA: U3/U6 promoters do have small speech cassettes, they can be produced with overlapping PCR or adaptor ligation using 300–600 bp.

Constitutive promoters are widely used, such as the 35S mosaic virus Cauliflower (CaMV) and A ubiquitin. Callus-based transformation methods may be used to track *Cas9* expression in dicots and monocots for extremely selective gene editing in various crops. *Cas9* gene expression cassettes can be delivered directly using biolistic and PEG-mediated transformation techniques [78]. The Agrobacterium-mediated transformation of the CRISPR/Cas9 method is a highly e-scientific approach to secure transformation of dicots and monocots. Although the *CRISPR/Cas9* gene-editing process has gained popularity in agriculture, it has some disadvantages as used in plant GE. This technology has several significant flaws, including non-specific targets, HR inefficiency, PAM series constriction, cargo-vector frequency, and other issues.

The APOBEC1 cytidine deaminase enzyme can accept ssDNA but not dsDNA as a substrate [79]. BE1 was made by combining *APOBEC1* with dead *Cas9* from *Streptococcus pyogenes* [80]. The merger of *APOBEC1* to *dCas9* provides the cytosine base editor with a large specific molarity target site, enhancing its effectiveness. The second kind of base editor is the adenine base editor, which transforms an A•T base pair to a G•C base pair. Since it can reverse pathogenic point mutations, this base editor is critical. To alter the base-pairing preferences of adenine, the exocyclic amine may be deaminated. This base editor is in charge of converting adenosine to inosine by deamination [81].

1.7.4.9 CRISPR/CAS AS NOVEL GENOME EDITING (GE) TOOL

The CRISPR/Cas mechanism has advanced to the point that several instances of transcriptional control targeted regulation in model plants exist, illustrating the system's evolutionary importance. The designer seed is made using Cas9 nuclease-mediated selective GE, which is also commonly used in animals and plants for gene substitution, transcription control, etc.

The CRISPR/Cas process, which is highly hereditary, can be used to add a particular change to a plant genome [82]. In terms of generating a null segregant line free of the transgenic insert, both techniques outperform traditional breeding. Though CRISPR-edited plants may be excluded from existing GMO legislation in this situation, this technology raises many new questions about public approval and control of genetically modified (GM) crop plants [82(b)].

Direct transfection of Cas9 and guide RNAs into plant protoplasts following by single-cell plant regeneration was found to be efficient for GE in rice and tobacco, but the efficiency was poor, and full plant regeneration from protoplasts is reportedly not possible for many crop species [83]. Agrobacterium-mediated T-DNA transformation, accompanied by callus activation and organogenic plant regeneration (or floral dip transformation in Arabidopsis) is a beneficial relationship for plants. In this case, gRNA-Cas9 transmitted via T-DNA can incorporate into the genome and function properly in somatic tissues in addition to functioning temporarily during callus formation [84].

Ornamental trees, cereals, cash crops, fruits, and less commercially important plant species are among the plants studied. These methods have been used as proof of concept in basic science and in the majority of gene function studies. CRISPR-systems, which are developed from bacteria, can be used for targeted gene editing. Current CRISPR-Systems have experienced several modifications, including – (i) enhancing precision; (ii) modifying the adjacent protospacer motif; (iii) modifying the ability to cut DNA; and (iv) merging them with additional proteins. Other conversion techniques based on *Agrobacterium tumefaciens* or *Rhizobium rhizogenes* have already been accessible, and new methods are on their path to the crop sector. As a result, foreign DNA was also unable to incorporate into the host genome [85].

There are two questions to remember whenever it comes to DNA-free transformation: Plant regeneration from a cell or tissue to a cell that is wall-free I plant cell wall delivery (ii) plant regeneration from cell to wall-free cells, we were able to prevent editing the plant cell wall layer by using isolated protoplasts and single plant cells with enzymatic cell wall digestion. PEG-mediated fusion can easily target protoplasts. Scientists found RNA or an RNP complex in PEG vesicles and used it to combine protoplasts. The architecture only allows for around 10% average editing, that is less than DNA-based frameworks [86].

Chlamydomonas reinhardtii was successfully converted using electroporation and RNPs. Petunia, tobacco, lettuce, and soy are all available, despite the lack of practical protocols for potato regeneration. There is a dire need that DNA-free GE methods must be updated. Protoplast microinjection has been used to deliver DNA to oil palms since 1983 [87], but it has not yet been tested for RNP delivery. In wheat, planta particle bombardment (iPB) was used to regenerate immature cells, with mature plant tissue as the target [88]. In wheat, planta particle bombardment (iPB) has been used to regenerate immature cells, with mature plant tissue as the target [88]. The development of a novel method for converting immature plant tissue has been made. Infiltration of cell penetration peptides into mature plant tissue is a novel method of converting mature plant tissue.

The original and enormous GE potential in plants that is DNA-free has yet to be realized. Plants are yet to put modern methods of single-cell transition and propagation, such as microfluidic or sonicating deformation cells, and heavy light beams, to the test [89]. Engineered nucleases are being used to make proteins, and the science of GE is still developing. Both were studied, but in robust transformation schemes, the DNA-free method is favored.

1.8 GENE DRIVE TECHNOLOGIES

Gene drive technologies can disseminate engineered genes within a population more rapidly than mere genetic inheritance [90]. Selfish genetic elements use a technique in nature to generate several copies around the entire genome to increase their chances of being inherited [91]. CRISPR/Cas9 is a bacteria-derived endonuclease program that utilizes a 20 base pairs complementary gRNA to sever the targeted DNA sequence. CRISPR/Cas9 enables any traditional molecular biology laboratory to incorporate this technology with minimal preparation and specialist facilities [59(g)]. In contrast to all previous technologies of gene drive, CRISPR/Cas9 is regarded as both less costly and reliable. The use of CRISPR-based gene drive technologies allows for the rapid dissemination of DNA cassette into the targeted species. A gene which codes for Cas9, a gene which codes for gRNA that target a particular site in the genome. The construct, which is located in a particular position in the genome, has the ability to copy and paste itself, enabling it to propagate within a population. A normal allele has a 50% chance of passing down to the next generation, while a gene drive cassette does have a 90% chance of being passed down thanks to its capability to replicate itself across the genome.

Gene drive produces the precise mutations that it was designed to cut, resulting in the desired DNA sequence. Within 16–20 generations, a tiny proportion of people in the community could unleash the whole gene drive cassette attack [90]. The gene drive cassette can be programmed to take a cargo gene into a specified genome to regulate metabolism and physiology, or it could be programmed to knock out a gene at a specific position inside the genome. The species would become obsolete if the male and female unique genes used for reproduction are replaced by gene drive [90]. It seems that destroying a population is better than modifying its metabolism. The field of gene drive research and applications is increasingly growing. CRISPR/Cas9 GE technology has only been around for 3–4 years. In 2015, the first study on gene drive species was published. Over US $140 million have been funded by various organizations around the globe in gene drive research to improve seed production and control disease vectors over the last two years. Biotech firms recently reached a licensing agreement with Bayer, Dupont, and Monsanto to use CRISPR/Cas9 engineering. Prime editing and base editing are relatively a fresh arrival in the toolkit of GE to tackle the off targets issue.

1.9 BASE EDITING

Base editing is a technique involving irreversibly changing a single base in the whole genome. It can be very useful for manipulating single base mutation related properties or diseases, without producing unwanted double stand breaks in the DNA. Scientists converted cytosine (C) into thymine (T) by first using cytidine deaminase to convert C into uracil (U). Consequent changing in the complementary strand through DNA repair mechanism resulted in the final conversion of G:U into A:T. There are some reports of off targeting caused by Cytosine based base editing. However, to further increase the efficiency of base editing, various amendments including addition of uracil N-glycosylase (UGI), Gam protein and POBEC3A have been made so far. Also, to increase the target sites, various variants of Cas proteins have been used with base editing machinery. The unintended modification in RNA by base editing complex was also reported, which can be reduced by using deaminases.

1.10 PRIME EDITING

The newest member in the family of GE tools is prime editing. It was reported back in 2019. It not only overcomes the off-targeting issue of base editing but also it can perform all 12 kinds of base-to-base conversions, including 4 transition mutations and 8 transversion mutations [92]. The complex consists of a prime editing guide RNA (peg-RNA), a reverse transcriptase domain (RT domain) and a Cas9 protein. The Cas9 in it does not produces a double strand break (DSB) at the target areas instead it produces a nick in a single strand, thus avoiding the complex aftermath of DSBs. The peg-RNA contains the sequence needed to be introduced at the target place. Upon attachment of prime editing complex at targeted sequence, Cas9 produces a nick in the PAM containing DNA strand and then primer binding site (PBS) of peg-RNA hybridizes with it and RT produces the desired DNA strand using the sequence provided by peg-RNA [92].

1.11 FUTURE PERSPECTIVES

Genome editing (GE) technology has been progressing with an exponential pace. Although, it was started with a lot of controversies and concerns, but, with the advent of time, things are improving with better understanding and concepts. From MNs to CRISPR, GE technology has been improved in terms of specificity, ease of designing and cloning, cost-effectiveness, time-consumption, and safe application. CRISPR/Cas9 has been considered one of the most powerful tools for GE of various important crops, because of its high efficiency, relatively low cost, and ease of use compared with other GE techniques, such as ZFNs and TALENs. CRISPR/Cas9 has begun to revolutionize biological research, as the method of choice for targeting specific genome sequences in simple or complex organisms. Although significant progress has been made to increase its efficiency and target specificity, more work remains to be done to improve it.

Plant GE technology based on type II CRISPR/Cas system depends upon *Agrobacterium tumefaciens* or direct gene transfer, using cultured plant tissues. However, in many crops or other plant species, genetic transformation and/or regeneration from tissue culture is not efficient; it takes a long time to select and characterize mutants and can generate somaclonal variation, which creates additional mutations. Therefore, to overcome this problem, the use of tissue culture-free GE systems, such as ribonucleoproteins (RNPs), viral delivery, and nanoparticle systems provide alternatives that can accelerate the GE process. Tissue culture-free GE systems are likely to be easier, cheaper, and less expertise-intensive. At the same time, it can increase the efficiency of CRISPR/Cas and reduce the time required to generate edited plants. Countries like USA has approved no-regulations for genome edited organism with small, targeted deletions. Moreover, brainstorming session for devising new regulation policy have been organized in many countries. Along with traditional CRISPR tools, several other tools of CRISPR, such as base-editing and prime editing, have been developed, which as more specific and precise for GE, to address concerns associated with traditional CRISPR technology.

A number of concerns, such as off-targeting and biosafety, are still under consideration. Scientists are working to make this technology safer and more acceptable for the community especially in case of GE in animals and humans. The future of GE is bright and vivid with the fact that billions of dollars industry investments are involved in GE technology to provide solutions to the prevailing problems in a variety of fields from agriculture to medical sciences.

1.12 CONCLUSION

Genome editing (GE) technology has great potential for revolutionizing crop production worldwide. Although, ZFNs and TALENs are routinely used in plant research that uses HDR-based mutations, these tools still produce unwanted results and lead to HDR-altered alleles. Additionally, these technologies can potentially be used to create fusion proteins containing domains other than nucleases. TALE array repeats can be used in order to induce epigenetic modifications in specific genomic regions to induce stable and heritable mutations. Lack of sufficient genetic data set to address the sequence specificities for different GE tools is the biggest limitation for target prediction, and a larger effort is necessary to address this problem. However, the presence of DBD is enough to predict the target to little extent. The CRISPR/Cas9 system can be very useful for post-transcriptional control of gene expression. ABE-mediated GE will be very useful for generating point mutations/deletions with high accuracy and less indels. Chromosome engineering and the synthetic plant genome approach can be used as a potential tool for DSB-mediated GE in future. These genetic devices can be integrated into the genetic circuit to switch on and off to a particular trait or pathways. Introduction of a "genome editing" database that harbor experimental references, as well as in-silico prediction data of model organisms could be of particular interest. From the database, a researcher could be able to find suitable GE tools for a complex gene.

KEYWORDS

- **clustered regularly interspaced short palindromic repeats**
- **database**
- **gene expression**
- **genetic devices**
- **genome editing**
- **meganucleases**
- **ribonucleoproteins**
- **transcription activator-like effector nucleases**
- **zinc finger nucleases**

REFERENCES

1. Ahmad, S., Wei, X., Sheng, Z., Hu, P., & Tang, S., (2020). CRISPR/Cas9 for development of disease resistance in plants: Recent progress, limitations, and future prospects. *Briefings in Functional Genomics, 19*(1), 26–39.
2. Voytas, D. F., (2013). Plant genome engineering with sequence-specific nucleases. *Annu. Rev. Plant Biol., 64*, 327–350.
3. Smith, N., & Owocki, S. P., (2006). On the role of continuum-driven eruptions in the evolution of very massive stars and population III stars. *The Astrophysical Journal Letters, 645*(1), L45.
4. Arnould, S., Delenda, C., Grizot, S., Desseaux, C., Paques, F., Silva, G., & Smith, J., (2011). The I-CreI meganuclease and its engineered derivatives: Applications from cell modification to gene therapy. *Protein Eng. Des. Sel., 24*(1, 2), 27–31.
5. Argast, G. M., Stephens, K. M., Emond, M. J., & Monnat, Jr. R. J., (1998). I-PpoI and I-CreI homing site sequence degeneracy determined by random mutagenesis and sequential *in vitro* enrichment. *J. Mol. Biol., 280*(3), 345–353.
6. Chandrasegaran, S., & Carroll, D., (2016). Origins of programmable nucleases for genome engineering. *J. Mol. Biol., 428*(5), 963–989.
7. Ramirez, A. A., Bénard, S., Giroir-Fendler, A., Jones, J. P., & Heitz, M., (2008). Kinetics of microbial growth and biodegradation of methanol and toluene in biofilters and an analysis of the energetic indicators. *J. Biotechnol., 138*(3, 4), 88–95.
8. Maeder, M. L., Thibodeau-Beganny, S., Osiak, A., Wright, D. A., Anthony, R. M., Eichtinger, M., Jiang, T., et al., (2008). Rapid "open-source" engineering of customized zinc-finger nucleases for highly efficient gene modification. *Mol. Cell, 31*(2), 294–301.
9. Townsend, A. K., Clark, A. B., McGowan, K. J., Buckles, E. L., Miller, A. D., & Lovette, I. J., (2009). Disease-mediated inbreeding depression in a large, open population of cooperative crows. *Proceedings of the Royal Society B: Biological Sciences, 276*(1664), 2057–2064.
10. Minteer, B. A., & Miller, T. R., (2011). The new conservation debate: Ethical foundations, strategic trade-offs, and policy opportunities. *Biol. Conserv., 144*(3), 945–947.
11. Baccarani, M., Deininger, M. W., Rosti, G., Hochhaus, A., Soverini, S., Apperley, J. F., Cervantes, F., et al., (2013). European LeukemiaNet recommendations for the management of chronic myeloid leukemia: 2013. *Blood, The Journal of the American Society of Hematology, 122*(6), 872–884.
12. (a) Ginevra, M. C., Annovazzi, C., Santilli, S., Di Maggio, I., & Camussi, E., (2018). Breadth of vocational interests: The role of career adaptability and future orientation. *The Career Development Quarterly, 66*(3), 233–245; (b) Holkers, M., Maggio, I., Henriques, S. F., Janssen, J. M., Cathomen, T., & Gonçalves, M. A., (2014). Adenoviral vector DNA for accurate genome editing with engineered nucleases. *Nat. Methods, 11*(10), 1051–1057.
13. De Luca, G., Dudek, D., Sardella, G., Marino, P., Chevalier, B., & Zijlstra, F., (2008). Adjunctive manual thrombectomy improves myocardial perfusion and mortality in patients undergoing primary percutaneous coronary intervention for ST-elevation myocardial infarction: A meta-analysis of randomized trials. *Eur. Heart J., 29*(24), 3002–3010.

14. Jinek, M., Chylinski, K., Fonfara, I., Hauer, M., Doudna, J. A., & Charpentier, E., (2012). A programmable dual-RNA–guided DNA endonuclease in adaptive bacterial immunity. *Science, 337*(6096), 816–821.
15. (a) Barrangou, R., Fremaux, C., Deveau, H., Richards, M., Boyaval, P., Moineau, S., Romero, D. A., & Horvath, P., (2007). CRISPR provides acquired resistance against viruses in prokaryotes. *Science, 315*(5819), 1709–1712; (b) Makarova, K., Slesarev, A., Wolf, Y., Sorokin, A., Mirkin, B., Koonin, E., Pavlov, A., et al., (2006). Comparative genomics of the lactic acid bacteria. *Proceedings of the National Academy of Sciences, 103*(42), 15611–15616; (c) Mulder, N. J., Apweiler, R., Attwood, T. K., Bairoch, A., Bateman, A., Binns, D., Bradley, P., et al., (2005). InterPro, progress and status in 2005. *Nucleic Acids Res., 33*(suppl_1), D201–D205; (d) Mojica, F. J., García-Martínez, J., & Soria, E., (2005). Intervening sequences of regularly spaced prokaryotic repeats derive from foreign genetic elements. *J. Mol. Evol., 60*(2), 174–182.
16. Ishino, Y., Krupovic, M., & Forterre, P., (2018). History of CRISPR-Cas from encounter with a mysterious repeated sequence to genome editing technology. *J. Bacteriol., 200*(7), e00580–17.
17. Oliver, S. G., Van, D. A. Q., Agostoni-Carbone, M., Aigle, M., Alberghina, L., Alexandraki, D., Antoine, G., et al., (1992). The complete DNA sequence of yeast chromosome III. *Nature, 357*(6373), 38–46.
18. Chevalier, B. S., & Stoddard, B. L., (2001). Homing endonucleases: Structural and functional insight into the catalysts of intron/intein mobility. *Nucleic Acids Res., 29*(18), 3757–3774.
19. (a) Kim, Y. D., & Morr, C. V., (1996). Microencapsulation properties of gum arabic and several food proteins: Spray-dried orange oil emulsion particles. *J. Agric. Food. Chem., 44*(5), 1314–1320; (b) Urnov, F. D., Rebar, E. J., Holmes, M. C., Zhang, H. S., & Gregory, P. D., (2010). Genome editing with engineered zinc finger nucleases. *Nature Reviews Genetics, 11*(9), 636–646; (c) Porteus, M. H., & Carroll, D., (2005). Gene targeting using zinc finger nucleases. *Nat. Biotechnol., 23*(8), 967–973.
20. Kim, H. J., Lee, H. J., Kim, H., Cho, S. W., & Kim, J. S., (2009). Targeted genome editing in human cells with zinc finger nucleases constructed via modular assembly. *Genome Res., 19*(7), 1279–1288.
21. Pavletich, N. P., & Pabo, C. O., (1991). Zinc finger-DNA recognition: Crystal structure of a Zif268-DNA complex at 2.1 A. *Science, 252*(5007), 809–817.
22. Wolfe, S. A., Nekludova, L., & Pabo, C. O., (2000). DNA recognition by Cys2His2 zinc finger proteins. *Annu. Rev. Biophys. Biomol. Struct., 29*(1), 183–212.
23. (a) Choo, Y., & Klug, A., (1994). Selection of DNA binding sites for zinc fingers using rationally randomized DNA reveals coded interactions. *Proceedings of the National Academy of Sciences, 91*(23), 11168–11172; (b) Jamieson, A. C., Kim, S. H., & Wells, J. A., (1994). *In vitro* selection of zinc fingers with altered DNA-binding specificity. *Biochemistry, 33*(19), 5689–5695.
24. Beerli, R. R., Segal, D. J., Dreier, B., & Barbas, C. F., (1998). Toward controlling gene expression at will: Specific regulation of the erbB-2/HER-2 promoter by using polydactyl zinc finger proteins constructed from modular building blocks. *Proceedings of the National Academy of Sciences, 95*(25), 14628–14633.
25. (a) Greisman, H. A., & Pabo, C. O., (1997). A general strategy for selecting high-affinity zinc finger proteins for diverse DNA target sites. *Science, 275*(5300), 657–661; (b) Isalan, M., Klug, A., & Choo, Y., (2001). A rapid, generally applicable method to

engineer zinc fingers illustrated by targeting the HIV-1 promoter. *Nat. Biotechnol., 19*(7), 656–660; (c) Hurt, J. A., Thibodeau, S. A., Hirsh, A. S., Pabo, C. O., & Joung, J. K., (2003). Highly specific zinc finger proteins obtained by directed domain shuffling and cell-based selection. *Proceedings of the National Academy of Sciences, 100*(21), 12271–12276; (d) Magnenat, L., Blancafort, P., Barbas, III. C. F., (2004). *In vivo* selection of combinatorial libraries and designed affinity maturation of polydactyl zinc finger transcription factors for ICAM-1 provides new insights into gene regulation. *J. Mol. Biol., 341*(3), 635–649.

26. (a) Bhakta, M. S., Henry, I. M., Ousterout, D. G., Das, K. T., Lockwood, S. H., Meckler, J. F., Wallen, M. C., et al., (2013). Highly active zinc-finger nucleases by extended modular assembly. *Genome Res., 23*(3), 530–538; (b) Sander, J. D., Dahlborg, E. J., Goodwin, M. J., Cade, L., Zhang, F., Cifuentes, D., Curtin, S. J., et al., (2011). Selection-free zinc-finger-nuclease engineering by context-dependent assembly (CoDA). *Nat. Methods, 8*(1), 67–69; (c) Gupta, S. K., Rai, A. K., Kanwar, S. S., & Sharma, T. R., (2012). *Comparative Analysis of Zinc Finger Proteins Involved in Plant Disease Resistance, 7*(8), e42578. https://doi.org/10.1371/journal.pone.0042578.
27. (a) Gabriel, R., Lombardo, A., Arens, A., Miller, J. C., Genovese, P., Kaeppel, C., Nowrouzi, A., et al., (2011). An unbiased genome-wide analysis of zinc-finger nuclease specificity. *Nat. Biotechnol., 29*(9), 816–823; (b) Pattanayak, V., Ramirez, C. L., Joung, J. K., & Liu, D. R., (2011). Revealing off-target cleavage specificities of zinc-finger nucleases by *in vitro* selection. *Nat. Methods, 8*(9), 765–770.
28. (a) Miller, J. C., Holmes, M. C., Wang, J., Guschin, D. Y., Lee, Y. L., Rupniewski, I., Beausejour, C. M., et al., (2007). An improved zinc-finger nuclease architecture for highly specific genome editing. *Nat. Biotechnol., 25*(7), 778–785; (b) Doyon, Y., Vo, T. D., Mendel, M. C., Greenberg, S. G., Wang, J., Xia, D. F., Miller, J. C., et al., (2011). Enhancing zinc-finger-nuclease activity with improved obligate heterodimeric architectures. *Nat. Methods, 8*(1), 74–79.
29. Guo, J., Gaj, T., & Barbas, III. C. F., (2010). Directed evolution of an enhanced and highly efficient FokI cleavage domain for zinc finger nucleases. *J. Mol. Biol., 400*(1), 96–107.
30. Liu, J., Gaj, T., Wallen, M. C., & Barbas, III. C. F., (2015). Improved cell-penetrating zinc-finger nuclease proteins for precision genome engineering. *Molecular Therapy-Nucleic Acids, 4*, e232.
31. Gaj, T., Guo, J., Kato, Y., Sirk, S. J., & Barbas, C. F., (2012). Targeted gene knockout by direct delivery of zinc-finger nuclease proteins. *Nat. Methods, 9*(8), 805–807.
32. (a) Kim, E., Kim, S., Kim, D. H., Choi, B. S., Choi, I. Y., & Kim, J. S., (2012). Precision genome engineering with programmable DNA-nicking enzymes. *Genome Res., 22*(7), 1327–1333; (b) Ramirez, C. L., Certo, M. T., Mussolino, C., Goodwin, M. J., Cradick, T. J., McCaffrey, A. P., Cathomen, T., et al., (2012). Engineered zinc finger nickases induce homology-directed repair with reduced mutagenic effects. *Nucleic Acids Res., 40*(12), 5560–5568; (c) Wang, J., Friedman, G., Doyon, Y., Wang, N. S., Li, C. J., Miller, J. C., Hua, K. L., et al., (2012). Targeted gene addition to a predetermined site in the human genome using a ZFN-based nicking enzyme. *Genome Res., 22*(7), 1316–1326.
33. Smith, A. M., Takeuchi, R., Pellenz, S., Davis, L., Maizels, N., Monnat, R. J., & Stoddard, B. L., (2009). Generation of a nicking enzyme that stimulates site-specific gene conversion from the I-AniI LAGLIDADG homing endonuclease. *Proceedings of the National Academy of Sciences, 106*(13), 5099–5104.

34. Tebas, P., Stein, D., Tang, W. W., Frank, I., Wang, S. Q., Lee, G., Spratt, S. K., et al., (2014). Gene editing of CCR5 in autologous CD4 T cells of persons infected with HIV. *N. Engl. J. Med., 370*(10), 901–910.
35. Sharma, R., Anguela, X. M., Doyon, Y., Wechsler, T., DeKelver, R. C., Sproul, S., Paschon, D. E., et al., (2015). *In vivo* genome editing of the albumin locus as a platform for protein replacement therapy. *Blood, The Journal of the American Society of Hematology, 126*(15), 1777–1784.
36. (a) Boch, J., Scholze, H., Schornack, S., Landgraf, A., Hahn, S., Kay, S., Lahaye, T., et al., (2009). Breaking the code of DNA binding specificity of TAL-type III effectors. *Science, 326*(5959), 1509–1512; (b) Moscou, M. J., & Bogdanove, A. J., (2009). A simple cipher governs DNA recognition by TAL effectors. *Science, 326*(5959), 1501–1501.
37. (a) Christian, M., Cermak, T., Doyle, E. L., Schmidt, C., Zhang, F., Hummel, A., Bogdanove, A. J., & Voytas, D. F., (2010). Targeting DNA double-strand breaks with TAL effector nucleases. *Genetics, 186*(2), 757–761; (b) Miller, J. C., Tan, S., Qiao, G., Barlow, K. A., Wang, J., Xia, D. F., Meng, X., et al., (2011). A TALE nuclease architecture for efficient genome editing. *Nat. Biotechnol., 29*(2), 143–148.
38. Guilinger, J. P., Thompson, D. B., & Liu, D. R., (2014). Fusion of catalytically inactive Cas9 to FokI nuclease improves the specificity of genome modification. *Nat. Biotechnol., 32*(6), 577–582.
39. (a) Liu, J., Li, C., Yu, Z., Huang, P., Wu, H., Wei, C., Zhu, N., et al., (2012). Efficient and specific modifications of the *Drosophila* genome by means of an easy TALEN strategy. *Journal of Genetics and Genomics, 39*(5), 209–215; (b) Mak, A. N. S., Bradley, P., Cernadas, R. A., Bogdanove, A. J., & Stoddard, B. L., (2012). The crystal structure of TAL effector PthXo1 bound to its DNA target. *Science, 335*(6069), 716–719; (c) Deng, D., Yan, C., Pan, X., Mahfouz, M., Wang, J., Zhu, J. K., Shi, Y., & Yan, N., (2012). Structural basis for sequence-specific recognition of DNA by TAL effectors. *Science, 335*(6069), 720–723.
40. (a) Cong, L., Zhou, R., Kuo, Y. C., Cunniff, M., & Zhang, F., (2012). Comprehensive interrogation of natural TALE DNA-binding modules and transcriptional repressor domains. *Nature Communications, 3*(1), 1–6; (b) Streubel, J., Blücher, C., Landgraf, A., & Boch, J., (2012). TAL effector RVD specificities and efficiencies. *Nat. Biotechnol., 30*(7), 593–595.
41. Cermak, T., Doyle, E. L., Christian, M., Wang, L., Zhang, Y., Schmidt, C., Baller, J. A., et al., (2011). Efficient design and assembly of custom TALEN and other TAL effector-based constructs for DNA targeting. *Nucleic Acids Res., 39*(12), e82-e82.
42. Reyon, D., Tsai, S. Q., Khayter, C., Foden, J. A., Sander, J. D., & Joung, J. K., (2012). FLASH assembly of TALENs for high-throughput genome editing. *Nat. Biotechnol., 30*(5), 460–465.
43. Briggs, A. W., Rios, X., Chari, R., Yang, L., Zhang, F., Mali, P., & Church, G. M., (2012). Iterative capped assembly: Rapid and scalable synthesis of repeat-module DNA such as TAL effectors from individual monomers. *Nucleic Acids Res., 40*(15), e117–e117.
44. Schmid-Burgk, J. L., Schmidt, T., Kaiser, V., Höning, K., & Hornung, V., (2013). A ligation-independent cloning technique for high-throughput assembly of transcription activator–like effector genes. *Nat. Biotechnol., 31*(1), 76–81.
45. (a) Guilinger, J. P., Pattanayak, V., Reyon, D., Tsai, S. Q., Sander, J. D., Joung, J. K., & Liu, D. R., (2014). Broad specificity profiling of TALENs results in engineered nucleases with improved DNA-cleavage specificity. *Nat. Methods, 11*(4), 429–435; (b)

Juillerat, A., Pessereau, C., Dubois, G., Guyot, V., Maréchal, A., Valton, J., Daboussi, F., et al., (2015). Optimized tuning of TALEN specificity using non-conventional RVDs. *Sci. Rep., 5*(1), 1–7.
46. Hubbard, B. P., Badran, A. H., Zuris, J. A., Guilinger, J. P., Davis, K. M., Chen, L., Tsai, S. Q., et al., (2015). Continuous directed evolution of DNA-binding proteins to improve TALEN specificity. *Nat. Methods, 12*(10), 939–942.
47. Mussolino, C., Alzubi, J., Fine, E. J., Morbitzer, R., Cradick, T. J., Lahaye, T., Bao, G., & Cathomen, T., (2014). TALENs facilitate targeted genome editing in human cells with high specificity and low cytotoxicity. *Nucleic Acids Res., 42*(10), 6762–6773.
48. Meckler, J. F., Bhakta, M. S., Kim, M. S., Ovadia, R., Habrian, C. H., Zykovich, A., Yu, A., et al., (2013). Quantitative analysis of TALE–DNA interactions suggests polarity effects. *Nucleic Acids Res., 41*(7), 4118–4128.
49. Liu, H., Chen, Y., Niu, Y., Zhang, K., Kang, Y., Ge, W., Liu, X., et al., (2014). TALEN-mediated gene mutagenesis in rhesus and cynomolgus monkeys. *Cell Stem Cell, 14*(3), 323–328.
50. (a) Mahiny, A. J., Dewerth, A., Mays, L. E., Alkhaled, M., Mothes, B., Malaeksefat, E., Loretz, B., et al., (2015). *In vivo* genome editing using nuclease-encoding mRNA corrects SP-B deficiency. *Nat. Biotechnol., 33*(6), 584–586; (b) Mock, U., Machowicz, R., Hauber, I., Horn, S., Abramowski, P., Berdien, B., Hauber, J., & Fehse, B., (2015). mRNA transfection of a novel TAL effector nuclease (TALEN) facilitates efficient knockout of HIV co-receptor CCR5. *Nucleic Acids Res., 43*(11), 5560–5571.
51. Kim, Y., Kweon, J., Kim, A., Chon, J. K., Yoo, J. Y., Kim, H. J., Kim, S., et al., (2013). A library of TAL effector nucleases spanning the human genome. *Nat. Biotechnol., 31*(3), 251–258.
52. Bolotin, A., Quinquis, B., Sorokin, A., & Ehrlich, S. D., (2005). Clustered regularly interspaced short palindrome repeats (CRISPRs) have spacers of extrachromosomal origin. *Microbiology, 151*(8), 2551–2561.
53. Marraffini, L. A., & Sontheimer, E. J., (2008). CRISPR interference limits horizontal gene transfer in staphylococci by targeting DNA. *Science, 322*(5909), 1843–1845.
54. Cotten, S. R., Ford, G., Ford, S., & Hale, T. M., (2012). Internet use and depression among older adults. *Comput. Human Behav., 28*(2), 496–499.
55. Deveau, H., Barrangou, R., Garneau, J. E., Labonté, J., Fremaux, C., Boyaval, P., Romero, D. A., et al., (2008). Phage response to CRISPR-encoded resistance in *Streptococcus thermophilus*. *J. Bacteriol., 190*(4), 1390–1400.
56. Marraffini, L. A., & Sontheimer, E. J., (2010). CRISPR interference: RNA-directed adaptive immunity in bacteria and archaea. *Nature Reviews Genetics, 11*(3), 181–190.
57. Garneau, J. E., Dupuis, M. È., Villion, M., Romero, D. A., Barrangou, R., Boyaval, P., Fremaux, C., et al., (2010). The CRISPR/Cas bacterial immune system cleaves bacteriophage and plasmid DNA. *Nature, 468*(7320), 67–71.
58. Deltcheva, E., Chylinski, K., Sharma, C. M., Gonzales, K., Chao, Y., Pirzada, Z. A., Eckert, M. R., et al., (2011). CRISPR RNA maturation by trans-encoded small RNA and host factor RNase III. *Nature, 471*(7340), 602–607.
59. (a) Baltimore, D., Berg, P., Botchan, M., Carroll, D., Charo, R. A., Church, G., Corn, J. E., et al., (2015). A prudent path forward for genomic engineering and germline gene modification. *Science, 348*(6230), 36–38; (b) Barrangou, R., & Marraffini, L. A., (2014). CRISPR-Cas systems: Prokaryotes upgrade to adaptive immunity. *Mol. Cell, 54*(2), 234–244; (c) Hsu, P. D., Lander, E. S., & Zhang, F., (2014). Development and

applications of CRISPR-Cas9 for genome engineering. *Cell, 157*(6), 1262–1278; (d) Lochbaum, M., Zazo, R., Kazak, Ç. Z., Wright, T., Graham, K. A., & Konttinen, N., (2016). A meta-analytic review of achievement goal orientation correlates in competitive sport: A follow-up to Lochbaum et al., (2016). *Kinesiology, 48*(2), 159–173; (e) Van, D. O. J., Westra, E. R., Jackson, R. N., & Wiedenheft, B., (2014). Unravelling the structural and mechanistic basis of CRISPR–Cas systems. *Nature Reviews Microbiology, 12*(7), 479–492; (f) Samai, P., Pyenson, N., Jiang, W., Goldberg, G. W., Hatoum-Aslan, A., & Marraffini, L. A., (2015). Co-transcriptional DNA and RNA cleavage during type III CRISPR-Cas immunity. *Cell, 161*(5), 1164–1174; (g) Sander, J. D., & Joung, J. K., (2014). CRISPR-Cas systems for editing, regulating, and targeting genomes. *Nat. Biotechnol., 32*(4), 347–355.

60. Jansen, R., Embden, J. D. V., Gaastra, W., & Schouls, L. M., (2002). Identification of genes that are associated with DNA repeats in prokaryotes. *Mol. Microbiol., 43*(6), 1565–1575.
61. East-Seletsky, A., O'Connell, M. R., Knight, S. C., Burstein, D., Cate, J. H., Tjian, R., & Doudna, J. A., (2016). Two distinct RNase activities of CRISPR-C2c2 enable guide-RNA processing and RNA detection. *Nature, 538*(7624), 270–273.
62. Shmakov, S., Smargon, A., Scott, D., Cox, D., Pyzocha, N., Yan, W., Abudayyeh, O. O., et al., (2017). Diversity and evolution of class 2 CRISPR–Cas systems. *Nature Reviews Microbiology, 15*(3), 169–182.
63. Burstein, D., Harrington, L. B., Strutt, S. C., Probst, A. J., Anantharaman, K., Thomas, B. C., Doudna, J. A., & Banfield, J. F., (2017). New CRISPR–Cas systems from uncultivated microbes. *Nature, 542*(7640), 237–241.
64. Kazlauskiene, M., Tamulaitis, G., Kostiuk, G., Venclovas, Č., & Siksnys, V., (2016). Spatiotemporal control of type III-A CRISPR-Cas immunity: Coupling DNA degradation with the target RNA recognition. *Mol. Cell, 62*(2), 295–306.
65. Makarova, K. S., Wolf, Y. I., Alkhnbashi, O. S., Costa, F., Shah, S. A., Saunders, S. J., Barrangou, R., et al., (2015). An updated evolutionary classification of CRISPR–Cas systems. *Nature Reviews Microbiology, 13*(11), 722–736.
66. Garrett, B., (2011). *Convicting the Innocent.* Harvard University Press.
67. Makarova, K. S., & Koonin, E. V., (2015). Annotation and classification of CRISPR-Cas systems. *CRISPR*, 47–75.
68. Koonin, E. V., & Makarova, K. S., (2017). Mobile genetic elements and evolution of CRISPR-Cas systems: All the way there and back. *Genome Biol. Evol., 9*(10), 2812–2825.
69. Schiml, S., Fauser, F., & Puchta, H., (2014). The CRISPR/C as system can be used as nuclease for in planta gene targeting and as paired nickases for directed mutagenesis in A rabidopsis resulting in heritable progeny. *The Plant Journal, 80*(6), 1139–1150.
70. Nishimasu, H., Ran, F. A., Hsu, P. D., Konermann, S., Shehata, S. I., Dohmae, N., Ishitani, R., et al., (2014). Crystal structure of Cas9 in complex with guide RNA and target DNA. *Cell, 156*(5), 935–949.
71. Bikson, M., Grossman, P., Thomas, C., Zannou, A. L., Jiang, J., Adnan, T., Mourdoukoutas, A. P., et al., (2016). Safety of transcranial direct current stimulation: Evidence based update 2016. *Brain Stimulation, 9*(5), 641–661.
72. Abudayyeh, O. O., Gootenberg, J. S., Konermann, S., Joung, J., Slaymaker, I. M., Cox, D. B., Shmakov, S., et al., (2016). C2c2 is a single-component programmable RNA-guided RNA-targeting CRISPR effector. *Science, 353*(6299).

73. Zetsche, B., Heidenreich, M., Mohanraju, P., Fedorova, I., Kneppers, J., DeGennaro, E. M., Winblad, N., et al., (2017). Multiplex gene editing by CRISPR–Cpf1 using a single crRNA array. *Nat. Biotechnol., 35*(1), 31–34.
74. Shmakov, S., Abudayyeh, O. O., Makarova, K. S., Wolf, Y. I., Gootenberg, J. S., Semenova, E., Minakhin, L., et al., (2015). Discovery and functional characterization of diverse class 2 CRISPR-Cas systems. *Mol. Cell, 60*(3), 385–397.
75. Jasin, M., & Rothstein, R., (2013). Repair of strand breaks by homologous recombination. *Cold Spring Harb. Perspect. Biol., 5*(11), a012740.
76. Cong, L., Ran, F. A., Cox, D., Lin, S., Barretto, R., Habib, N., Hsu, P. D., et al., (2013). Multiplex genome engineering using CRISPR/Cas systems. *Science, 339*(6121), 819–823.
77. Wu, Y., Liang, D., Wang, Y., Bai, M., Tang, W., Bao, S., Yan, Z., et al., (2013). Correction of a genetic disease in mouse via use of CRISPR-Cas9. *Cell Stem Cell, 13*(6), 659–662.
78. Scheben, A., Wolter, F., Batley, J., Puchta, H., & Edwards, D., (2017). Towards CRISPR/Cas crops–bringing together genomics and genome editing. *New Phytol., 216*(3), 682–698.
79. Harris, R. S., Petersen-Mahrt, S. K., & Neuberger, M. S., (2002). RNA editing enzyme APOBEC1 and some of its homologs can act as DNA mutators. *Mol. Cell, 10*(5), 1247–1253.
80. Komor, A. C., Badran, A. H., & Liu, D. R., (2017). CRISPR-based technologies for the manipulation of eukaryotic genomes. *Cell, 168*(1, 2), 20–36.
81. Gaudelli, N. M., Komor, A. C., Rees, H. A., Packer, M. S., Badran, A. H., Bryson, D. I., & Liu, D. R., (2017). Programmable base editing of A• T to G• C in genomic DNA without DNA cleavage. *Nature, 551*(7681), 464–471.
82. (a) Mahfouz, M. M., Piatek, A., & Stewart, Jr. C. N., (2014). Genome engineering via TALENs and CRISPR/Cas9 systems: Challenges and perspectives. *Plant Biotechnol. J., 12*(8), 1006–1014; (b) Kanchiswamy, C. N., Malnoy, M., & Maffei, M. E., (2015). Chemical diversity of microbial volatiles and their potential for plant growth and productivity. *Frontiers in Plant Science, 6*, 151.
83. Nielsen, K. L., Godfrey, P. A., Stegger, M., Andersen, P. S., Feldgarden, M., & Frimodt-Møller, N., (2014). Selection of unique *Escherichia coli* clones by random amplified polymorphic DNA (RAPD): Evaluation by whole genome sequencing. *J. Microbiol. Methods, 103*, 101–103.
84. Bridgen, M. P., Van, H. W., & Eeckhaut, T., (2018). Plant tissue culture techniques for breeding. In: *Ornamental Crops* (pp. 127–144). Springer.
85. Metje-Sprink, J., Menz, J., Modrzejewski, D., & Sprink, T., (2019). DNA-free genome editing: Past, present, and future. *Frontiers in Plant Science, 9*, 1957.
86. Svitashev, S., Schwartz, C., Lenderts, B., Young, J. K., & Cigan, A. M., (2016). Genome editing in maize directed by CRISPR–Cas9 ribonucleoprotein complexes. *Nature Communications, 7*(1), 1–7.
87. Griesbach, R. J., (1983). Protoplast microinjection. *Plant Mol. Biol. Rep., 1*(4), 32–37.
88. Bowman, J. L., Kohchi, T., Yamato, K. T., Jenkins, J., Shu, S., Ishizaki, K., Yamaoka, S., et al., (2017). Insights into land plant evolution garnered from the *Marchantia polymorpha* genome. *Cell, 171*(2), 287–304. e15.
89. Schlicher, R. K., Radhakrishna, H., Tolentino, T. P., Apkarian, R. P., Zarnitsyn, V., & Prausnitz, M. R., (2006). Mechanism of intracellular delivery by acoustic cavitation. *Ultrasound Med. Biol., 32*(6), 915–924.

90. Burt, A., (2003). Site-specific selfish genes as tools for the control and genetic engineering of natural populations. *Proc. R. Soc. Lond. B. Biol. Sci., 270*(1518), 921–928.
91. Werren, J. H., (2011). Selfish genetic elements, genetic conflict, and evolutionary innovation. *Proceedings of the National Academy of Sciences, 108*(2), 10863–10870.
92. Anzalone, A. V., Randolph, P. B., Davis, J. R., Sousa, A. A., Koblan, L. W., Levy, J. M., Chen, P. J., et al., (2019). Search-and-replace genome editing without double-strand breaks or donor DNA. *Nature, 576*(7785), 149–157.

CHAPTER 2

APPLICATION OF CRISPR-CAS SYSTEM IN PLANT BREEDING TECHNOLOGIES

BRAULIO EDGAR HERRERA-CABRERA,[1]
VÍCTOR MANUEL CETINA ALCALÁ,[2]
LUIS GERMÁN LÓPEZ-VALDEZ,[3] RAFAEL SALGADO-GARCIGLIA,[4]
LETICIA MÓNICA SÁNCHEZ-HERRERA,[5]
VÍCTOR MANUEL OCAÑO-HIGUERA,[6]
JORGE MONTIEL-MONTOYA,[7] and
HEBERT JAIR BARRALES-CUREÑO[8]

[1]Colegio de Postgraduados, Campus Puebla, Boulevard Forjadores, Puebla, Mexico

[2]Colegio de Postgraduados, Campus Montecillo, Texcoco, Mexico

[3]Laboratorio de Productos Naturales, Universidad Autónoma Chapingo, Texcoco, State of Mexico, Mexico

[4]Instituto de Investigaciones Químico-Biológicas, Universidad Michoacana de San Nicolás de Hidalgo, Morelia, Michoacán, Mexico

[5]Universidad Autónoma de Nayarit, Tepic, Nayarit, Mexico

[6]Universidad de Sonora, Hermosillo, Sonora, Mexico

[7]Centro Interdisciplinario de Investigación Para el Desarrollo Integral Regional, CIIDIR, Instituto Politécnico Nacional Unidad Sinaloa, Guasave, Sinaloa, Mexico

[8]Instituto Tecnológico de Estudios Superiores de Zamora, Ingeniería en Innovación Agrícola Sustentable, Zamora, Michoacan, Mexico

CRISPR-Based Gene Editing: A Solution for Global Food Security. Shakeel Ahmad, Asif Ali Khan, and Muhammad Arshad Javed (Eds.)

ABSTRACT

Crops that address the challenges of climate change and the growing demand for safe food for humans and animals, as well as the various industrial applications of plants, require versatile, cheap, efficient, and fast genetic breeding techniques. The manipulation of gene function is a basic tool in obtaining genetically improved crops. The CRISPR/Cas (clustered regularly interspaced short palindromic repeats) is a tool for gene editing (*CRISPR/Cas*) is a new, simple, fast, less expensive method of assembly and offers the capacity to multiplex or mutate multiple genes simultaneously, using multiple RNA guides (gRNA) and a single Cas nuclease. In addition, the CRISPR/Cas system is based on the recognition of the white sequence through complementary RNA. In genetic improvement, CRISPR/Cas genome editing (GE) is used in plants to increase yield, improve plant architecture, induce resistance to pathogenic microorganisms, and increase the nutritional and quality value for the consumer, as well as ensure adaptation to drought and tolerance to disease and abiotic stress. Both CRISPR-Cas9 and CRISPR-Cpf1 (the CRISPR system from the bacterium *Francisella tularensis*) are the two systems best studied and most commonly used in plants, but new versions are being investigated. The number of resistant plants obtained with CRISPR/Cas technology increases annually. This chapter describes the use of certain sophisticated applications of the CRISPR/Cas technology system in plant improvement and discusses the potentialities and limitations of its use in the genetic breeding of plants of great international, agricultural importance.

2.1 INTRODUCTION

Actually, the CRISPR/Cas9 technology has been widely studied and used to edit plant genomes. CRISPR/Cas refers to the system formed by the CRISPR machinery and Cas proteins, which confers adaptive immunity against exogenous elements (such as viruses) in bacteria and archaea [1, 2]. The CRISPR structure consists of palindromic repetitions (23 to 55 bases) separated (21 to 72 bases). Cas9 is a large protein (approximately 160 kDa and 1,368 aa) [3, 4] consisting of two primary lobes: a recognition lobe (REC) and a nuclease action lobe (NUC) [5, 6]. The REC lobe is comprised of three domains (-helix bridge, REC1, and REC2), and the NUC lobe contains an additional three domains (RuvC, HNH, and protospacer-adjacent motif (PAM) interaction (PI)). The binding of a CRISPR RNA:trans-activating CRISPR RNA (crRNA:tracrRNA) complex with Cas9 initiates a conformational change in Cas9, forming a groove between the two lobes and allowing the entry of the double-stranded target DNA [5]. Cas proteins are nucleases that cleave DNA at specific positions, directed by RNA sequences [6]. The use of the CRISPR/Cas9-based genomic editing system in plants requires specific elements, which has both advantages and limitations. On the one hand, the CRISPR/Cas-based editing system enables the transformation of plant cells and tissues using available methods, such as those used by bacteria in the genus *Agrobacterium* or biobalistics. On the other hand, these methods do not work in some species and plant varieties, preventing the introduction of the elements required to identify and modify cellular DNA. Cas9 and the RNA that guides this protein must be introduced to cells in the form of protoplasts (cells without a cell wall) or through biolistic methods, which limits the use of this technology to those species with the capacity to regenerate entire fertile plants from protoplasts [7]. A second limitation is associated with the precision of the CRISPR/Cas method. Effects have been reported outside of the intended site of action, which are mentioned to as off-target effects. Experience shows that off-target effects can occur and are difficult to predict. Therefore, proper controls must be performed to ensure that the genome is being modified with the expected level of precision. Good candidate genes must be identified for editing, including precise information regarding the effects of interest that gene mutations are expected to produce. The use of genomic-editing techniques is also restricted by existing regulations regarding the approval of these techniques in various countries for use in plants or animals that depend on the – (i) gene endowment; (ii) use in food production; or (iii) the potential for dissemination throughout the environment. Cas9 nuclease has used to successfully alter genes in plant biological models, such as Nicotiana tabacum, Arabidopsis thaliana, rice, wheat, corn, sorghum, tomato, and orange, among other crops [8–12]. This chapter describes the use of certain sophisticated applications of the CRISPR/Cas technology system in plant improvement and discusses the potentialities and limitations of its use in the genetic improvement of plants of great international, agricultural importance.

2.2 CRISPR/CAS9 METHOD

The use of CRISPR/Cas9 technology allows for precise editing and modification to the genome and can be used in agricultural crops to produce precise improvements to certain characteristics. The efficiency of the CRISPR/Cas construct not only lies in the model of the guide and the nuclease used, but also in the chosen transformation method.

In several reports describing gene editing in plant cells, *Agrobacterium tumefaciens* (*Rhizobium radiobacter*) and biolistic transfer are used to carry the vector with Cas9 and its single guide (sg)RNA into the nucleus. The Cas9-RNA complex adheres to the site defined by the guide (g)RNA, and mutations are produced. Therefore, it is likely that the CRISPR/Cas9 construct will adhere to the plant genome and subsequently, the possibility of producing undesirable changes increases, for example, the integration of transgenes or the presence of out-of-range mutations. Once inside the receptor cell, CRISPR/Cas9 is degraded. The resulting fragments insert themselves in desired or undesired sites to repair double-stranded DNA (dsDNA) breaks, resulting in the introduction of small insertions and deletions (indels) or balanced chromosomal rearrangements. To eliminate these side effects of CRISPR/Cas9-ARNg genome editing (GE) (e.g., indel formation), an approximate strategy is to use the Cas9 protein and a pre-assembled gRNA. In this technique, a ribonucleoprotein (RNP) is formed instead of a plasmid encoder. Kim et al. [13] demonstrated in human cells that the RNP complex acts by cutting the chromosomal DNA after transfection [13]. The RNP is then damaged by proteases, thus reducing the risk of indel introduction into the genome. This new technology can also be used for the advanced analysis of genetic expression in plants and allow functional analysis of genes involved in molecular patterns associated with disease resistance. The interpretation of this data can be used for the genetic improvement of crops. Site-specific nucleases (SSN) cut specific DNA sequences, resulting in a targeted mutation or correction. CRISPR/Cas uses an RNA molecule to guide the nuclease to its target (RNA guided endonuclease-RGEN). Therefore, in order to target a specific sequence, redesign of the nuclease is not necessary, and it is sufficient to simply add the appropriate RNA guide. Furthermore, Cas nucleases are naturally occurring and unlike other nucleases such as Fok1 they are not restricted to a single genus but have been identified in many prokaryotes (both bacteria and archaea).

Woo et al. [14] analyzed pre-assemblies of the Cas9 protein and gRNA in an RNP complex in Arabidopsis, tobacco, rice, and lettuce protoplasts, in the presence of PEG. A T7E1 endonuclease was used to measure the mutation frequency in the transfected cells, and indels were found to occur in the expected positions [14]. Using the brassinosteroid insensitivity (BRI1) gene in Arabidopsis, demonstrated that mutations occurred 24 hours after transfection, identifying that the RNP induces mutations before the cell can complete the cell division cycle. Comparable results were derived with the *phytochrome B* gene (*PHYB*).

2.3 APPLICATIONS IN PLANT BREEDING TECHNIQUES

2.3.1 SOMATIC HYBRIDS

The production of somatic hybrids by fusing two cells from different species is a technique that has been used by developers of transgenic products. In oilseed rape and tobacco, andro-sterile plants have been obtained thanks to the introduction of a chimeric gene composed of two parts: one that is only expressed in the anther tissue surrounding the pollen grains and another that encodes the synthesis of an enzyme that destroys the RNA in the cells of this tissue. This procedure will make it easier to obtain commercial hybrids. It is important to obtain and characterize the edited product. There are two features for full characterization of the products, including DNA sequence level information to confirm: (i) successful editing; and (ii) absence of DNA insertion from the plasmids used in the transformation. The ability of CRISPR-Cas9, combined with its efficacy in rice, enabled clonal breeding (F1 hybrids) while conserving heterozygosity [15]. Three meiotic genes allow the replacement of meiosis with mitosis-like cell division, obtaining tetraploid seeds and diploid gametes [15].

2.3.2 GENETIC BREEDING

The use of CRISPR-Cas9 as a breeding tool is used in conventional back-cross breeding. This requires multiple generations, e.g., a cross between a stress resistant (low yield potential) line A plant crossed with a stress susceptible (good production potential) line B (100% B) plant to generate stress tolerance genes incorporated into elite germplasm with high yield potential. The CRISPR-Cas technology is based on an enzyme that provides highly effective endonuclease activity and programmable selectivity. There are two routes of starch synthesis in the normal corn kernel: straight-chain polysaccharide: amylose and branched-chain polysaccharide: amylopectin. There are two types of corn starch: normal: 75% amylopectin + 25% amylose. While waxy has high amylopectin greater than 97% amylopectin. To be a high value commercial product: specialty enters the high value niche market, with food and industrial applications derived from physicochemical properties: high viscosity at high temperature, clearer, and more stable gels at low temperature, good stability in freeze/thaw cycle. The corn gene *Wx1* encodes granule-bound starch synthase. The current *Wx1* allele comes from a corn variety discovered in 1908, commercially grown since 1940 and sold by Pioneer since 1980.

DuPont Pioneer is the leading supplier of Waxy corn seed in the USA. The problems of current waxy corn are – delay for introgression of the *Wx1* allele in the current elite lines. Some yield losses, partly due to linkage drag. The solution is to disrupt the *Wx1* gene using CRISPR-Cas technology by editing the elite lines (Figure 2.1). However, the mechanism is illustrated in Figure 2.2.

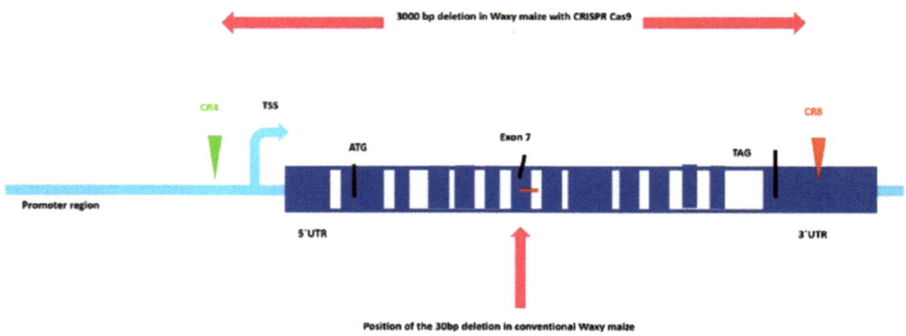

FIGURE 2.1 Cr4 and Cr8-RNAs guide two double cuts in the DNA strand. The gen is delimited, via CRISPR-Cas, the gen is interrupted with approximately 100% amylopectin and waxy phenotype.

2.3.3 APOMIXIS

Apomixis is a type of asexual plant reproduction that generates clonal offspring through seeds [16]. Xie et al. [16] mention apomixis insertion in *O. sativa* by mutating *OsMATL*, *OsOSD1*, *OsREC8*, and *OsSPO11-1*, using CRISPR/Cas9 technology. The quadruple mutant formed a meiosis to mitosis transformation producing clonal diploid gametes. With mutated *OsMATL*, which results in haploid generation in plants, the quadruple mutant generates apomictic diploid. The quadruple mutant was named AOP (Apomictic Offspring Producer) because of the obtention of apomictic offspring. Apomixis is a trait that allows flowering plants to produce seeds asexually. The results of the studies reveal a simple form to produce apomixis to maintain hybrid vigor in *O. sativa*. In addition, this focus can also work for other crops that have equivalent genes, such as maize, wheat, barley, and millet, increasing the navel of seeds with hybrid vigor and these seeds can reach farmer's fields. In addition, identifying the genes that control the molecular mechanism of apomictic breeding is critical to understanding its genetic architecture [17].

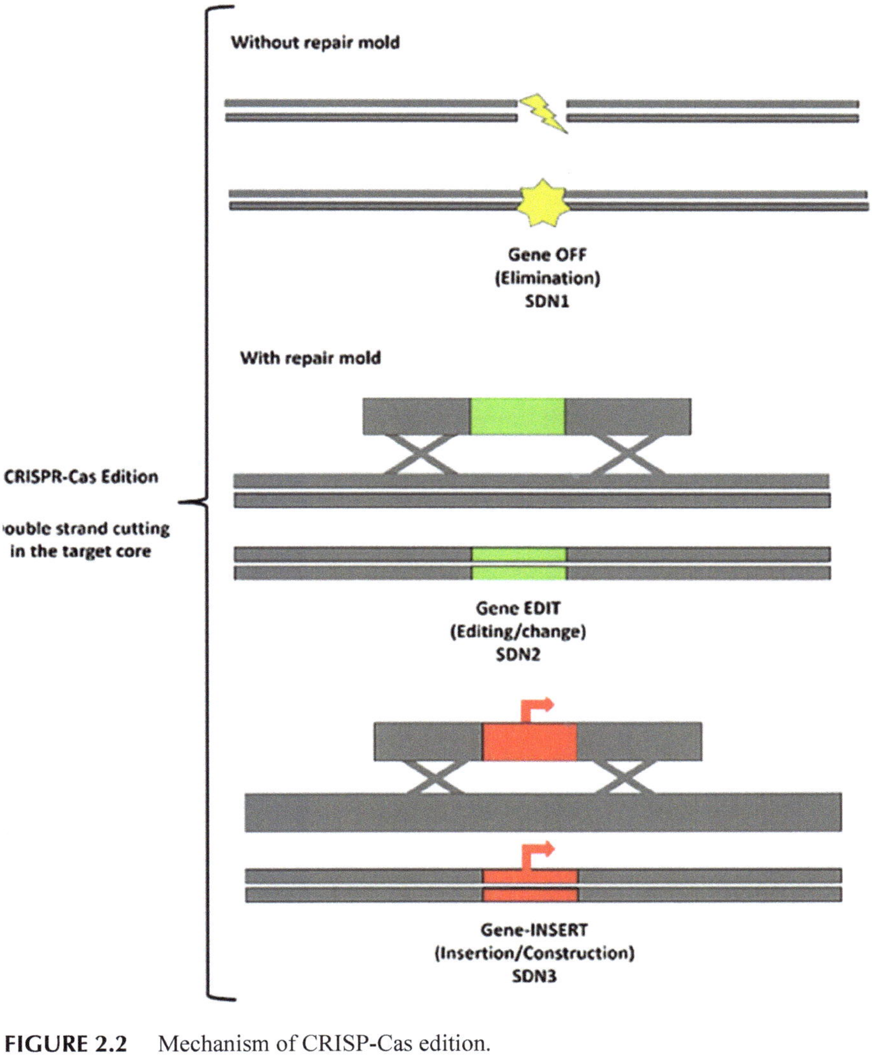

FIGURE 2.2 Mechanism of CRISP-Cas edition.

2.3.4 *DOMESTICATION*

Domestication is a selection process leading to adaptation/acclimatization of plants for agriculture. Domestication and plant breeding alters crop architecture, at the phenotypic and genomic level. It results in the transformation of formerly wild species into elite, high-yielding cultivars

[18]. There are other locally important crops, such as sorghum, groundnut, chickpea, cassava, and potato, known as "orphan crops," with the characteristic of addressing food security issues [19]. With CRISPR-Cas9 technology, desirable traits are obtained in traditional orphan crops in a limited time. Zsögön et al. [20, 21] indicate that GE of wild tomatoes not only improved fruit yield and size, but also enhanced the nutritional value of the fruit. They targeted a set of six genes [self-programming, fruit weight, and lycopene beta cyclase] applying a multiplexed CRISPR/Cas9 to create a new variety with desirable characteristics. De novo domestication involves identifying and introgressing the genes/mutations that guide to the successful domestication and adaptation of crop varieties. Lemmon et al. generated de novo domestication of an orphan Solanaceae crop, *Physalis pruinosa*. Lemmon et al. demonstrated that CRISPR-mediated editing of genes to improve plant architecture, fruit size, and flower production had the potential to accelerate domestication of these productivity traits [22, 23].

2.3.5 HAPLOID INDUCTION

The goal of crop breeding is to generate pure lines with multiple desired traits. Double haploid (DH) and CRISPR/Cas9 GE are two powerful technologies used in crop breeding. Wang et al. [15] created a haploid inducer-mediated genome editing (IMGE) approach, which uses a haploid maize inducer variety carrying a CRISPR/Cas9 cassette targeting an agronomic characteristic to pollinate an elite maize inbred line and generate genome-edited haploids in soil. Homozygous pure DH lines with the expected trait improvement could be generated in two generations, thus avoiding the lengthy crossing and back-crossing procedure used in conventional breeding to integrate an expected trait into elite of value backgrounds [24].

Khanday et al. [25] mentioned that triple knock-out of *BBM1*, *BBM2*, and *BBM3* genes originated embryo abortion and arrest, phenotypes that were fully rescued by male-transmitted BBM1. So, the requirement for fertilization in embryogenesis is determined by genome transmission of pluripotency elements. When GE to replace the mitosis by meiosis (MiMe) phenotype is combined with BBM1 expression in the egg, progeny of the clones can be reached that retain parental heterozygosity throughout the genome. The asexual propagation trait is heritable through generations of clones [25].

2.3.6 IMPROVED STRESS TOLERANCE

2.3.6.1 DROUGHT STRESS

In Arabidopsis thaliana, the open stomata 2 (OST2) was used as a target gene, and delivery method/main strategy was used Agrobacterium-mediated transformation/truncated gRNA (trugRNA) in the CRISPR/Cas9 system was used to produce site-directed (–+) modifications [26]. For *miR169a* target gene, the main delivery method/strategy was *Agrobacterium*-mediated transformation/double-sgRNA/Cas9-mediated targeted elimination to generate null mutations [76]. In *Arabidopsis thaliana*, vacuolar H+-pyrophosphatase (AVP1) was used as a target gene, and delivery method/main strategy was used *Agrobacterium*-mediated transformation of CRISPR/Cas9 for activation to generate overexpression of the target gene [27]. In Arabidopsis, ABA-responsive element-binding protein 1 (*AREB1*) was used as a target gene, and delivery method/main strategy was used Agrobacterium-mediated transformation/CRISPR activation (CRISPRa) method to increase target gene expression [28].

2.3.6.2 SALINITY TOLERANCE

In rice, for salinity tolerance breeding, rice type-B response regulator (*OsRR22*) was used as a target gene, and *Agrobacterium*-mediated transformation/CRISPR/Cas9 system was used to generate knockdown plants [30]. The BBM gene resides to the apetala response factor 2/ethylene response factor (AP2/ERF) DNA-binding domain (DBD) family. The AP2/ERF gene family is categorized into the ERF-like category (with a single AP2 domain), and the AP2-like category (with two AP2 domains similar to each other and separated by a linker region). In *P. squamulatum*, apomixis is directed through a physically large, hemizygous, non-recombinant chromosomal region, the ASGR [31]. Several copies of the *PsASGR-BBML* genes are utilized candidate genes that promote parthenogenesis due to their linkage between Pennisetum/Cenchrus apomictic species [32, 33]. Conner et al. [33] analyzed a mutant of the *CcASGR-BBML* gene in *C. ciliaris* that eventually lost its ability to undergo parthenogenesis; in addition, ASGR-BBML has also been found in Brassica and *A. thaliana* [34, 35]. They successfully reduced *PsASGR-BBML* expression in apomictic F1 RNAi transgenic plants and endorsed the key role of *PsASGR-BBML* in producing parthenogenesis in *P. squamulatum*, which can lead to haploid induction to vigorously obtain homozygous lines for genetic breeding [25].

2.4 CRISPR APPLICATIONS IN PLANTS

CRISPR-Cas9 is an efficient system to perform functional analysis of genes in plants, due to the design of vectors that include the Cas9 gene with codons optimized for its expression in plants. With this method it is possible to work with haploid, diploid, tetraploid species [36]. Table 2.1 lists the main applications of CRISPR-Cas technology in agriculture.

TABLE 2.1 Applications of CRISPR-Cas Technology in Agriculture

Culture	Disease Resistance	Yield and Stability	Drought Resistance	Output Trait (Quality)	Maturity
Zea mays					
Soja					
Canola					
Oryza sativa					
Wheat					
Helianthus					

2.4.1 SOLANUM LYCOPERSICUM

The CRISPR/Cas9 technique has been used in recent studies to improve the post-harvest life and certain organoleptic characteristics of model fruits, such as tomatoes. Recently, a group from the Centro Nacional de Biotecnología in Madrid reported the editing of the gene *SlJAZ2* gene in tomato, which is related to jasmonate. This edition confers protection to the plant against pathogenic fungi through a mechanism associated with stomatal closure. Flowering has also been manipulated by researchers at several institutions, including Cold Spring Harbor, USA, in the United States [43]. Harbor, USA, in tomato (*Solanum lycopersicum*) plants. Using CRISPR-Cas9, the researchers generated mutations in the generated mutations in the flowering suppressor *SELF-PRUNING5G* (*SP5G*) to manipulate photoperiod response in manipulate the response to photoperiod in tomato. The CRISPR-Cas9-driven mutations caused rapid flowering and improved the compact growth habit of tomatoes, resulted in rapid flower production [44].

By generating mutations in the gene responsible for the suppression of flowering, known as SP5G, the compact growth of tomatoes in the field was induced, with rapid flowering that produced an early yield [45]. The microorganisms *Xanthomonas gardneri* Sg153, *X. perforans* Xp4b, *P. capsici* LT1534, and *P. syringae* DC3000 were used in the culture of *S.*

lycopersicum, the modified gene was *SIDMR6*, with modification type: 6. The Cas version used was undetermined, the Cas9/sgRNA promoters used were 2 x CaMV35S/AtU6-26, the delivery method was transfer of cotyledonous segments with *Agrobacterium tumefasciens* and mutation frequency undetermined [50]. In the *Solanum lycopersicum* culture, the microorganism *Oidium neolycopersici* was used, the modified gene was SIMlo1, with the type of modification: deletion in ORF. The Cas version used were codons optimized for humans, the Cas9/sgRNA promoters used were CaMV35S/AtU6, the delivery method was transfer of cotyledonous segments with Agrobacterium tumefaciens and mutation frequency not determined [49].

2.4.2 SOYBEAN

Cai et al. [77] generated soybean plants with late flowering due to the induction of mutations using the CRISPR-based technology in a gene related to the photoperiod of flowering, and the resulting variety was characterized by a high yield and a larger vegetative size [46].

2.4.3 RICE

In rice, Syngenta used the CRISPR-based technology to induce deletions in dense and erect panicle 1 (*DEP1*) and gene dense. This process resulted in plants that produced a higher density of erect panicles, producing plants with lower heights and higher yields [47]. Rice has been modified to be resistant to radioactive contamination [48], providing a great opportunity to grow rice on the large tracts of agricultural land that were contaminated by the accidents at the Chernobyl and Fukushima nuclear power plants.

A group from the Institute of Integral Plant Biology, in Montpellier, in collaboration with a team from the University of Tokyo, investigated an alternative based on the CRISPR/Cas9 genetic editing technique. They developed rice plants that do not have the capacity to absorb radioactive cesium Cs^+ from the soil. This strategy offers the ability to generate food crops without the risk of radioactive contamination in areas with radioactive soil [49]. The use of the CRISPR/Cas9 genome-editing tool allows for the specific targeting of genes associated with characteristics that are desirable to alter. Therefore, candidate genes have been explored for later use as target genes [50].

In rice (*Oryza sativa*), the microorganism *Xanthomonas oryzae* was used, the modified genes were *OsSWEET11* and *OsSWEET14*, with the type of modification: insertion and deletion. The Cas version used were SpCas9 and SpCas9 codons optimized for rice, the Cas9/sgRNA promoters used were CaMV35s/AtU6, the delivery technique was protoplast transfer with PEG (polyethylene glycol) and the mutation frequency was not available. In rice (*Oryza sativa*) the microorganism *Magnaporthe oryzae* was used, the modified gene was *OsERF922*, with the type of modification: insertion and deletion. The Cas version used was the optimized SpCas9 codons for plants, the Cas9/sgRNA promoters used were Ubiquitin 1 from maize/U6 from rice, the delivery method was embryogenic callus transfer with Agrobacterium tumefaciens, and the mutation frequency was 42.0% [51].

2.4.4 COFFEE

Souto et al. [52] identified candidate genes concomitant with drought-tolerance properties in coffee by comparing the molecular responses to this factor in two commercial cultivars of Coffee arabica L.: Rubi (susceptible var.) and IAPAR59 (tolerant var.) [52]. Both coffee varieties were grown in the field in controlled conditions of either irrigation or drought, and 38 candidate genes were analyzed by performing RNA pyrosequencing on their apices. Under drought conditions, a positive regulatory effect was observed in the IAPAR59 cultivar, which was not observed in the Rubi cultivar due to differences in the genes encoding *non-specific lipid-transfer proteins* (*nsLTPs*), *CaSTK1* (a serine/threonine kinase), *CaMAS1* (involved in abscisic acid biosynthesis), *CaSLP1* (a stellacyanin-like protein involved in plant development) and *CaSAMT1* (S-adenosyl methionine dependent methyltransferase [53–55].

2.4.5 WHEAT

Blumeria graminis f. sp. Tritici, an obligate biotrophic fungus, causes powdery mildew, a serious disease that reduces crop yields in temperate zones. CRISPR/Cas9 technology was used to edit hexaploid wheat (*Triticum aestivum* L.) into three homoalleles of resistance to wheat powdery mildew, where CRISPR/Cas9 generated mutations on the *TaMLO-A1* allele. However, Patel & Mishra [58] did not describe the degree of resistance this type of mutation confers [56].

Liang et al. [59], in two wheat varieties, used the purified protein Cas9 pre-assembly complex and the RNA guide in the RNP complex to edit the *TaGW2* and *TaGASR7* genes [57]. The main steps to generate the RNP-CRISPR/Cas9 complex were as follows. Over 7–9 weeks, preparation, functional validation, RNP binding, and bombardment preparation, plant regeneration, and mutant identification were conducted. A reduction in mutation frequency was observed, compared to the method using DNA. This finding is mainly explained by the speed of degradation of the RNPs (i.e., the short half-life). As expected, the plants regenerated by this process did not contain any transgene.

Zhang et al. [60] demonstrated the production of transgene-free mutants. This was done for several wheat genes with CRISPR/Cas9 DNA (TEECCDNA) or in the form of RNA (TECCRNA) in hexaploid wheat (*Triticum aestivum*) and durum wheat (*T. turgidum*, tetraploid) [58]. The genes analyzed were: *TaGASR7-A1*, *B1*, and *D1* (involved in the control of weight and grain); *DEP1* (control of inflorescence architecture); *NAC2* (branching of tillers); *pin1* (formation of auxin-induced adventitious roots); and *TaLOX2* (grain development). The mutated plants regenerated with high efficiency, with reduced integration of transgenes in the genome of T_0 plants and without marker-assisted selection. The use of RNA technology (TECCRNA) will allow the generation of free plants in the introduction of nucleic acids, since RNA molecules are not integrated into nuclear DNA. By using RNA, the mutation frequency was lower than with DNA.

Wang et al. [78] obtained a strain of oidium-resistant wheat. The strain lacks genes that produce proteins which suppress defenses against powdery mildew. The researchers removed all copies of the genes from the hexaploid genome of wheat. The strain promises to reduce or eliminate the extensive use of plaguicides to control the plant diseases. Gao used transcription activator-like effector nuclease (TALENs) and CRISPR gene editing systems to add or change other genes; however, field trials are not yet available [59]. Researchers studied type-A gliadins, which are gluten proteins considered to be the main proteins responsible for the immune disorder caused by wheat. The researchers created small fragments of genetic material from the grain, which direct the Cas9 protein, acting as molecular scissors gliadin A genes. The modified wheat reduced the immune response by up to 85% in the laboratory [60, 61].

To obtain gluten-free wheat, CRISPR/Cas9 is used to eliminate or reduce the content of the immunogenic fraction of gluten. Sánchez León et al. [61] obtained a high mutation frequency using CRISPR/Cas9 of the alpha-gliadin gene family in baking and durum wheat. This provided wheat lines with an 85% reduction in immunoreactivity compared to control wheat. These low-gluten wheat lines represent a breakthrough in the use of this technology to edit complex gene families in wheat. Furthermore, they are the first step in the development of wheat lines devoid of gliadin genes [34].

The main challenge is to exploit the total capacity of CRISPR/Cas technology to precisely modify gliadin genes, suppressing their immunogenic capacity but maintaining their functionality and organoleptic characteristics. Biotechnology, and in particular new genome-editing tools such as CRISPR/Cas, offer hitherto unknown potential for the removal of specific epitopes in gliadin genes. RNAi and CRISPR/Cas technologies are not only applicable to wheat, but also to other cereals containing gluten, such as barley and rye [34]. One example of gene editing is the development of wheat plants with a low gluten protein content, which is responsible for various disorders of the digestive system. This research was successful in introducing mutations in a specific and simultaneous way, resulting in plants with low gluten content simultaneously, resulting in plants with a low gluten content and a decrease in immunoreactivity of 85% [61].

2.4.6 *CITRULLUS* LANATUS

Researchers at the Beijing Plant Germplasm Breeding Laboratory, led by Shouwei, used CRISPR-Cas9 to edit the *ClPDS* gene, the watermelon (*Citrullus lanatus*), phytoene desaturase (PDS) to get the albino phenotype. All edited watermelons showed a full albino or mosaic phenotype albino or mosaic. This analysis benefited as a proof of utility for the use of the CRISPR-Cas9 method in watermelon genetic breeding [63].

2.4.7 *ARABIDOPSIS THALIANA*

Regeneration efficiency of genetically modified (GM) *A. thaliana* plants has been reported to be higher than 46% with germ lines containing small insertions or deletions [14]. Yang et al. [64] compared two promoters for the correct edition of the genome mediated by the transformation with Agrobacterium in Arabidopsis. The YAO promoter (expressed in tissues with active cell division, such as the root meristem) and 35S of CaMV, a strong constitutive promoter, were compared. High mutation rates were obtained in the transformations with the YAO promoter and lower percentages in those containing 35S. Thus, it is necessary to select an efficient promoter, even if the transformation is by *A. tumefaciens* [64].

Ma et al. [65] effectively inserted into Arabidopsis different guides in a single vector with cloning via the Golden Gate technique or via the Gibson assembly. Each of the guides specific to SCEI, its parallels, and the chelatase gene were cloned independently using the Golden Gate technique [65].

2.4.8 TOBACCO

Pharmacological glycoproteins transport N-glycans with typical residues, which affect protein activity. Two enzymes are important for the addition of plant-specific glycans: (1,3)-fucosyltransferase and (1,2)-xylyltransferase. Researchers from the Catholic University of Leuven, attempted to delate four *FucT* genes and two *XylT* genes in *N. tabacum* BY-2 suspension cells using CRISPR/Cas9. Six *FucTs* and Three *XylT* sgRNAs were delineated to conserved regions. After the transformation of Nicotiana tabacum BY-2, edited genome lines were obtained, and protein complements were analyzed.

Three lines showed a significant reduction of (1,2)-xylose and (1,3)-fucose, while two lines lacked them completely, indicating complete genetic inactivation. The KO lines showed no morphological change and grew similarly to the wild types. One KO line was replaced with genes encoding a human IgG2 antibody. The level of IgG2 expression was as high as in an unedited control transformer [66]. Scientists from the Technical University of Dortmund in Germany obtained tobacco plants with 99.7% less nicotine. Using CRISPR gene editing technology, six enzymes that produce the addictive stimulant were inactivated. The edited plant biosynthesized 0.04 milligrams of nicotine per gram [66].

In the culture of *Nicotiana benthamiana* with the microorganism bean yellow dwarf virus (*BeYDV*), the modified gene was *BeYDV*, with the type of modification: insertion and deletion. The Cas version used was optimized SpCas9 codons for plants, the Cas9/sgRNA promoters used were 35SPPDK/AtU6, the delivery method was co-infiltration of leaves with Agrobacterium tumefaciens and mutation frequency 0.03–70.01% [44]. In the culture of *Nicotiana benthamiana* with the microorganism TYLCV (tomato yellow leaf curl virus), and beet curly top virus (BCTV), the modified gene was TYLCV-IR intergenics regions, regions RCA, with the type of modification: insertion and deletion. The Cas version used was optimized codons SpCas9 for plants, the Cas9/sgRNA promoters used were CaMV35S/promoter of the PEBV, the delivery method was transfection with sgRNA virus in overexpressing plants Cas9 and mutation frequency 42% [45].

2.4.9 CUCUMBER

CVYV (cucumber vein yellowing virus), *ZYMV* (zucchini yellow mosaic virus), and *PRSV* (papaya ring spot virus) were used in *Cucumis sativus* L. culture, the modified gene was F4E, with the type of modification: deletion in Exon 1. The Cas version used was optimized codons SpCas9 for plants, the Cas9/sgRNA promoters used were CaMV35S/AtU6, the delivery method was cotyledon transfer without embryo with Agrobacterium tumefaciens and mutation frequency not determined [46, 47].

2.4.10 LETTUCE

Researchers at the University of California worked on improving the ability of multiple varieties of lettuce to tolerate rising temperatures. The scientists identified the gene responsible for heat sensitivity in lettuce and used CRISPR/Cas9 technology to access DNA, silence the gene in question and obtain lettuce plants that could withstand temperatures of more than 11°C [67].

Lettuce plants have the ability to germinate at higher temperatures, and the edited plants showed identical characteristics to current lettuce varieties. The new lettuce varieties released by CRISPR/Cas9 are in their experimental stage. Therefore, it was demonstrated that CRISPR/Cas9 may be used as a plant-breeding tool to protect crops from threats including diseases, degraded soil conditions and climate change [68]. In lettuce plants with an edited *BIN2* gene, mono-allergic mutants were found in 5.7% of tripe with bi-allergic mutations. The edited plants produced seeds, and the mutated character was demonstrated in the progeny [69].

2.4.11 CITRUS

The microorganism *Xanthomonas citri* subsp. Citri was used in Citrus sinensis culture, the modified gene was *CsLOBI*, with the type of modification: deletion. The Cas version used were codons optimized for plants, the Cas9/sgRNA promoters used were 35SPPDK/AtU6-1, the delivery method was epicotyl transfer with Agrobacterium tumefaciens and mutation frequency 11.5–64.7% [48].

2.4.12 CORN

Through CRISPR-Cas9, Svitashev et al. [70, 71], in corn, reported the release of the *ALS* gene to grant resistance to the chlorsulfuron herbicide [70, 71]. They succeeded in mutating the *ALS2* gene by bombarding immature embryos with gold particles from a 127-nucleotide single-stranded DNA oligo-repair template. Calluses resistant to chlorsulfuron were obtained. Also, homozygous plants were regenerated for the desired mutation (selecting only in chlorsulfuron) and were self-pollinating. The segregated seeds were sprayed with the herbicide, and the plants with the edited ALS alleles showed resistance to the selection factor [72].

Gene blocking (gene inactivation or deactivation) consists of suppressing the expression of a specific gene in an organism, replacing the original gene at its locus. The *waxy* gene (*Wx1*) codes for starch synthase, which is required for amylose synthesis in endosperm. Traditional corn contains amylopectin (75%) and amylose (25%). Waxy corn obtained by plant breeding contains only amylopectin. Due to the low yields of the waxy variety, *Wx1* in pure corn lines was edited by CRISPR-Cas9 to produce waxy corn in higher quantities.

Genome modification was achieved by bombarding particles of immature totally embryos with gold sub-particles with six different plasmids, which encode, for the two gRNAs, the Cas9 gene of *S. pyogenes*, the *WUSCHEL* and *ODP2* pluripotency genes of maize, and the NPTII selection marker for kanamycin and G418 resistance. In sum, *Wx1* deletions were generated in more than 10 different elite maize pure lines, allowing direct production of CRISPR-Cas9 derived hybrids with wax production [73]. The waxy corn seed mutants obtained had a similar appearance to candle wax because of the amylopectin content, giving greater uniformity and better texture to the starch for use as an ingredient in various food products, which have important applications in the textile, adhesive, and paper manufacturing industries.

Drought tolerance is a desirable characteristic in commercial corn. The *ARGOS8* gene negatively regulates the response of ethylene, which is important in drought tolerance. Naturally, the expression of *ARGOS8* is lower and absent in non-reproductive tissues; however, when ARGOS8 is over-expressed, the response of ethylene is negatively regulated. Shi et al. [29] carried out GE using CRISPR-Cas9 to alter the expression of *ARGOS8* in maize by replacing promoters [74]. The native promoter was replaced with maize's GOS2 promoter, promoting the constitutive expression of the SU1 transcription factor. The exchange of promoters was achieved by bombarding gold microparticles with a DNA repair mold (GOS2 flanked by two DNA fragments of 400 base pairs). These match the sequences adjacent to the Cas9 cleavage sites at the ARGOS8 locus. Five carrier plasmids were also added: each of the two gRNAs, a Streptococcus pyogenes Cas9 encoding gene, the maize pluripotency genes (*WUSCHEL* and *ODP2*) and the *PMI* gene as a selection marker. Drought tolerance was achieved by placing the modified lines in regeneration medium with the inclusion of mannose 6-phosphate, which is toxic to corn; with the expression of PMI, mannose-6-phosphate is converted to fructose-6-phosphate, which is not toxic to corn.

The result enabled 172 seedlings to grow in the selective agent (23 were positive in at least one binding, three were positive for both bindings, and two produced fertile plants). One of these plants was free of foreign DNA (without containing the genes used in the transformation). The plant edited in *ARGOS8* was evaluated in the field by crossing it with an innate tester. This hybrid was then compared with a similar wild-type hybrid in environments with drought stress and with sufficient irrigation. The previous agronomic characterization of the transgenic plants *GOS2-ARGOS8* was carried out. The edited hybrids behaved similarly to the wild type in non-limiting conditions of irrigation, but they registered a higher yield when they were cultivated under drought stress.

Transgenic maize grown in the United States of America includes resistance to two or more herbicides and multiple *cry* genes of Bacillus thuringiensis, which provide resistance to several insect pests. It is common to find that such a commercial product contains five different transgenic traits. These traits were generated by random transformation and are found throughout the genome.

To create a corn CTL, genomes are edited by CRISPR-Cas9 technology, so that transgenes are targeted in genomic regions selected by homologous recombination (HR). Chilcoat et al. [41] improved the efficiency of this process in maize by inserting specific integration sites of the SSI type in certain locations of the genome, based on the FLP/FRT recombinase system [47]. The target locations were selected to be close to each other but sufficiently distant (0.1 centimorgan) to be genetically separable, and far from the endogenous genes. In maize, which reproduces sexually, transgenes (Cas9 and gRNA) and pluripotency genes (*ODP2* and *WUSCHEL*) were removed from the plant genome after seed formation [24, 35]. However, the researchers mention that this strategy is not useful for vegetatively propagated plants, as it takes years for the plants to reach sexual maturity [75].

Svitashev et al. [71] conducted the *ALS2* gene editing of maize using CRISPR-Cas9, using the RNP complex [43, 44]. The gene was modified by replacing the amino acid proline with serine at position 165, and this change gave maize full resistance to chlorsulfuron. A 127-nucleotide single-stranded oligo was used to repair the DNA and was co-bombarded with Cas9-ALS-ARNg-complex RNP and pluripotency genes (*ODP2* and *WUSCHEL*) from maize. As a result, plants resistant to chlorsulfuron were regenerated. The mutation frequencies found when using DNA were similar to those found when using the RNP complex, demonstrating the efficiency and high rate of the RNP complex cutting activity. Using this procedure, it is possible to generate non-transgenic mutagenesis (i.e., mutagenesis free of foreign DNA and selection markers).

Char et al. [62] generated mutations in two maize genes encoding argonaute 18 and dihydroflavonol 4-reductase [62]. The vector used principal components of Cas9 and the U6 promoter tracrRNA, and, by double cloning, the double-stranded fragment was inserted as a guide. Then, the transformation process was carried out with Agrobacterium tumefaciens. Mutants were obtained with indels caused by the predominant repair mechanism in plants, NHEJ, with the transformation by the bacterium being effective.

For example, there are studies in grain breeding such as corn (drought resistance), rice (higher density and erectable panicles, and lower-height plants, which leads to a higher yield), wheat (gluten-free wheat indicated for patients with coeliac disease), and soybeans; horticultural crops such as tomato (resistance to high temperatures due to the inactivation of a single gene); potato, cucumber, and watermelon; and woody species such as citrus, papaya, and apple, among others. The mutation of susceptibility genes, which facilitate plant pathogen compatibility, could provide a more durable and broad-spectrum resistance than that afforded by resistance genes. Global crop productivity is limited by adverse environmental conditions such as water deficits, high temperatures, high salinity, and the low nutritional quality of soils. However, the use of CRISPR/Cas in the production of varieties tolerant to biotic stress has potentialities not yet exploited by genetic improvement.

2.5 CONCLUSIONS

New GE technologies are a powerful new method in the genetic yield of plants. New genomic editing technologies are adapted for use in plant breeding programs, as good candidate genes are necessary for modification. Questions remain as to the predictability of effects in different regions of the genome and about the technical conditions for carrying out the introduction of components of the modification process in various species. These new methodologies will produce new varieties of plants that perhaps in the future we will find in our fields of cultivation, whose background is the ease and potential that CRISPR/Cas has to edit or modulate the DNA, making it *in vivo* and in any type of organism. Gene-editing methods have the potential to grant new responses to the main challenges to meet the growing demand for food, increasingly sustainable, under changing climatic conditions and obtain a high production of non-GM plants. The advantages of the CRISPR gene-editing technique are accuracy, speed, economy, and predictability. Mutations from natural genetic variation can be transferred to other varieties of the same species. Therefore, the use of CRISPR in modern agriculture is a new breeding method, as it produces identical results to those that would be produced by conventional methods. Most studies are directed at essential crops with the goal of improving characteristics such as yield, plant architecture and disease tolerance. Such applications can help to preserve the natural genetic diversity (biodiversity) of species, which is the basis of any breeding process and whose loss over time is one of the problems of modern agriculture. By avoiding the use of recombinant DNA, plants produced by this system should be liberated from GMO legislation.

In the future, the CRISPR tool will make it possible to delete genes, alter expression, combine genes of interest, and perform precise GE. In plant breeding it is possible to discover the introduction of genetic diversity, identify responsible genes, validate research and in terms of efficiency accumulate several genetic traits, break the imbalance, break the linkage, and also eliminate negative traits, optimize existing traits.

KEYWORDS

- ***Agrobacterium tumefaciens***
- **clustered regularly interspaced short palindromic repeats**
- **drought stress**
- **endonucleases**
- **genomic editing**
- **haploid induction**
- **ribonucleoprotein complex**

REFERENCES

1. Barrangou, R., & Marraffini, L., (2014). CRISPR-Cas systems: Prokaryotes upgrade to adaptive immunity. *Mol. Cell, 54*, 234–244.
2. Soon, H., Heung, J., Jin, K., & Ko, K., (2014). N-Glycosylation modification of plant-derived virus-like particles: An application in vaccines. *Biomed Res. Int.,* 1–8. doi: 10.1155/2014/249519.

3. Gasiunas, G., Sinkunas, T., & Siksnys, V., (2014). Molecular mechanisms of CRISPR mediated microbial immunity. *Cell. Mol. Life Sci., 71*, 449–465.
4. Mali, P., Esvelt, K., & Church, G., (2013). Cas9 as a versatile tool for engineering biology. Nat. *Methods, 10*, 957–963. doi: 10.1038/nmeth.2649.
5. Nishimasu, H., Ann, F., Hsu, P., Konermann, S., Shehata, S., Dohmae, N., Ishitani, R., et al., (2014). Crystal structure of Cas9 in complex with guide RNA and target DNA. *Cell, 156*, 935–949. doi: 10.1016/j.cell.2014.02.001.
6. Jinek, M., Jiang, F., Taylor, D., Sternberg, S., Kaya, E., Ma, E., Anders, C., et al., (2014). Structures of Cas9 endonucleases reveal RNA-mediated conformational activation. *Science, 343*, 1–28.
7. Lei, Y., Lu, L., Liu, Y., Li, S., Xing, F., & Chen, L., (2014). CRISPR-P: A web tool for synthetic single-guide RNA design of CRISPR-System in plants. *Mol. Plant, 7*, 1494–1496. doi: 10.1093/mp/ssu044.
8. Sander, J., & Joung, K., (2014). CRISPR-Cas systems for genome editing, regulation and targeting. *Nat. Biotechnol., 32*, 347–355.
9. Jiang, W., Zhou, H., Bi, H., Fromm, M., Yang, B., & Weeks, P., (2013). Demonstration of CRISPR/Cas9/sgRNA-mediated targeted gene modification in *Arabidopsis*, tobacco, sorghum, and rice. *Nucleic Acids Res., 41*, 1–12. doi:10.1093/nar/gkt780.
10. Belhaj, K., Chaparro-García, A., Kamoun, S., Patrón, N., & Nekrasov, V., (2015). Editing plant genomes with CRISPR/Cas9, *Curr. Opin. Biotech., 32*, 76–84. doi: 10.1016/j.copbio.2014.11.007.
11. Shan, Q., Wang, Y., Li, J., & Gao, C., (2014). Genome editing in rice and wheat using the CRISPR/Cas system. *Nat. Protoc., 9*, 2395–2410. doi: 10.1038/nprot.2014.157.
12. Endo, M., Mikami, M., & Toki, S., (2015). Multigene knockout utilizing off-target mutations of the CRISPR/Cas9 system in rice. *Plant Cell Physiol., 56*, 41–47. doi: 10.1093/pcp/pcu154.
13. Kim, S., Kim, D., Woo, S. C., Kim, J., & Jin-Soo, K., (2014). Highly efficient RNA-guided genome editing in human cells via delivery of purified Cas9 ribonucleoproteins. *Genome Res., 24*, 1012–1019. doi: 10.1101/gr.171322.113.
14. Woo, J. W., Kim, J. K., Won, S. I., Corvalán, C., Cho, S. W., Kim, H., Kim, S. G., et al., (2015). DNA-free genome editing in plants with preassembled CRISPR-Cas9 ribonucleoproteins. *Nat. Biotechnol., 33*, 1162–1178. doi:10.1038/nbt.3389.
15. Wang, C., Liu, Q., Shen, Y., Hua, Y., Wang, J., Lin, J., et al., (2019). Clonal seeds from hybridrice by simultaneous genome engineering of meiosis and fertilization genes. *Nat Biotechnol.* https://doi.org/10.1038/s41587-018-0003-0.
16. Xie, E., Li, Y., Tang, D., Lv, Y., Shen, Y., & Cheng, Z., (2019). A strategy for generating rice apomixis by gene editing. *J. Integr. Plant Biol., 61*(8), 911–916.
17. Janaiah, A., Hossain, M., & Husain, M., (2002). Hybrid rice for tomorrow's food security: Can the Chinese miracle be replicated in other countries? *Outlook Agri., 31*, 2333.
18. Vaughan, D. A., et al., (2007). From crop domestication to super-domestication. *Ann. Bot., 100*, 893–901.
19. Varshney, R. K., et al., (2012). Can genomics boost productivity of orphan crops? *Nat. Biotechnol., 30,* 1172–1176.
20. Li, T., et al., (2018). Domestication of wild tomato is accelerated by genome editing. *Nat. Biotechnol., 36*, 1160–1163.

21. Zsögön, A., et al., (2018). De novo domestication of wild tomato using genome editing. *Nat. Biotechnol., 36*, 1211.
22. Lemmon, Z. H., et al., (2018). Rapid improvement of domestication traits in an orphan crop by genome editing. *Nat. Plants, 4*, 766–770.
23. Zuhab, K. M., Shan-e-Ali, Z. S., Amin, I., & Mansoor, S. (2007). A CRISPR way for fast-forward crop domestication. *Trends in Plant Science*, 1–3.
24. Wang, B., Zhu, L., Zhao, B., Zhao, Y., Xie, Y., Zhang, Z., Li, Y., et al. (2019). Development of a haplid-inducer mediated genome editing system for accelerating maize breeding. *Molecular Plant,* 597–602.
25. Khanday, I., Skinner, D., Yang, B., Mercier, R., & Sundaresan, V., (2019). A male-expressed rice embryogenic trigger redirected for asexual propagation through seeds. *Nature, 565*, 9195.
26. Osakabe, Y., Watanabe, T., Sugano, S. S., Ueta, R., Ishihara, R., Shinozaki, K., & Osakabe, K., (2016). Optimization of CRISPR/Cas9 genome editing to modify abiotic stress responses in plants. *Scientific Reports, 6*, 26685.
27. Park, J. J., Dempewolf, E., Zhang, W., & Wang, Z. Y., (2017). RNA-guided transcriptional activation via CRISPR/dCas9 mimics overexpression phenotypes in *Arabidopsis*. *PLoS One, 12*, e0179410.
28. Roca, P. J. F., Gillet, F. X., Ribeiro, T. P., Bournaud, C., Lourenço-Tessutti, I. T., Noriega, D. D., Melo, B. P., et al., (2019). Improved drought stress tolerance in *Arabidopsis* by CRISPR/dCas9 fusion with a histone acetyl transferase. *Scientific Reports, 9*, 8080.
29. Shi, J., Gao, H., Wang, H., Lafitte, H. R., Archibald, R. L., Yang, M., Hakimi, S. M., et al., (2017). ARGOS8 variants generated by CRISPR-Cas9 improve maize grain yield under field drought stress conditions. *Plant Biotechnology Journal, 15*, 207–216.
30. Zhang, A., Liu, Y., Wang, F., Li, T., Chen, Z., Kong, D., Bi, J., et al., (2019). Enhanced rice salinity tolerance via CRISPR/Cas9-targeted mutagenesis of the OsRR22 gene. *Molecular Breeding, 39,* 47.
31. Ozias-Akins, P., & Van, D. P. J., (2007). Mendelian genetics of apomixis in plants. *Annu. Rev. Genet., 41*, 50937.
32. Conner, J. A., Goel, S., Gunawan, G., Cordonnier-Pratt, M. M., Johnson, V. E., Chun, L., Wang, H., Pratt, L. H., Mullet, J. E., DeBarry, J., et al., (2008). Sequence analysis of bacterial artificial chromosome clones from the apospory-specific genomic region of *Pennisetum* and *Cenchrus*. *Plant Physiol., 147*, 1396411.
33. Conner, J. A., Gunawan, G., & Ozias-Akins, P., (2013). Recombination within the apospory specific genomic region leads to the uncoupling of apomixis components in *Cenchrus ciliaris*. *Planta, 238*, 5163.
34. Conner, J. A., Mookkan, M., Huo, H., Chae, K., & Ozias-Akins, P., (2015). A parthenogenesis gene of apomict origin elicits embryo formation from unfertilized eggs in a sexual plant. *Proc. Natl. Acad. Sci., 112*, 11205–11210.
35. Boutilier, K., Offringa, R., Sharma, V. K., Kieft, H., Ouellet, T., Zhang, L., Hattori, J., Liu, C. M., Van, L. A. A., Miki, B. L., et al., (2002). Ectopic expression of BABY BOOM 68S triggers a conversion from vegetative to embryonic growth. *Plant Cell, 14*, 1737–1749.
36. Wang, S., Zhang, S., Wang, W., Xiong, X., Meng, F., & Cui, X., (2015). Efficient targeted mutagenesis in potato by the CRISPR/Cas9 system. *Plant Cell Rep., 34*, 1473–1476. doi: 10.1007/s00299-015-1816-7.

37. Wang, H. X., Li, M., Lee, C. M., Chakraborty, S., Kim, H. W., Bao, G., & Leong, K. W., (2017). CRISPR/Cas9-based genome editing for disease modeling and therapy: Challenges and opportunities for nonviral delivery. *Chem. Rev., 117*(15), 9874–9906. doi: 10.1021/acs.chemrev.6b00799.
38. Zhang, H., Zhang, J., Wei, P., Zhang, B., Gou, F., Feng, Z., Mao, Y., et al., (2014). The CRISPR/Cas9 system produces specific and homozygous targeted gene editing in rice in one generation. *Plant Biotechnol. J., 12*(6), 797–807. doi: 10.1111/pbi.12200.
39. Mao, T., Zhu, M., Sheng, Z., et al., (2021). Effects of grain shape genes editing on appearance quality of erect-panicle Geng/Japonica rice. *Rice, 14*, 74. https://doi.org/10.1186/s12284-021-00517-5.
40. Cai, Y., Li, C., Xiujie, L., Chen, G., Shi, S., Cunxiang, W., Bingjun, J., Tianfu, H., & Wensheng, H., (2018). CRISPR/Cas9-mediated targeted mutagenesis of GmFT2a delays flowering time in soya bean. *Plant Biotechnology Journal, 16*(1), 176–185. https://doi.org/10.1111/pbi.12758.
41. Chilcoat, D., Liu, Z. B., & Sander, J., (2017). Use of CRISPR/Cas9 for crop improvement in maize and soybean. *Progress Molecular Biol. Trans. Sci., 149*, 27–46.
42. Peng, A., Chen, S., Lei, T., Xu, L., & He, Y., (2017). Engineering canker-resistant plants through CRISPR/Cas9-targeted editing of the susceptibility gene CsLOB1 promoter in citrus. *Plant Biotechnol. J., 15*, 1509–1519. doi: 10.1111/pbi.12733.
43. Ortigosa, A., Gimenez-Ibanez, S., Nathalie, L., & Roberto, S., (2019). Design of a bacterial speck resistant tomato by CRISPR/Cas9-mediated editing of SlJAZ2. *Plant Biotechnology Journal, 17*(3), 665–673. https://doi.org/10.1111/pbi.13006.
44. Soyk, S., Niels, A M., Soon, J. P., Inga, S., Ke, J., Ryosuke, H., Lei, Z., et al., (2017). Variation in the flowering gene SELF PRUNING 5G promotes day-neutrality and early yield in tomato. *Nature Genetics, 49*(1), 162–168. https://doi.org/10.1038/ng.3733.
45. Soyk, S., Müller, N. A., Park, S. J., Schmalenbach, I., Jiang, K., Hayama, R., Zhang, L., et al., (2017). Variation in the flowering gene SELF PRUNING 5G promotes day-neutrality and early yield in tomato. *Nat. Genet., 49*, 162–168. doi: 10.1038/ng.3733.
46. Cai, Y., Chen, L., Liu, X., Guo, C., Sun, S., Wu, C., Jiang, B., et al., (2018). CRISPR/Cas9-mediated targeted mutagenesis of GmFT2a delays flowering time in soya bean. *Plant Biotechnol. J., 16*, 176–185. doi: 10.1111/pbi.12758.
47. Wang, Y., Geng, L., Yuan, M., Wei, J., Jin, C., Li, M., Yu, K., et al., (2017). Deletion of a target gene in indica rice via CRISPR/Cas9. *Plant Cell Rep., 36*, 1333–1343. doi: 10.1007/s00299-017-2158-4.
48. Nieves-Cordones, M., Mohamed, S., Tanoi, K., Kobayashi, N. I., Takagi, K., Vernet, A., Guiderdoni, E., et al., (2017). Production of low-Cs+ rice plants by inactivation of the K+ transporter OsHAK1 with the CRISPR-Cas system. *Plant J., 92*, 43–56.
49. Nieves-Cordones, M., Mohamed, S., Tanoi, K., Kobayashi, N. I., Takagi, K., Vernet, A., Guiderdoni, E., et al., (2017). Production of low-Cs+ rice plants by inactivation of the K + transporter OsHAK1 with the CRISPR-Cas system. *Plant J., 92*, 43–56. doi: 10.1111/tpj.13632.
50. Liu, H., Ding, Y., Zhou, Y., Jin, W., Xie, K., & Chen, L. L., (2017). CRISPR-P 2.0: An improved CRISPR-Cas9 tool for genome editing in plants. *Mol. Plant, 10*, 530–532. doi: 10.1016/j.molp.2017.01.003.

51. Wang, F., Wang, C., Liu, P., Lei, C., Hao, W., Gao, Y., & Zhao, K., (2016). Enhanced rice blast resistance by CRISPR/Cas9-targeted mutagenesis of the ERF transcription factor gene OsERF922. *PloS One, 11*, 1–18. doi: 10.1371/journal.pone.0154027.
52. Souto, L., De-Araújo, F., Gomes, N., Duarte, K., Oliveira, R., Alekcevetch, J. C., Guitton, M., et al., (2016). Identification of candidate genes for drought tolerance in coffee by high-throughput sequencing in the shoot apex of different *Coffea arabica* cultivars. *BMC Plant Biol., 16*, 1–18. doi: 10.1186/s12870-016-0777-5.
53. Baltes, N. J., Hummel, A. W., Konecna, E., Cegan, R., Bruns, A. N., Bisaro, D. M., & Voytas, D. F., (2015). Conferring resistance to geminiviruses with the CRISPRCas prokaryotic immune system. *Nature Plants, 1*, 1–7. doi: 10.1038/nplants.2015.145" target="_blank">10.1038/nplants.2015.145.
54. Ali, Z., Abulfaraj, A., Idris, A., Ali, S., Tashkandi, M., & Mahfouz, M. M., (2015). CRISPR/Cas9-mediated viral interference in plants. *Genome Biol., 16*, 1–11. doi: 10.1186/s13059-015-0799-6.
55. Chandrasekaran, J., Brumin, M., Wolf, D., Leibman, D., Klap, C., & Pearlsman, M., (2016). Development of broad virus resistance in non-transgenic cucumber using CRISPR/Cas9 technology. *Mol. Plant Pathol., 17*, 1140–1153. doi: 10.1111/mpp.12375.
56. Nekrasov, V., Wang, C., Win, J., Lanz, C., Weigel, D., & Kamoun, S., (2017). Rapid generation of a transgene-free powdery mildew resistant tomato by genome deletion. *Sci. Rep., 7*, 1–6. doi: 10.1038/s41598-017-00578-x.
57. De Toledo, D. P., Brail, Q., Dahlbeck, D., & Staskawicz, B., (2016). CRISPR-Cas9 mediated mutagenesis of a DMR6 ortholog in tomato confers broad-spectrum disease resistance. *bioRxiv., 1*, 1–23. doi: 10.1101/064824.
58. Patel, J., & Mishra, A., (2019). Genome editing: Advances and prospects. *Plant Biotechnology: Progress in Genomic Era, 1*, 147–174.
59. Liang, Z., Chen, K., Li, T., Zhang, Y., Wang, Y., Zhao, Q., & Gao, C., (2017). Efficient DNA-free genome editing of bread wheat using CRISPR/Cas9 ribonucleoprotein complexes. *Nat. Commun., 8*, 1–5.
60. Zhang, Y., Liang, Z., Zong, Y., Wang, Y., Liu, J., Chen, K., & Gao, C., (2016). Efficient and transgene-free genome editing in wheat through transient expression of CRISPR/Cas9 DNA or RNA. *Nat. Commun., 7*, 1–8.
61. Sánchez-León, S., Gil-Humanes, J., Carmen, V. O., María, J. G., Carolina, S., Daniel, F. V., & Francisco, B., (2018). Low-gluten, nontransgenic wheat engineered with CRISPR/Cas9. *Plant Biotechnology Journal, 16*(4), 902–910. https://doi.org/10.1111/pbi.12837.
62. Char, S., Neelakandan, K., Nahampun, H., Frame, B., Main, M., Spalding, M. H., Becraft, P. W., et al., (2017). An Agrobacterium-delivered CRISPR/Cas9 system for high-frequency targeted mutagenesis in maize. *Plant Biotechnol. J., 15*, 257–268. doi: 10.1111/pbi.12611.
63. Tian, S., Linjian, J., Qiang, G., Jie, Z., Mei, Z., Haiying, Z., Yi, R., et al., (2017). Efficient CRISPR/Cas9-based gene knockout in watermelon. *Plant Cell Reports, 36*(3), 399–406. https://doi.org/10.1007/s00299-016-2089-5.
64. Yang, W., Wu, Y., Li, H., Wei, S., Yan, L., Xie, Q., & Hu, R., (2015). High-efficiency genome editing in *Arabidopsis* using YAO promoter-driven CRISPR/Cas9 system. *Mol. Plant, 8*, 1820–1823. doi: 10.1016/j.molp.2015.10.004.
65. Ma, X., Zhang, Q., Zhu, Q., Liu, W., Chen, Y., Qiu, R., Wang, B et al., (2015). A robust CRISPR/Cas9 system for convenient, high-efficiency multiplex genome editing in monocot and dicot plants. *Mol. Plant, 8*, 1274–1284. doi: 10.1016/j.molp.2015.04.007.

66. Schachtsiek, J., & Stehle, F., (2019). Nicotine-free, nontransgenic tobacco (Nicotiana tabacum L.) edited by CRISPR-Cas9. Plant *Biotechnol. J., 17*, 2228–2230. doi: 10.1111/pbi.13193.
67. Agricultural Marketing Resource Center. *Lettuce*. https://www.agmrc.org/commodities-products/vegetables/lettuce (accessed on 16 December 2023).
68. Bertier, L. D., Ron, M., Huo, H., Bradford, K. J., Britt, A. B., & Michelmore, R. W., (2018). High-resolution analysis of the efficiency, heritability, and editing outcomes of CRISPR/Cas9-induced modifications of nced4 in lettuce (*Lactuca* sativa). *G3, 8*, 1513–1521. doi: 10.1534/g3.117.300396.
69. Mou, B., (2011). Mutations in lettuce improvement. *Int. J. Plant Genomics, 2011*, 1–7. doi: 10.1155/2011/723518.
70. Svitashev, S., Young, J., Schwartz, C., Gao, H., Falco, S. C., & Cigan, A. M., (2015). Targeted mutagenesis, precise gene editing and site-specific gene insertion in maize using Cas9 and guide RNA. *Plant Physiol., 169*, 931–945.
71. Svitashev, S., Schwartz, C., Lenderts, B., Young, J. K., & Cigan, A. M., (2016). Genome editing in maize directed by CRISPRCas9 ribonucleoprotein complexes. *Nat. Commun., 7*, 1–7.
72. Cabrera, J. L., Valencia, E., & Trejo, D. L., (2019). Genetic modifications of corn. In: Serna-Saldivar, S. O., (ed.), *Corn. 3rd* (p. 4385). AACC International Press: Ed.; Oxford, UK.
73. Zhang, J., Zhang, H., Botella, J. R., & Zhu, J. K., (2018). Generation of new glutinous rice by CRISPR/Cas9targeted mutagenesis of the waxy gene in elite rice varieties. *J. Integrative Plant Biol., 60*, 369–375.
74. Shi, J., Gao, H., Wang, H., Lafitte, H. R., Archibald, R. L., Yang, M., & Habben, J. E., (2017). ARGOS 8 variants generated by CRISPRCas9 improve maize grain yield under field drought stress conditions. *Plant Biotechnol., 152*, 207–216.
75. Artlip, T. S., Wisniewski, M. E., Arora, R., & Norelli, J. L., (2016). An apple rootstock overexpressing a peach CBF gene alters growth and flowering in the scion but does not impact cold hardiness or dormancy. *Hort. Res., 3*, 1–9.
76. Zhao, Y., Zhang, C., Liu, W., Gao, W., Liu, C., Song, G., Li, W. X., Mao, L., Chen, B., Xu, Y., Li, X., & Xie, C. (2016). An alternative strategy for targeted gene replacement in plants using a dual-sgRNA/Cas9 design. *Sci Rep.* 2016 Apr 1; *6*, 23890. doi: 10.1038/srep23890.
77. Cai, Y., Chen, L., Liu, X., Guo, C., Sun, S., Wu, C., Jiang, B., Han, T., & Hou, W. (2018). CRISPR/Cas9-mediated targeted mutagenesis of GmFT2a delays flowering time in soya bean. *Plant Biotechnol J. 16*(1), 176–185. doi: 10.1111/pbi.12758.
78. Wang, Y., Cheng, X., Shan, Q., Zhang, Y., Liu, J., Gao, C., & Qiu, J. L. (2014). Simultaneous editing of three homoeoalleles in hexaploid bread wheat confers heritable resistance to powdery mildew. *Nat Biotechnol 32,* 947–951

CHAPTER 3

STRATEGIES FOR DELIVERING CRISPR-CAS REAGENTS INTO PLANTS

LUAN LUONG CHU[1] and NGUYEN QUANG HUY[2]

[1] *National Key Laboratory of Enzyme and Protein Technology, University of Science, Vietnam National University, Hanoi (VNU), 334 Nguyen Trai, Thanh Xuan, Hanoi 10000, Vietnam*

[2]*Faculty of Biology, University of Science, Vietnam National University, Hanoi (VNU), 334 Nguyen Trai, Thanh Xuan, Hanoi 10000, Vietnam*

ABSTRACT

The clustered regularly interspaced short palindromic repeat (CRISPR)-associated nuclease (CRISPR-Cas) systems are archaeal and bacterial adaptive immunity that provide a promising genome-editing tool in the field of molecular biology. This technology is rapidly being used in plant for developing transgene-free mutants, improving crops, and securing global foods. Delivering CRISPR-Cas reagents into plant cells is the most important factor for high editing efficiency. Currently, there are many approaches for delivering CRISPR-Cas reagents into various plants, including *Agrobacterium*-mediated transformation, viral vector delivery, polyethylene glycol (PEG)-mediated delivery, and particle bombardment delivery. This chapter covers the various strategies for delivering CRISPR-Cas component with advantages and disadvantages for genome editing (GE) in plant cells. Moreover, the application of nanoparticle and pollen magnetofection-mediated delivery are also discussed in the present chapter.

CRISPR-Based Gene Editing: A Solution for Global Food Security. Shakeel Ahmad, Asif Ali Khan, and Muhammad Arshad Javed (Eds.)

3.1 INTRODUCTION

Plants play an essential role in the planet and human life. Plants not only supply oxygen for planet though photosynthesis process but also provide the various essential resource as water, food, fruits, and medicine [1]. Along with the rising population in many parts of the world, the demand for food and energy is highly increasing. Plant engineering and breeding are the most important approaches to achieve a greater yield of crops [2–4]. Many biotools have been developed to plant genetic engineering, such as synthetic promoter, expression vectors, zinc-finger proteins (ZFPs), transcription activator-like effectors (TALEs), TALE nucleases (TALENs), RNA interferences, CRISPR-Cas systems [5–9]. These tools exhibit the ability to transcriptional regulation of endogenous and/or heterologous gene expression in the plant cell. As an expected result, many crops have been reported to improve the yield through genetic engineering, such as maize, wheat, soybean [10–13].

CRISPR-Cas system was discovered from the bacterial immune system in 1987. It has rapidly developed for genome editing (GE) of plants, animals, and microbes. CRISPR-Cas systems have been becoming a powerful tool for crop improvement in recent years [14–16]. According to the component of Cas protein, CRISPR systems are categorized into two classes: class 1 and class 2. Class 1 system, include type I (Cas3), III (Cas10), and IV, need multiple Cas protein to carry out the function. By contrast, class 2 system, consist of type II (Cas9), V (Cas12), and VI (Cas13), require only a single Cas protein to be functional [17]. Among these CRISPR-Cas systems, type II CRISPR-Cas9 consists of Cas9 nuclease from *Streptococcus pyogenes* and the single-guide RNA (sgRNA) [18]. In this system, and histidine-asparagine-histidine (HNH) nuclease domain of Cas9 cutting the targeting strand, while a RuvC-like domain cleaving the non-targeting strand. Cas9 cleavages site-specific double-stranded break (DSB) through recognizing NGG PAM sequences (Figure 3.1). DSB-based CRISPR-Cas systems can be repaired through the nonhomologous end-joining (NHEJ) pathway or the homology-directed repair (HDR) system [19]. Along with Cas9, CRISPR-Cas12a (Cpf1) in the class 2 has been exhibited the significant yield in the various crops [20]. The simple and highly efficiency in genomic engineering are two major reasons for the extensive application of CRISPR-Cas system. Many factors have been demonstrated effective on the editing efficiency in plant cells, such as the hosts, the expression level of *Cas* gene and gRNA, the delivery methods, and temperatures [21]. Strategies for delivering

CRISPR-Cas component into plants are one of the most important factors to improve the efficiency of mutant frequencies. In this chapter, we focus to review the different delivery CRISPR-Cas system approaches with advantages and disadvantages for GE in various plants (Figure 3.1). Moreover, several two methods, nanoparticle, and pollen magnetofection-mediated delivery system, have suggested as promising approach in the near future.

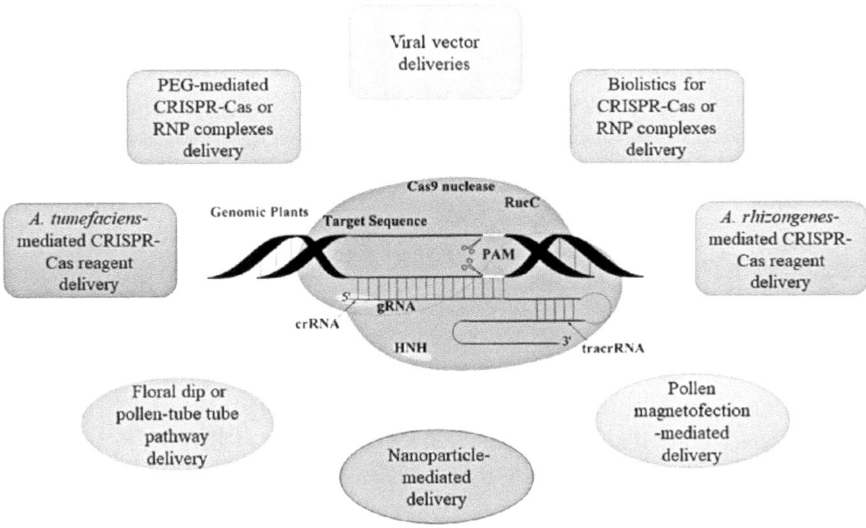

FIGURE 3.1 CRISPR-Cas9 system, traditional (rectangle) and future perspective (oval) methods for delivering of CRISPR-Cas reagents. [HNH: An endonuclease domain named for characteristic histidine and asparagine residues; RuC: An endonuclease domain named for a protein involved in DNA repair; gRNA: Single guide RNA; crRNA: CRISPR RNA; tracrRNA: Trans-activating crRNA; and PAM: Protospacer-adjacent motif].

3.2 *AGROBACTERIUM*-MEDIATED CRISPR-CAS9 SYSTEM DELIVERY

Since T-DNA binary vector systems are known as the greatest technology to generate genetically modified (GM) plants, they have been used to deliver CRISPR-Cas system into plan cells The different parts of the plants are used for delivering CRISPR-Cas reagents, including callus, leaf, and floral organs [22, 23]. *Agrobacterium*-mediated CRISPR-Cas reagents have been used to efficiently edited genome for more than 20 plant species. These plants include monocot and dicot crops [24]. *Agrobacterium* strains, including *Agrobacterium tumefaciens* and *Agrobacterium rhizongenes*, are the most preferred strains for genetic transformation due to their compatibility or interaction with the plant genome. Furthermore, both strains have a great effect on the efficiency of the biotransformation process and or foreign protein expression [25].

A. tumefaciens is known as a gram-negative, soil bacterium that causes crown gall, and can transfer T-DNA into the plant chromosome for expression [26]. Using *A. tumefaciens* to deliver CRISPR-Cas reagent into plant cells has been successfully reported in both monocotyledon and dicotyledon plants. Monocotyledon plants include *Zea mays*; *Oryza sativa*; *Sorghum bicolor*, while the different dicotyledon plants consist of *Arabidopsis thaliana*; *Solanum lycopersicum; Vitis vinifera; Manihot esculenta; Populus tremula; Brassica napus; Citrullus lanatus; Nicotiana benthamiana; N. tabacum* [27–42] (Table 3.1). Transformed plants are generally targeted to the callus, leaf, and floral organs [21]. In the first report of plant genomics editing, transgenic *A. thaliana* and *N. benthamiana* achieved through *A. tumefaciens* leaf infiltration. Phytoene desaturase (*PDS3*) gene in carotenoid biosynthesis and chlorophyll oxidation were mutated by using the plant codon optimized *SpCas9* (*pcoCas9*) and gRNA. A 2.7% and 4.8% mutagenesis frequency of *AtPDS3* and *NbPDS* mutation were obtained, respectively. Notably, this CRISPR-Cas9 system also was used to edit the *Arabidopsis* activated C kinase 1 (*AtRACK1b* and *AtRACK1c*) with mutagenesis frequency around 2.5–2.7% [27]. Along with the development of CRISPR-Cas9 system, using *A. tumefaciens* carrying a construction for editing of the genome have been increased the efficient multiplex editing of both moncotyle-don and dicotyledon. For example, 100% efficiency of acetolactate synthase 1 mutation was obtained in callus *S. tuberosum* by *A. tumefaciens*-mediated transformation [31]. In *Z. mays*, *A. tumefaciens* carrying a CRISPR-Cas9 mutated *ZmAgo18a* and *ZmAgo18b* with 74% and 70% mutation frequency, respectively [33]. Interestingly, Cas9 is the ubiquitous nuclease for GE, however, there is a number of a variant of Cas9 in constructing of an effective platform for CRISPR-Cas application. Firstly, a nickase Cas9 mutant, which cuts only one strand of DNA, is used to suppress the off-target mutant in *O. sativa* L. cv. Nipponbare. As result, the highest average mutation frequency of DMC1A gene was obtained up to 96.8% using scutellum-derived calli [34]. Since dead Cas9 nuclease (dCas9) has been shown to the ability to bind without cleaving DNA, a CRISPR-dCas9 system was carried out a base-editing with watermelon actolactate synthase (*ALS*) gene using *A. tumefaciens*-mediate biotransformation. In this case, the base-editing efficiency could be achieved 23% at T_0 generation [1]. Recently,

Cas12a (Cpf1), a Cas nuclease from *Prevotella* and *Francisella* 1 bacteria, has been widely applied for *Agrobacterium*-mediated GE [42–44]. Interestingly, both Cas9 and Cpf1 for *Agrobacterium* biotransformation could be targeted in the maize *glossy2* genes. While editing efficiency in the case of Cas9-edited T_0 plants was 90–100%, the results showed 0–60% efficiency frequencies in apply of Cpf1-edited T_0 plants. The reasons that the crRNAs of Cpf1 are shorter than Cas9, the formation the secondary structures are undesirable. Therefore, the efficiency frequency of mutant is followed decrease in the case of maize [43].

Like *A. tumefaciens*, *A. rhizongenes* is a soil bacterium that has capability of transferring new genetic material to plant cells [45]. *A. rhizongenes*-based transformation for CRISPR-Cas reagents exhibited many the advantages in compared with *A. tumefaciens*-based transformation. Firstly, *A. rhizongenes* causes hairy roots, which is induced from the infected site. The editing event occurs because of transferring of the alien DNA into the plant genome through the root-inducing (Ri) plasmid. Ri plasmid harboring *Cas* gene and a cassette for gRNA provide a significant approach to improving the efficiency of potential transformants. The functionality of constructed Ri plasmid could be verified through the phenomenon of hairy root development [48]. Secondly, this method is a simple and rapid approach. While *A. tumefaciens*-based transformation requires 4–6 months, Ri plasmid transgenic roots generate transgenic plants within 1–1.5 months [47]. Moreover, there is a different mechanism for the import of single-stranded DNA (ssDNA; T-strands) into plant cell between *A. rhizongenes* and *A. tumefaciens*. *A. tumefaciens* virD2 and virE2 are known as nuclear import of T-strands into plant cells. However, GALLAS gene, an *A. tumefaciens* virE2 mutant contain nuclear localization sequence, is used to instead of virE2 in *A. rhizongenes*. GALLAS full length shows the ability of a strand transferase involved in conjugation through ATP-biding and helicase motifs, whereas VirE2-T-strand complex moves T-strand without ATP hydrolysis. Therefore, *A. tumefaciens* and *A. rhizongenes* showed the different import of T-strand on the efficiency and quality of transgenic event [48, 49]. Since it has been successful knocked-out for genomic editing in *Glycine max* L. with more than 95% of mutatated hairy root samples, using of *A. rhizongenes* to deliver CRISPR-Cas reagents into plant cells has been applied to *Medicago truncatula, Lotus japonicas, Salvia miltiorrhiza, Barassica carinata,* and *P. tomentosa* [50–57] (Table 3.1). For example, *A. rhizongenes*-mediated GE of *P. tomentosa* Carr. showed 51.7% mutation frequency of *PDS* gene [51]. In another case, use of *A. rhizongenes* to deliver CRISPR-Cas reagents for the knock-out of *SmCPS1* (diterpene synthase) in *S. miltiorrhiza* was reported with 42.3% of mutation frequency [55]. Interestingly, the inactivation of a rosmarinic acid synthase gene for the water-soluble phenolic acid biosynthetic pathway in hairy roots of *S. miltiorrhiza* was reported to 50% of efficiency through using *A. rhizongenes*-mediated biotransformation [57].

TABLE 3.1 *Agrobacterium*-Mediated CRISPR-Cas Reagent Delivery in Plant Species

Plant Species	CRISPR-Cas Vector	Targeted Genes	Editing Efficiency	References
A. tumefaciens-Mediated CRISPR-Cas Reagent Delivery				
Watermelon	pBSE90	*ALS* (acetolactate synthase).	NA	[1]
A. thaliana	pUC119-RCS	*AtPDS3, AtRACK1b, AtRACK1c.*	2.5–2.7%	[27]
N. benthamiana		*NbPDS*	4.8%	
A. thaliana	pCAMBIA1300	*BRI1, GAI, JAZ1*	30–84%	[28]
O. sativa		*ROC5, SPP, YSA*	4.7–75%	
T. aestivum	pBI121	*inox, pds*	11.2–22.3%	[29]
N. benthamiana		*pds*	1.8–23.8%	
N. tabacum	pORE	*NtPDS, NtPDR6*	81.8%, 87.5%	[30]
S. tuberosum L.	pMDC32	*StALS1* (acetolactate synthase 1).	87–100%	[31]
Cucumis sativus L.	pRCS	*elF4E* and *elF(iso)4E* (eukaryotic translation initiation factor 4E).	NA	[32]
O. sativa L. cv. Nipponbare	pZH_OsU6gRNA_MMCas9-Nickase	*DMC1A* (a conserved recombinase).	6.2–96.8%	[33]
Z. mays	pMCG1005	*ZmAgo18a* and *ZmAgo18b*; *a1* (dihydroflavonol 4-reductase), or *a4* (anthocyaninless).	74% and 70%; 15–49%	[34]
Banana cv. Rasthali	pRGEB31	*RAS-PDS*	59%	[35]
Kiwi fruit	pYLsgRNA-AtU6-1, pYLCRISPR/Cas9$_{P35S\text{-}N}$	*AcPDS*	8.3–91.67%	[36]
Marchantia polymorpha	pMpGE013, pMpGE014	*MpARF1* (auxin response factor 1), *MpNOP1* (U-box E3 ubiquitin ligase).	75%, 87.5%	[37]

TABLE 3.1 *(Continued)*

Plant Species	CRISPR-Cas Vector	Targeted Genes	Editing Efficiency	References
S. lycopericum cv. AC.	pYLCRISPR/Cas9	*SGR1* (silencing of stay-green 1), *LCY-E* (lycopene Ɛ-cyclase), *Blc* (β-lycopene cyclase), *LCY-B1* (lycopene β-cyclase 1), *LCY-B2* (lycopene β-cyclase 2).	0–95.83%	[38]
O. sativa	VK005	*ISA1* (*Isoamylase* 1)	NA	[39]
Citrus sinesis L. Osbeck	pCas9-GN	*CsWRKY22* (plant immunity response to citrus canker).	68.2–85.7%	[40]
S. lycopericum L.	pENTR-CCD8sgRNA:pMR290/Cas9	*CCD8* (carotenoid cleavage dioxygenase 8).	71.4%	[41]
Z. mays	pYPQ166, pYPQ210 CRISPR-Cas9, LbCas12a	*glossy2* (*gl2*)	90–100%, 0–60%	[43]
Banana cultivar (cv.) Grand Naine	pRGEB31	LCY_{ε} (lycopene epsilon-cyclase).	NA	[44]
A. rhizongenes-Mediated CRISPR-Cas Reagent Delivery				
G. max	P201N	*GFP* transgene, *Glyma01g38150*, *Glyma11g07220*	> 95%	[50]
P. tomentosa Carr.	pYLCRISPR/Cas9	*PtoPDS*	51.7%	[51]
G. max	pMDC23-AtU6-26 pro	*GS1* (glutamine synthase), *CHI20* (chalcone-flavanone isomerase).	NA	[52]
M. truncatula		*GUS* (β-glucuronidase)		
L. japonicus	pCAMBIA1300	*LjLb1*, *LjLb2*, *LjLb3* (symbiotic leghemoglobin genes); *LjSYMRK* (symbiosis receptor kinase).	35%	[53]
M. truncatula	pFGC5941	*MtPDS*	10.35%	[54]
S. miltiorrhiza	AtU6-26SK, pCambia1300	*SmCPS1* (diterpene synthase).	42.3%	[55]
B. carinata	CRISPR+35S::GFP (V112)	*BcFLA1* (Fasciclin-like arabinogalactan protein 1).	–	[56]
S. miltiorrhiza	pAt-RAS-Cas9, pOs-RAS-Cas9	*SmRAS* (rosmarinic acid synthase).	50%	[57]

3.3 VIRAL VECTOR DELIVERIES

Due to virus shows the ability of injection and replication their genetic material into host cells, it has been highly considerate as a promising delivery system. Viruses have been widely used in animal and plant cells as a vector system to delivery GE reagents, such as ZFNs, TALENs, RGENs, and CRISPR-Cas. In plants, family Germiniviridae and Virgaviridae could be effected to both monocotyledon and dicotyledon. Although plant DNA and RNA viruses have great effect on achieving higher transformation efficiency, the delivery of the CRISPR-Cas system into plant cells for gene knock-out, knock-in (KI), and/or silencing is currently limited. The main reason that DNA and positive-strand RNA viruses are unable to carry a large CRISPR-Cas system (> 4.5 kb). Therefore, it was firstly processed to use viruses for the delivery of sgRNA into transgenic plants expressing the Cas9 gene, which is delivered by *Agrobacterium*-mediated transformation. Germiniviridae based gRNA delivery for CRISPR-Cas9 system has been successfully carried out in white, rice, potato, and tomato. The viruses have been used to achieve high-efficiency gene targeting in plant as the bean yellow dwarf virus (BeYDV), wheat dwarf virus (WDV), and maize streak virus (MSV) [28, 58, 59]. In another case, cabbage leaf curl virus (CaLCuV) has been applied to induce gene silencing and miRNA expression vector in cabbage, cauliflower, *Arabidopsis*, as well as *N. benthamiana* [60]. Unlike Germiniviridae, family Virgaviridae infects more than 400 plant species through single-strand RNA viruses. For example, tobacco rattle virus (TRV) is one of the most ubiquitous species for virus-induced gene silencing, which infects the growing points in *N. benthamiana*. RNA2 genome of TRV can be modified to carry 2–3 kb alien DNA. Although TRV facilitate cloning and delivering one or more sgRNAs, it is still inefficient for Cas9 gene. In order to the editing *PDS* gene in *N. benthamiana*, TRV virus harboring sgRNA under PEBV promoter was delivered by agro-infiltration. Then, it was reconstituted in the leaves of *N. benthamiana* expressing Cas9. As result, the efficiency mutagenesis from three leaves collected at different time points was achieved with 43 to 61% in compared with control sample [61]. Interestingly, Sonchus yellow net rhabdovirus (SYNV) has been successfully employed to delivery of the entire CRISPR-Cas9 into *N. benthamiana* for single, multiplex mutagenesis and chromosome deletion with high frequency. SYNV is known as a negative-strand RNA virus, which can be modified to carry a large DNA alien as Cas9. A SYNV tgtRNA (tRNA-gRNA-tRNA)-Cas9 cassette was used to carry out a knock-out of three endogenous tobacco genes, *PDS*, RNA-dependent RNA polymerase 6 (*RDR6*), and suppressor of gene silencing 3 (*SGS3*). The efficiency mutation was obtained from 40–90%

in each gene [62]. This research provides great promise for delivery of the entire CRISPR-Cas system into the various plant cells, even plant rhabdoviruses only exhibit the mutation efficiency in somatic cells and requires isolation and culture of an individual genome-edited plant.

3.4 POLYETHYLENE GLYCOL (PEG)-MEDIATED CRISPR-CAS REAGENT DELIVERY

Plasmid integration or an episomal plasmid harboring Cas9 and gRNA sequences can be delivered into plant cells through a polyethylene glycol (PEG) transformation of protoplasts or biolistic particles. According to tailored for purpose, a specific approach will be selected to target the high-efficiency delivery and expression of CRISPR-Cas9 system. While the transformation of protoplasts approach is major used for rapid testing the activity of CRISPR-Cas in the model as well as the various plants, biolistic particles are primarily applied for regeneration of transformed tissues or embryo plants. The PGE-mediated CRISPR-Cas9 reagent was the earliest described in *O. sativa*. Three selected gRNAs were directed Cas9 nuclease to mutate a mitogen-activated protein kinase (MAPK) with the mutation efficiency of 3–8% [63]. Interestingly, the mutation efficiency has reached to 100% in transgenic *C. lanatus* line through using PGE-mediated CRISPR-Cas system [64]. Moreover, PGE-mediated CRISPR-Cas9 system was used for a directed GE of *Z. mays*, *G. max*, *Physcomitrella patens*, and *Triticum aestivum* [11, 65–68] (Table 3.2). Recently, CRISPR-Cas12a have been optimized successfully delivery of the *PDS* gene into banana protoplasts. The editing efficiency could be achieved up 88.89% [69].

On another hand, Cas and gRNA of ribonucleoprotein (RNP) complexes can also be delivered into plant cells through the transformation of protoplasts. Due to the plant cell walls are removed by enzymatic digestion, plant protoplasts could be used to transfection for RNP complex in the various plants, as *A. thaliana,* rice, lettuce, grapevine, apple, and tobacco [70, 71, 74, 75]. For example, a successful Cas9-gRNA RNP complex editing of brassinosteroid insensitive (*BRI*) and phytochrome B (*PHYB*) genes was carried out in *A. thaliana* and *Lactuca sativa* using PEG-mediated delivery. The editing frequency obtained up to 46% [70]. There is also research on the transient transformation of protoplasts of Cabbage by Cas9-gRNA RNP complex. The editing frequency of *FRI* and *PDS* in *B. oleracea* was 2.25 and 1.33%, respectively, while 12.58% was the efficiency mutation of FRI in *Brassica rapa* [72]. Recently, RNP complex, which is make from Cpf1 and only crRNA, has been used to edit FAD2-1A and FAD2-1B in *G. max*. The editing frequency was obtained at 11.7% and 1.2%, respectively [73].

TABLE 3.2 Transformation of Protoplasts for CRISPR-Cas or RNP Complexes into the Different Plants

Plant Name	CRISPR-Cas Vector or RNP Complexes	Targeted Genes	Editing Efficiency	References
CRISPR-Cas Vector				
T. aestivum	pJIT163-2NLSCas9	*TaDREB2* (dehydration responsive element binding protein 2), *TaERF3* (ethylene responsive factor 3).	50%	[11]
O. sativa	pRGE3, pRGE6	*OsMPK5* (a mitogen-activated protein kinase).	3–8%	[63]
Z. mays	pU3-gRNA	*ZmIPK* (myo-inositol polyphosphate kinase).	16.4–19.1%	[65]
G. max	pCas9-GmU6-sgRNA	*Glyma06g14180*	14.7%	[66]
		Glyma08g02290	20.2%	
		Glyma12g37050	17.9%	
	pCas9-AtU6-sgRNA	*Glyma06g14180*	6.6%	
		Glyma08g02290	3.2%	
		Glyma12g37050	9.7%	
C. lanatus (Thunb.) Matsum and Nakai.	PHSN1 or PHSN2, PHSE3	*PDS*	100%	[64]
Physcomitrella patens	psgRNA, pAct-Cas9, PpAPT-KO4, PpAPT-KO7	*APT* (adenine phosphoribosyl transferase).	2.2–3.2%	[67]
T. aestivum	pCR8-U6-gRNA	(*EPSPS*) 5 enolpyruvylshikimate-3-phosphate synthases.	0–23.3%	[68]
Banana	CRISPR/Cas9-PDS	*PDS*	7.14%	[69]
	CRISPR/Cas12a-PDS		88.89%	
RNP Complexes				
A. thaliana	CRISPR RNA-guided endonucleases (RGENs).	*BRI1* (brassinosteroid insensitive 1).	46%	[70]
		PHYB (phytochrome B)		
Lactuca sativa		*BIN2* (brassinosteroid insensitive 2).		

TABLE 3.2 *(Continued)*

Plant Name	CRISPR-Cas Vector or RNP Complexes	Targeted Genes	Editing Efficiency	References
Grapvine	Purified Cas9 protein + sgRNA mix.	*MLO-7* (mildew locus O genetic factors).	0.1%	[71]
Apple		*DIPM-1* (DspE-interacting proteins of *Malus*).	6.7%	
		DIPM-2	3.3%	
		DIPM-4	6.9%	
G. max var. William 82.	Cpf1 with *in vitro* target-specific crRNAs.	*FAD2-1A* (fatty acid desaturase 2-1A; *Glyma10g42470*).	11.7%	[73]
		FAD2-1B (Glyma20g24530)	1.2%	
Cabbage (*B. oleracea* var. *capitata f. alba*)	The purified Cas9 enzyme and *in vitro* synthesized sgRNAs.	*FRI*	2.25%	[72]
		PDS	1.33%	
Cabbage (*B. rapa* subsp. *pekinensis*)		*FRI*	12.58%	
Petunia cv. mirage rose	Cas9 and *in vitro* transcribed sgRNA.	*ACO* (1-amino cyclopropane-1-carboxylate OXIDASE).	31.5%	[74]
Banana	CRISPR/Cas9 RNP-PDS complex (RNPs).	*PDS*	0.92%	[75]

3.5 BIOLISTIC DELIVERY OF CRISPR-CAS REAGENTS

Biolistic transformation is a method for the delivery of CRISPR-Cas system and/or Cas-gRNA RNP complex into plant cells by high-speed particle bombardment. CRISPR-Cas system and/or Cas-gRNA RNP complex are coated by tungsten, silver, and gold particles. Then, these nucleic acid-coated particles are transferred into the cell by a "gene gun" or "biolistic gun." Owing to the high efficiency, gold particle bombardment offers a ubiquitous platform for the delivery of CRISPR-Cas system and/or Cas-gRNA RNP complex. Gold particle bombardment has been used for GE in *O. sativa*, *T. aestivum*, *Z. mays*, *G. max*, and *Hordeum vulgare* (Table 3.3) [75–85]. For example, maize immature embryo cells were gold particle bombardment with a plasmid harboring Cas9 gene and gRNA intended to edit five different genomic regions: the liguleless 1 (*LIG1*) gene, male fertility genes (*Ms26* and *Ms45*), and acetolactate synthase (*ALS*) genes (*ALS1* and *ALS2*). The highest percentage of mutant reads at *LIG1*, *Ms26*, *Ms45*, and *ALS* were 4.61, 3.11, 1.87, and 2.23%, respectively [69–76]. In another case, the use of CRISPR-Cas9 RNP-coated gold for biolistic transformation showed the editing efficiency in *Z. mays*, *O. sativa*, *T. aestivum* (Table 3.3) [83–85]. For example, biolistic bombardment of Cas9 RNP complexes was delivered into the *T. aestivum* immature embryos to edit *GW2* at three homoeologs (*TaGW2*-B1, -D1, and -A1) with the mutation frequencies of 3.8, 4.1, 4.4% [84].

TABLE 3.3 Biolistics for CRISPR-Cas Reagents into the Different Plants

Plant Name	CRISPR-Cas Vector or RNP Complexes	Targeted Genes	Editing Efficiency	References
CRISPR-Cas Vector				
O. sativa *T. aestivum* L.	pOsU3-sgRNA, pJIT163-2NLSCas9, pAct1-HPT	*PDS* *DEP1* *LOX2*	NA	[75]
Z. mays	pSB11-Ubi:Cas9, U6:gRNA	Liguleless 1 (*LIG1*)	0.2–4.61%	[76]
		Male fertility genes (*Ms26*)	0.13–3.11%	
		Male fertility genes (*Ms45*)	0.47–1.87%	
		Acetolactate synthase (*ALS1, ALS2*)	1.35–2.23%	
G. max	QC810 and RTW830	*DD20*	59%	[77]
	QC799 and RTW831	*DD43*	76%	
O. sativa	pCXUN-Cas9-gRNA1-gRNA2	Acetolactate synthase 1(*ALS1*)	–	[78]
T. aestivum	pGE-sgRNA	*TaGASR7*	2.6–61.9%	[79]
		TaDEP1	0–53.8%	
		TaNAC2	2.0–75%	
		TaPIN1	1–62.5%	
		TaLOX2	9.5–86.8%	
		TdGASR7	0–58.3%	
		TaGW2	0–100%	

TABLE 3.3 *(Continued)*

Plant Name	CRISPR-Cas Vector or RNP Complexes	Targeted Genes	Editing Efficiency	References
Hordeum vulgare	pcasENTRY	*ENGase* (endo-N-acetyl-b-D-glucosaminidase).	78%	[80]
T. aestivum	pE(R4-R3)ZmUbi_ OsCas9_ver3, pTAKN-sg-GR7, and pUba-GFP	*TaGASR7*	5.2%	[81]
O. sativa sp. *japonica*	pCambia1300-CRIPSR-B	*SSU-crtI* (the chloroplast transit peptide from the pea RUBISCO small subunit and the *Erwinia uredovora* carotenoid desaturase).	NA	[82]
		ZmPsy (a maize phytoene synthase).		
RNP Complexes				
Z. mays	RNP complexes.	*LIG*	0.9–9.7%	[83]
		MS26	0.3–2.4%	
		MS45	0.5–3.7%	
T. aestivum L.	CRISPR-Cas9 RNP	*TaGW2-A1*, *-B1*, and *-D1*	3.8, 4.1, 4.4%	[84]
O. sativa L. cv. Kitaake	FnCpf1 and LbCpf1, crRNA, and repairing template.	*OsCOA1* (chlorophyllide-a oxygenase).	32% and 10%	[85]

Interestingly, transiently expressing CRISPR/Cas9 DNA (TECCDNA), a simple and efficient method, has been developed to edit the genome plant with golden particle biolistic. In this study, homozygous mutants of *TaGASR7*, *TaGW2*, and *TaLOX2* in hexaploid bread wheat and *TdGASR7* in tetraploid durum wheat were obtained with no detectable transgenes in the T_0 generation. The percentage of the transgene-free mutant of TaGW2 could be achieved up to 100% [79]. Moreover, a Cpf1 nuclease from *Francisella novicida* and *Lachnospiraceae bacterium* ND2006 has been applied to insert *OsCOA1* (Chlorophyllide-a oxygenase) into the genome of *O. sativa* via homology-directed repair. DNA of plasmid harboring FnCpf1, crRNA, and repairing template were co-precipitated in gold microprojectiles. As result, the frequency of OsCOA1 was obtained with 8%. Note that the use of Cpf1 for biolistic transformation has a great significant effect on improving the editing efficiency [85].

3.6 DELIVERING CRISPR-CAS REAGENTS WITH ADVANTAGES AND DISADVANTAGES

In general, the delivery of CRISPR-Cas reagents can be classified into two major methods: direct and indirect. *Agrobacterium* mediated transformation and viral vector delivery are indirect methods. The indirect methods are introduced CRISPR-Cas reagents into plant cells by using a plasmid system. Besides, the direct methods include PEG-mediated transfection and biolistic bombardment. These methods deliver CRISPR-Cas reagents into the plant cells using chemical and/or physical factors (Figure 3.2). While *Agrobacterium*-mediated transformation add PEG-mediated transfection are commonly used for testing editing efficiency, biolistic bombardment is a promising approach to deliver CRISPR-Cas reagent into plant cells [86]. However, the development of delivery of CRISPR-Cas reagent is still requirement for the plant GE. Each method remains the major barrier for limiting the editing efficiency in plant cells. Firstly, *Agrobacterium*-mediated transformation has been widely used method of the various plant cells. The two main advantages of this method are highly editing efficiency and more cost-effective than other known methods. Another reason that the use of *Agrobacterium* would be advantageous is it would not require labor-intensive and sophisticated equipment. The ability to transfer of relatively large segment of DNA is also the dominant point of *Agrobacterium*-mediated transformation. It allows carrying a large CRISPR-Cas system (> 4.5 kb) into the plant cells [87, 88]. While other delivery methods as bioplastic bombardment introduces multiple copies, *Agrobacterium*-mediated transformation integrates a small number of copies of genes. As result, stable transgene integration is obtained with the high fertility of transgenic plants [77–80]. Although the method has been rapidly applied to the various plant species as a routinely transformed approach, the use of binary vector and remains of incorporates non-target DNA segments into the plant genome can cause a further improvement of *Agrobacterium*-mediated transformation. By contrast, no requirement of the CRISPR-Cas binary vector is the major strength of bombardment-mediated delivery. Therefore, there is no dependence on the size of DNA segments and multiples gene can be introduced into the plant chromosome *via* co-transformation. Due to it is not relative to microbe, the physical process could eliminate the inherent influence of bacteria as *A. tumefaciens* and *A. rhizogenes*. Moreover, Cas9/gRNA RNP complex can be directly and efficiently delivered into plant cells [89]. Bombardment-mediated delivery demonstrated that it not only removes random integration of CRISPR-Cas, but also reduce the effect of off-target into plant chromosome [81, 90]. Noticeably, one of the weakness disadvantages of PGE-mediated delivery is the formation of suspension cells and protoplasts isolation. Therefore, alternative methods, as viral delivery, have been highly developed in recent years. Quickly replication of reagents and stable inheritance of mutation in plant tissue culture leads to application of CRISPR-Cas system in the diversity plants [60–62].

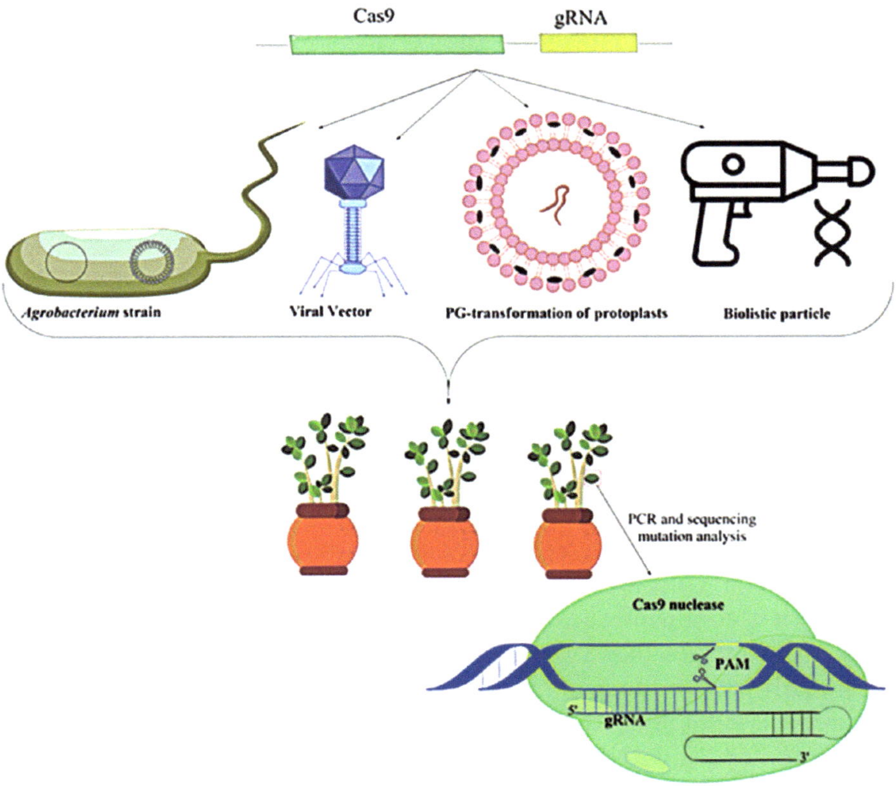

FIGURE 3.2 Schematic representation of delivering CRISPR-Cas reagents into plant cells and their genome editing processes.

3.7 FUTURE PERSPECTIVE

It is believed that various factor effects the efficiency of GE through CRISPR-Cas systems, such as a promoter strength, design of gRNA, and position of target site. However, strategies for delivering CRISPR-Cas reagent into plants are known as the most important point on getting the mutated plants with high efficiency. *Agrobacterium*-mediated delivery of CRISPR-Cas reagents have been successfully reported in *S. tuberosum* and *Z. mays* with 100% editing efficiency [28, 32], while the transformation of protoplasts have been achieved 100% editing efficiency in *C. lanatus* [64]. Moreover, 100% editing efficiency was reported using biolistic methods in *T. aestivum* [79]. Recently, several methods are a promising approach for the delivery of CRISPR-Cas reagents, such as the floral-dip or pollen-tube pathway method, the sequential floral-dip method. The major advantage of both approaches is not required the facility of culturing method. Furthermore, floral-dip-mediated delivery of CRISPR-Cas reagents is also known as straightforward and cost-effective methods. Although the floral-dip method has been successfully reported in various crops, floral-dip-mediated delivery method is still limited to few plants such as *Arabidopsis*, tomato, flax [91, 92]. Interestingly, nanoparticle and magnetofection technologies are being incorporated into many fields of agriculture and medicine. Both technologies have been demonstrated to increase nucleic acid delivery. Nanoparticle and pollen magnetofection-mediated delivery has been applied to carry out the CRISPR-Cas system and/or RNP complexes. The advantage of these approaches are they spend less time on the culture of plant tissue and transgenetic selection. Nanoparticles include magnetic (gold, silver, iron oxide) NPs, bio-inspired (calcium phosphate, chitosan, liposomes) NPs, carbon-based (single-walled carbon nanotubes (CNTs), multiwalled carbon nanoturbes, fullerenes) NPs, silicon-based (silica spheres, mesoporous silica, silicon carbide) NPs, and polymeric ((polyethylene-glycol, poly(lactic-co-glycolic acid), polyethylenimine) NPs [93]. NPs have some major advantages like a wide host range, noncytotoxic, and nonviral. It can easily cross biological membranes into monocot and dicot plant cells. Moreover, the ability to protect and release diverse biomolecules (DNA, RNA, protein, and RNP complex) is one of the most advantages of NPs to delivery in mammalian and plant systems [94, 95]. For example, silicon carbide whisker-mediated transformation has been successfully applied to produce the fertile transgenic maize, cotton, and rice [96, 97]. These advantages of nanoparticle-mediated delivery suggested that CRISPR-Cas and/or RNP complex along with nanoparticles are a promising approach to deliver into regenerative tissues. Recently, pollen magnetofection has been reported as a novel transformation platform to produce transgenic seeds without regeneration [98]. In this system, CRISPR-Cas reagents or RNP complexes combined with magnetic nanoparticles were delivered into pollen under the presence of a magnetic field. As a result, transgenic plants were successfully generated from transformed seeds. These approaches are promising technology owing to it can transform almost all crops and capable of multi-gene transformation [99].

3.8 CONCLUSION

In conclusion, strategies for delivering CRISPR-Cas reagents into plants have a great effect on the efficiency of GE. Although the traditional methods, such as *Agrobacterium*-mediated delivery, viral vector delivery, PEG-mediated delivery, and particle bombardment delivery, have been demonstrated the high efficiency in editing genome of crops, the novel approaches for delivering CRISPR-Cas reagents have highly required. The potential further methods, include nanoparticle mediated delivery and pollen magnetofection mediated delivery, not only improve to the high efficiency of GE but also reduce the time-consuming and laborious tissue culture. It is expected that the strategies for delivering CRISPR-Cas components would be improve the yield of crops.

KEYWORDS

- **agrobacterium-mediated transformation**
- **CRISPR-Cas**
- **magnetofection**
- **nanoparticle**
- **particle bombardment**
- **polyethylene glycol**
- **viral vector delivery**

REFERENCES

1. Tian, S., Jiang, L., Cui, X., Zhang, J., Guo, S., Li, M., Zhang, H., et al., (2018). Engineering herbicide-resistant watermelon variety through CRISPR/Cas9-mediated base-editing. *Plant Cell Rep., 37*, 1353–1356.
2. Aljanabi, S., (2001). Genomics and plant breeding. *Biotechnol. Annu. Rev., 7*, 195–238.
3. Hu, H., & Xiong, L., (2014). Genetic engineering and breeding of drought-resistant crops. *Annu. Rev. Plant Biol., 65*, 715–741.
4. Mao, T., Zhu, M., Sheng, Z., Shao, G., Jiao, G., Mawia, A. M., Ahmad, S., et al., (2021). Effects of grain shape genes editing on appearance quality of erect-panicle Geng/Japonica rice. *Rice (New York, N.Y.), 14*, 74.
5. Liu, W., Yuan, J. S., & Stewart, C. N. Jr., (2013). Advanced genetic tools for plant biotechnology. *Nat. Rev. Genet., 14*, 781–793.
6. Bailey-Serres, J., Parker, J. E., Ainsworth, E. A., Oldroyd, G. E. D., & Schroeder, J. I., (2019). Genetic strategies for improving crop yields. *Nature., 575*, 109–118.
7. Monsur, M. B., Shao, G., Lv, Y., Ahmad, S., Wei, X., Hu, P., & Tang, S., (2020). Base editing: The ever expanding clustered regularly interspaced short palindromic repeats (CRISPR) tool kit for precise genome editing in plants. *Genes., 11*, 466.
8. Ahmad, S., Shahzad, R., Jamil, S., Tabassum, J., Chaudhary, M. A. M., Atif, R. M., Iqbal, M. M., et al., (2021). Chapter 31 – regulatory aspects, risk assessment, and toxicity associated with RNAi and CRISPR methods. In nanobiotechnology for plant protection. *CRISPR and RNAi Systems,* 687–721. Elsevier.
9. Ahmad, S., Sheng, Z. H., Jalal, R. S., Tabassum, J., Farah, K., Ahmed, F. K., Hu, S., et al., (2021). Chapter 33 – CRISPR–Cas technology towards improvement of abiotic stress tolerance in plants. In nanobiotechnology for plant protection. *CRISPR and RNAi Systems,* 755–772. Elsevier.
10. Katoch, R., & Thakur, N., (2013). RNA interference: A promising technique for the improvement of traditional crops. *Int. J. Food Sci. Nutr., 64*, 248–259.
11. Kim, D., Alptekin, B., & Budak, H., (2018). CRISPR/Cas9 genome editing in wheat. *Funct. Integr. Genomics., 18*, 31–41.
12. Ahmad, S., Wei, X. J., Sheng, Z. H., Hu, P. S., & Tang, S. Q., (2020). CRISPR/Cas9 for development of disease resistance in plants: Recent progress, limitations, and future prospects. *Brief. Funct. Genomics, 19*, 26–39.
13. Tabassum, J., Ahmad, S., Hussain, B., Mawia, A. M., Zeb, A., & Ju, L., (2021). Applications and potential of genome-editing systems in rice improvement: Current and future perspectives. *Agron., 11*, 1359.
14. Manghwar, H., Lindsey, K., Zhang, X., & Jin, S., (2019). CRISPR/Cas system: Recent advances and future prospects for genome editing. *Trends Plant Sci., 24*, 1102–1125.
15. Bhat, M. A., Kumar, V., Wani, I. A., Bashir, H., Shah, A. A., Rahman, S., & Jan, A. T., (2020). The era of editing plant genomes using CRISPR/Cas: A critical appraisal. *J. Biotechnol., 324*, 34–60.
16. Ahmad, S., Tang, L., Shahzad, R., Mawia, A. M., Rao, G. S., Jamil, S., Wei, C., et al., (2021). CRISPR-based crop improvements: A way forward to achieve zero hunger. *J. Agric. Food Chem., 69*, 8307–8323.
17. Adli, M., (2018). The CRISPR tool kit for genome editing and beyond. *Nat. Commun., 9*, 1911.

18. Makarova, K. S., Wolf, Y. I., Iranzo, J., Shmakov, S. A., Alkhnbashi, O. S., Brous, S. J. J., Charpentier, E., et al., (2020). Evolutionary classification of CRISPR-Cas systems: A burst of class 2 and derived variants. *Nat. Rev. Microbiol., 18*, 67–83.
19. Cho, S., Shin, J., & Cho, B. K., (2018). Applications of CRISPR/Cas system to bacterial metabolic engineering. *Int. J. Mol. Sci., 19*, 1089.
20. Fiaz, S., Ahmad, S., Noor, M. A., Wang, X., Younas, A., Riaz, A., Riaz, A., & Ali, F., (2019). Applications of the CRISPR/Cas9 System for rice grain quality improvement: Perspectives and opportunities. *Int. J. Mol. Sci., 20*, 888.
21. Sandhya, D., Jogam, P., Allini, V. R., Abbagani, S., & Alok, A., (2020). The present and potential future methods for delivering CRISPR/Cas9 components in plants. *J. Genet. Eng. Biotechnol., 18*, 25.
22. Lee, L. Y., & Gelvin, S. B., (2008). T-DNA binary vectors and systems. *Plant Physiol., 146*, 325–332.
23. Montecillo, J. A. V., Chu, L. L., & Bae, H. H., (2020). CRISPR-Cas9 system for plant genome editing: Current approaches and emerging developments. *Agron., 10*, 1033.
24. Zhu, H., Li, C., & Gao, C., (2020). Applications of CRISPR-Cas in agriculture and plant biotechnology. *Nat. Rev. Mol. Cell Biol., 21*, 782.
25. Kuluev, B. R., Gumerova, G. R., Mikhaylova, E. V., Gerashchenkov, G. A., Rozhnova, N. A., Vershinina, Z. R., Khyazev, A. V., et al., (2019). Delivery of CRISPR/Cas components into higher plant cells for genome editing. *Russ. J. Plant Physiol., 66*, 694–706.
26. Gelvin, S. B., (2003). *Agrobacterium*-mediated plant transformation: The biology behind the "gene-jockeying" tool. *Microbiol Mol. Biol. Rev., 67*, 16–37.
27. Li, J. F., Norville, J. E., Aach, J., McCormack, M., Zhang, D., Bush, J., Church, G. M., & Sheen, J., (2013). Multiplex and homologous recombination-mediated genome editing in *Arabidopsis* and *Nicotiana benthamiana* using guide RNA and Cas9. *Nat. Biotechnol., 31*, 688–691.
28. Feng, Z., Zhang, B., Ding, W., Liu, X., Yang, D. L., Wei, P., Cao, F., et al., (2013). Efficient genome editing in plants using a CRISPR/Cas system. *Cell Res., 23*, 1229–1232.
29. Upadhyay, S. K., Kumar, J., Alok, A., & Tuli, R., (2013). RNA-guided genome editing for target gene mutations in wheat. *G3 (Bethesda), 3*, 2233–2238.
30. Gao, J., Wang, G., Ma, S., Xie, X., Wu, X., Zhang, X., Wu, Y., et al., (2015). CRISPR/Cas9-mediated targeted mutagenesis in *Nicotiana tabacum*. *Plant Mol. Biol., 87*, 99–110.
31. Butler, N. M., Atkins, P. A., Voytas, D. F., & Douches, D. S., (2015). Generation, and inheritance of targeted mutations in potato (*Solanum tuberosum* L.) using the CRISPR/Cas system. *PLoS One, 10*, e0144591.
32. Chandrasekaran, J., Brumin, M., Wolf, D., Leibman, D., Klap, C., Pearlsman, M., Sherman, A., et al., (2016). Development of broad virus resistance in non-transgenic cucumber using CRISPR/Cas9 technology. *Mol. Plant Pathol., 17*, 1140–1153.
33. Mikami, M., Toki, S., & Endo, M., (2016). Precision targeted mutagenesis via Cas9 paired nickases in rice. *Plant Cell Physiol., 57*, 1058–1068.
34. Char, S. N., Neelakandan, A. K., Nahampun, H., Frame, B., Main, M., Spalding, M. H., Becraft, P. W., et al., (2017). An *Agrobacterium*-delivered CRISPR/Cas9 system for high-frequency targeted mutagenesis in maize. *Plant Biotechnol. J., 15*, 257–268.

35. Kaur, N., Alok, A., Shivani, K. N., Pandey, P., Awasthi, P., & Tiwari, S., (2018). CRISPR/Cas9-mediated efficient editing in phytoene desaturase (PDS) demonstrates precise manipulation in banana cv. Rasthali genome. *Funct. Integr. Genomics., 18*, 89–99.
36. Wang, Z., Wang, S., Li, D., Zhang, Q., Li, L., Zhong, C., Liu, Y., & Huang, H., (2018). Optimized paired-sgRNA/Cas9 cloning and expression cassette triggers high-efficiency multiplex genome editing in kiwifruit. *Plant Biotechnol. J., 16*, 1424–1433.
37. Sugano, S. S., Nishihama, R., Shirakawa, M., Takagi, J., Matsuda, Y., Ishida, S., Shimada, T., et al., (2018). Efficient CRISPR/Cas9-based genome editing and its application to conditional genetic analysis in *Marchantia polymorpha. PLoS One, 13*, e0205117.
38. Li, X., Wang, Y., Chen, S., Tian, H., Fu, D., Zhu, B., Luo, Y., & Zhu, H., (2018). Lycopene is enriched in tomato fruit by CRISPR/Cas9-mediated multiplex genome editing. *Front. Plant Sci., 9*, 559.
39. Shufen, C., Yicong, C., Baobing, F., Guiai, J., Zhonghua, S., Ju, & Shaoqing, et al., (2019). Editing of rice isoamylase gene *ISA1* provides insights into its function in starch formation. *Rice Sci., 26*, 77–87.
40. Wang, L., Chen, S., Peng, A., Xie, Z., He, Y., & Zou, X., (2019). CRISPR/Cas9-mediated editing of *CsWRKY22* reduces susceptibility to *Xanthomonas citri* subsp. *citri* in wanjincheng orange (*Citrus sinensis* (L.) Osbeck). *Plant Biotechnol. Rep., 13*, 501–510.
41. Bari, V. K., Nassar, J. A., Kheredin, S. M., Gal-On, A., Ron, M., Britt, A., Steele, D et al., (2019). CRISPR/Cas9-mediated mutagenesis of carotenoid cleavage dioxygenase 8 in tomato provides resistance against the parasitic weed *Phelipanche aegyptiaca. Sci. Rep., 9*, 11438.
42. Bandyopadhyay, A., Kancharla, N., Javalkote, V. S., Dasgupta, S., & Brutnell, T. P., (2020). CRISPR-Cas12a (Cpf1): A versatile tool in the plant genome editing tool box for agricultural advancement. *Front. Plant Sci., 11*, 584151.
43. Lee, K., Zhang, Y., Kleinstiver, B. P., Guo, J. A., Aryee, M. J., Miller, J., Malzahn, A., et al., (2019). Activities and specificities of CRISPR/Cas9 and Cas12a nucleases for targeted mutagenesis in maize. *Plant Biotechnol. J., 17*, 362–372.
44. Kaur, N., Alok, A., Shivani, Kumar, P., Kaur, N., Awashi, P., Chaturvedi, S., et al., (2020). CRISPR/Cas9 directed editing of lycopene epsilon-cyclase modulates metabolic flux for β-carotene biosynthesis in banana fruit. *Metab. Eng., 59*, 76–86.
45. Kereszt, A., Li, D., Indrasumunar, A., Nguyen, C. D., Nontachaiyapoom, S., Kinkema, M., & Gresshoff, P. M., (2007). *Agrobacterium rhizogenes*-mediated transformation of soybean to study root biology. *Nat. Protoc., 2*, 948–952.
46. Jacobs, T. B., & Martin, G. B., (2016). High-throughput CRISPR vector construction and characterization of DNA modifications by generation of tomato hairy roots. *J. Vis. Exp., 110*, 53843.
47. Cui, M. L., Liu, C., Piao, C. L., & Liu, C. L., (2020). A stable *Agrobacterium rhizogenes*-mediated transformation of cotton (*Gossypium hirsutum* L.) and plant regeneration from transformed hairy root via embryogenesis. *Front. Plant Sci., 11*, 604255.
48. Li, X., Yang, Q., Peng, L., Tu, H., Lee, L. Y., Gelvin, S. B., & Pan, S. Q., (2020). *Agrobacterium*-delivered VirE2 interacts with host nucleoporin CG1 to facilitate the nuclear import of VirE2-coated T complex. *Proc. Natl. Acad. Sci. U S A., 117*, 26389–26397.

49. Xu, S., Lai, E., Zhao, L., Cai, Y., Ogutu, C., Cherono, S., Han, Y., & Zheng, B., (2020). Development of a fast and efficient root transgenic system for functional genomics and genetic engineering in peach. *Sci. Rep., 10*, 2836.
50. Jacobs, T. B., LaFayette, P. R., Schmitz, R. J., & Parrott, W. A., (2015). Targeted genome modifications in soybean with CRISPR/Cas9. *BMC Biotechnol., 15*, 16.
51. Fan, D., Liu, T., Li, C., Jiao, B., Li, S., Hou, Y., & Luo, K., (2015). Efficient CRISPR/Cas9-mediated targeted mutagenesis in populus in the first generation. *Sci. Rep., 5*, 12217.
52. Michno, J. M., Wang, X., Liu, J., Curtin, S. J., Kono, T. J., & Stupar, R. M., (2015). CRISPR/Cas mutagenesis of soybean and *Medicago truncatula us*ing a new web-tool and a modified Cas9 enzyme. *GM Crops Food., 6*, 243–252.
53. Wang, L., Wang, L., Tan, Q., Fan, Q., Zhu, H., Hong, Z., Zhang, Z., & Duanmu, D., (2016). Efficient inactivation of symbiotic nitrogen fixation related genes in *Lotus japonicus* using CRISPR-Cas9. *Front. Plant Sci., 7*, 1333.
54. Meng, Y., Hou, Y., Wang, H., Ji, R., Liu, B., Wen, J., Niu, L., & Lin, H., (2017). Targeted mutagenesis by CRISPR/Cas9 system in the model legume *Medicago truncatula. Plant Cell Rep., 36*, 371–374.
55. Li, B., Cui, G., Shen, G., Zhan, Z., Huang, L., Chen, J., & Qi, X., (2017). Targeted mutagenesis in the medicinal plant *Salvia miltiorrhiza. Sci. Rep., 7*, 43320.
56. Kirchner, T. W., Niehaus, M., Debener, T., Schenk, M. K., & Herde, M., (2017). Efficient generation of mutations mediated by CRISPR/Cas9 in the hairy root transformation system of *Brassica carinata. PLoS One, 12*, e0185429.
57. Zhou, Z., Tan, H., Li, Q., Chen, J., Gao, S., Wang, Y., Chen, W., & Zhang, L., (2018). CRISPR/Cas9-mediated efficient targeted mutagenesis of RAS in *Salvia miltiorrhiza. Phytochemistry, 148*, 63–70.
58. Gil-Humanes, J., Wang, Y., Liang, Z., Shan, Q., Ozuna, C. V., Sanchez-Leon, S., Baltes, N. J., et al., (2017). High-efficiency gene targeting in hexaploid wheat using DNA replicons and CRISPR/Cas9. *Plant J., 89*, 1251–1262.
59. Wang, M., Lu, Y., Botella, J. R., Mao, Y., Hua, K., & Zhu, J. K., (2017). Gene targeting by homology-directed repair in rice using a geminivirus-based CRISPR/Cas9 system. *Mol. Plant, 10*, 1007–1010.
60. Yin, K., Han, T., Liu, G., Chen, T., Wang, Y., Yu, A. Y. L., & Liu, Y., (2015). A geminivirus-based guide RNA delivery system for CRISPR/Cas9 mediated plant genome editing. *Sci. Rep., 5*, 14926.
61. Ali, Z., Abul-Faraj, A., Piatek, M., & Mahfouz, M. M., (2015). Activity and specificity of TRV-mediated gene editing in plants. *Plant Signal Behav., 10*, e1044191.
62. Ma, X., Zhang, X., Liu, H., & Li, Z., (2020). Highly efficient DNA-free plant genome editing using virally delivered CRISPR-Cas9. *Nat. Plants, 6*, 773–779.
63. Xie, K., & Yang, Y., (2013). RNA-guided genome editing in plants using a CRISPR-Cas system. *Mol. Plant, 6*, 1975–1983.
64. Tian, S., Jiang, L., Gao, Q., Zhang, J., Zong, M., Zhang, H., Ren, Y., et al., (2017). Efficient CRISPR/Cas9-based gene knockout in watermelon. *Plant Cell Rep., 36*, 399–406.
65. Liang, Z., Zhang, K., Chen, K., & Gao, C., (2014). Targeted mutagenesis in *Zea mays* using TALENs and the CRISPR/Cas system. *J. Genet. Genomics, 41*, 63–68.
66. Sun, X., Hu, Z., Chen, R., Jiang, Q., Song, G., Zhang, H., & Xi, Y., (2015). Targeted mutagenesis in soybean using the CRISPR-Cas9 system. *Sci. Rep., 5*, 10342.

67. Collonnier, C., Epert, A., Mara, K., Maclot, F., Guyon-Debast, A., Charlot, F., White, C., et al., (2017). CRISPR-Cas9-mediated efficient directed mutagenesis and RAD51-dependent and RAD51-independent gene targeting in the moss *Physcomitrella patens*. *Plant Biotechnol. J., 15*, 122–131.
68. Arndell, T., Sharma, N., Langridge, P., Baumann, U., Watson-Haigh, N. S., & Whitford, R., (2019). gRNA validation for wheat genome editing with the CRISPR-Cas9 system. *BMC Biotechnol., 19*, 71.
69. Wu, S., Zhu, H., Liu, J., Yang, Q., Shao, X., Bi, F., Hu, C., et al., (2020). Establishment of a PEG-mediated protoplast transformation system based on DNA and CRISPR/Cas9 ribonucleoprotein complexes for banana. *BMC Plant Biol., 20*, 425.
70. Woo, J. W., Kim, J., Kwon, S. I., Corvalan, C., Cho, S. W., Kim, H., Kim, S. G., et al., (2015). DNA-free genome editing in plants with preassembled CRISPR-Cas9 ribonucleoproteins. *Nat. Biotechnol., 33*, 1162–1164.
71. Malnoy, M., Viola, R., Jung, M. H., Koo, O. J., Kim, S., Kim, J. S., Velasco, R., & Nagamangala, K. C., (2016). DNA-free genetically edited grapevine and apple protoplast using CRISPR/Cas9 ribonucleoproteins. *Front. Plant Sci., 7*, 1904.
72. Murovec, J., Gucek, K., Bohanec, B., Avbelj, M., & Jerala, R., (2018). DNA-Free genome editing of *Brassica oleracea* and *B. rapa* protoplasts using CRISPR-Cas9 ribonucleoprotein complexes. *Front. Plant Sci., 9*, 1594.
73. Kim, H., Kim, S. T., Ryu, J., Kang, B. C., Kim, J. S., & Kim, S. G., (2017). CRISPR/Cpf1-mediated DNA-free plant genome editing. *Nat. Commun., 8*, 14406.
74. Xu, J., Kang, B. C., Naing, A. H., Bae, S. J., Kim, J. S., Kim, H., & Kim, C. K., (2020). CRISPR/Cas9-mediated editing of 1-aminocyclopropane-1-carboxylate oxidase1 enhances *Petunia flower* longevity. *Plant Biotechnol. J., 18*, 287–297.
75. Shan, Q., Wang, Y., Li, J., & Gao, C., (2014). Genome editing in rice and wheat using the CRISPR/Cas system. *Nat. Protoc., 9*, 2395–2410.
76. Svitashev, S., Young, J. K., Schwartz, C., Gao, H., Falco, S. C., & Cigan, A. M., (2015). Targeted mutagenesis, precise gene editing, and site-specific gene insertion in maize using Cas9 and guide RNA. *Plant Physiol., 169*, 931–945.
77. Li, Z., Liu, Z. B., Xing, A., Moon, B. P., Koellhoffer, J. P., Huang, L., Ward, R. T., et al., (2015). Cas9-Guide RNA directed genome editing in soybean. *Plant Physiol., 169*, 960–970.
78. Sun, Y., Zhang, X., Wu, C., He, Y., Ma, Y., Hou, H., Guo, X., et al., (2016). Engineering herbicide-resistant rice plants through CRISPR/Cas9-mediated homologous recombination of acetolactate synthase. *Mol. Plant, 9*, 628–631.
79. Zhang, Y., Liang, Z., Zong, Y., Wang, Y., Liu, J., Chen, K., Qiu, J. L., & Gao, C., (2016). Efficient and transgene-free genome editing in wheat through transient expression of CRISPR/Cas9 DNA or RNA. *Nat. Commun., 7*, 12617.
80. Kapusi, E., Corcuera-Gomez, M., Melnik, S., & Stoger, E., (2017). Heritable genomic fragment deletions and small indels in the putative ENGase gene induced by CRISPR/Cas9 in barley. *Front. Plant Sci., 8*, 540.
81. Hamada, H., Liu, Y., Nagira, Y., Miki, R., Taoka, N., & Imai, R., (2018). Biolistic-delivery-based transient CRISPR/Cas9 expression enables in planta genome editing in wheat. *Sci. Rep., 8*, 14422.
82. Dong, O. X., Yu, S., Jain, R., Zhang, N., Duong, P. Q., Butler, C., Li, Y., et al., (2020). Marker-free carotenoid-enriched rice generated through targeted gene insertion using CRISPR-Cas9. *Nat. Commun., 11*, 1178.

83. Svitashev, S., Schwartz, C., Lenderts, B., Young, J. K., & Mark, C. A., (2016). Genome editing in maize directed by CRISPR-Cas9 ribonucleoprotein complexes. *Nat. Commun., 7*, 13274.
84. Liang, Z., Chen, K., Li, T., Zhang, Y., Wang, Y., Zhao, Q., Liu, J., et al., (2017). Efficient DNA-free genome editing of bread wheat using CRISPR/Cas9 ribonucleoprotein complexes. *Nat. Commun., 8*, 14261.
85. Begemann, M. B., Gray, B. N., January, E., Gordon, G. C., He, Y., Liu, H., Wu, X., et al., (2017). Precise insertion and guided editing of higher plant genomes using Cpf1 CRISPR nucleases. *Sci. Rep., 7*, 11606.
86. Lino, C. A., Harper, J. C., Carney, J. P., & Timlin, J. A., (2018). Delivering CRISPR: A review of the challenges and approaches. *Drug Deliv., 25*, 1234–1257.
87. Song, G. Q., Prieto, H., & Orbovic, V., (2019). *Agrobacterium*-Mediated transformation of tree fruit crops: Methods, progress, and challenges. *Front. Plant Sci., 10*, 226.
88. Sabbadini, S., Capriotti, L., Molesini, B., Pandolfini, T., Navacchi, O., Limera, C., Ricci, A., & Mezzetti, B., (2019). Comparison of regeneration capacity and *Agrobacterium*-mediated cell transformation efficiency of different cultivars and rootstocks of *Vitis* spp. via *organogenesis*. *Sci. Rep., 9*, 582.
89. Altpeter, F., Baisakh, N., Beachy, R., Bock, R., Capell, T., Christou, P., Daniell, H., et al., (2005). Particle bombardment and the genetic enhancement of crops: Myths and realities. *Mol. Breeding, 15*, 305–327.
90. Liang, Z., Chen, K., Zhang, Y., Liu, J., Yin, K., Qiu, J. L., & Gao, C., (2018). Genome editing of bread wheat using biolistic delivery of CRISPR/Cas9 *in vitro* transcripts or ribonucleoproteins. *Nat. Protoc., 13*, 413–430.
91. Hu, D., Bent, A. F., Hou, X., & Li, Y., (2019). *Agrobacterium*-mediated vacuum infiltration and floral dip transformation of rapid-cycling *Brassica rapa*. *BMC Plant Biol., 19*, 246.
92. Bastaki, N. K., & Cullis, C. A., (2014). Floral-dip transformation of flax (*Linum usitatissimum*) to generate transgenic progenies with a high transformation rate. *J. Vis. Exp., 94*, 52189.
93. Cunningham, F. J., Goh, N. S., Demirer, G. S., Matos, J. L., & Landry, M. P., (2018). Nanoparticle-mediated delivery towards advancing plant genetic engineering. *Trends Biotechnol., 36*, 882–897.
94. Wilbie, D., Walther, J., & Mastrobattista, E., (2019). Delivery aspects of CRISPR/Cas for *in vivo* genome editing. *Acc. Chem. Res., 52*, 1555–1564.
95. Lv, Z., Jiang, R., Chen, J., & Chen, W., (2020). Nanoparticle-mediated gene transformation strategies for plant genetic engineering. *Plant J., 104*, 880–891.
96. Frame, B. R., Drayton, P. R., Bagnall, S. V., Lewnau, C. J., Bullock, W. P., Wilson, H. M., Dunwell, J. M., et al., (1994). Production of fertile transgenic maize plants by silicon carbide whisker-mediated transformation. *Plant J., 6*, 941–948.
97. Matsushita, J., Otani, M., Wakita, Y., Tanaka, O., & Shimada, T., (1999). Transgenic plant regeneration through silicon carbide whisker-mediated transformation of rice (*Oryza sativa* L.). *Breed. Sci., 49*, 21–26.
98. Zhao, X., Meng, Z., Wang, Y., Chen, W., Sun, C., Cui, B., Cui, J., et al., (2017). Pollen magnetofection for genetic modification with magnetic nanoparticles as gene carriers. *Nat. Plants, 3*, 956–964.

99. Lew, T. T. S., Sarojam, R., Jang, I. C., Park, B. S., Naqvi, N. I., Wong, M. H., Singh, G. P., et al., (2020). Species-independent analytical tools for next-generation agriculture. *Nat. Plants*, *6*, 1408–1417.

CHAPTER 4

HARNESSING THE POTENTIAL OF CRISPR TECHNOLOGIES ACROSS THE FOOD SUPPLY CHAIN

SAMEEN SHAHID,[1] FIDAA ASSLAM,[2] and AYESHA IMRAN[2]

[1]*Center of Applied Molecular Biology (CAMB), Punjab University, Lahore, Pakistan*

[2]*National Center of Excellence in Molecular Biology (CEMB), Punjab University, Lahore, Pakistan*

ABSTRACT

Over the past decade, the dawn of CRISPR-based technology has revolutionized genetics, and genome editing (GE) is now commonly applied for numerous medical and agricultural uses, such as addressing genetic abnormalities and enhancing the production of crops and livestock. There is also a tremendous opportunity for CRISPR-based technology to modify the genomic content of food microorganisms to regulate the structure and action of bacterial species in the food supply chain, from farming to customer goods. It is of huge economic significance to progress the food supply chain since it needs enhancing fermentation processes to increase the flavor and physical properties of food items, along with enhancing food quality and protection by the management of spoilage bacteria and pathogens. Here, we discuss the different technologies of CRISPR that can change bacterial functionalities and modify the composition of food microbial communities. We demonstrate how these applications could be used to control microbiota that contain disadvantaged and morbific bacteria in the food supply chain, as well as desirable cell culture and probiotics that improve health.

CRISPR-Based Gene Editing: A Solution for Global Food Security. Shakeel Ahmad, Asif Ali Khan, and Muhammad Arshad Javed (Eds.)

4.1 INTRODUCTION

Few decades ago, CRISPR has revolutionized genetics, and genome editing (GE) is now extensively used for a variety of health and agronomic purposes, including the correction of genetic diseases and crop and livestock breeding. CRISPR-based technologies have the ability to change the genetic content of food bacteria, allowing for better control of the composition and function of microbes all over the food supply chain, from farm to fork. Upgrading the food supply chain is critical for society since it entails optimizing fermentation processes to advance food eminence and safety by controlling spoilage bacteria and pathogens, as well as increasing food quality and safety by controlling spoilage bacteria and pathogens [1]. Figure 4.1 shows application of CRISPR/Cas that has revolutionized the world in food supply chain.

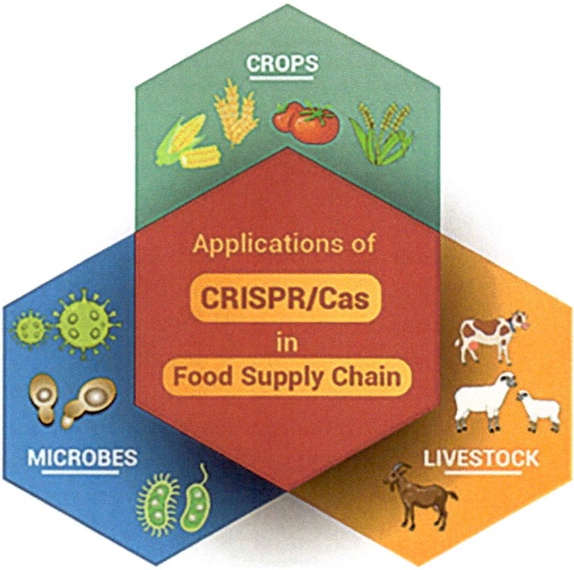

FIGURE 4.1 CRISPR/Cas based technologies and its application in food supply chain.

4.1.1 CROPS

Globally the population has been increasing rapidly and by 2050 it would reach up to 9.6 billion and along with it the demand of the staple crop will also increase [2]. Because the rate of increase in yields induced by the green revolution has been slowly decreasing and unfavorable climate change is projected to further restrict the development of plants, it is critical to grow cultivars that have more tolerance to harsh environments, higher yields, and better efficiency. The traditional crop breeding techniques, however, are labor-intensive, time-consuming, and complex. For this purpose, more efficient and time-saving breeding methods are required [3].

Agronomic crops are imperative because they are the largest crops from them 75–85% of the calories and 65–70% of the proteins are taken. However, plants are negatively impacted by the changing climate. Health and nutritional shortages arising from the epidemic of numerous abiotic and biotic stresses. To meet the 2050 target, models have predicted that a 2.4% annual rise in crop yield is required [4]. Unfortunately, according to a 2013 report that showed none of the world's top-produced crops are similar to this amount anywhere. This lack of increase in crop production has several contributing factors.

With rapid developments in sequencing technology, genomic knowledge is being accessible on an ever-increasing number of plant organisms, and GE systems are providing the ability to reliably edit genes and generate new prospects for crop improvement. The basic GE technique is using a sequence-specific nuclease (SSNs) to cause a double-strand break (DSB) of DNA at the target sites. Early SSNs have shown to be efficient for plant GE, like meganucleases (MNs) [5], zinc-finger nucleases [6], transcription activator-like effector nucleases (TALENs) [7] and clustered regularly interspaced short palindromic repeats (CRISPRs) but their architecture involves complex protein engineering, which restricts their applicability [8].

4.1.2 ADVANTAGE OF CRISPR IN PLANT BIOTECHNOLOGY

In order to benefit from CRISPR, in crop genetic modification the reagents of CRISPR are applied to plant cells in the form of DNA, RNA, or ribonucleoprotein (RNP) to cut plant DNA in a particular sequence. The plant cell has to restore the break in order to maintain genome integrity, and this leads to the incorporation of various forms of mutation in the targeted gene.

CRISPR-Cas9 is relatively inexpensive and easy to deal with. CRISPR-Cas9 technology has been commonly used to alter the genome of plants (more than 20 crops) to change many characteristics such as yield, promote growth and improve resistance to abiotic/biotic stresses. To date, the application of CRISPR technology in crop enhancement has concentrated on improving crop yields, efficiency and stress tolerance that could be accomplished through simple knockout of one or more genes that grant undesirable characteristics [9].

Knocking out *Gn1a*, *DEP1*, and *GS3* in rice, for instance, resulted in increased grain number, thick stiff panicles, and greater grain size [10]. *OsCYP71A1* gene disruption blocked serotonin biosynthesis and raised amounts of salicylic acid in rice plants considerably. This rice plants with higher salicylic content demonstrated high tolerance to plant harvesters and stem borers [11]. Rice blast is a devastating fungal disease that now in rice-growing countries lowers more than 20% of yields. Using CRISPR-Cas technology, Knocking-out *OsERF922*, an ethylene sensitive factor transcription factor gene [12], endows both blast and salinity tolerance in rice. *Xanthomonas oryzae pv. oryzae* causes bacterial blight in rice. Deactivation of the *OsSWEET13* promoter prompted the development of bacterial blight disease tolerant rice plants [13]. Disabling the waxy gene *Wx1* in maize resulted in high amylopectin content with enhanced protein content that has the potential to be marketed, there are special properties of this resulting rice phenotype that is best approach for use in the food industry; and destroying powdery mildew-resistant wheat and tomato developed by the *MLO allele* [14]. Recently, gene knockout controlled by CRISPR has been used to develop heterosis which is typically lost due to genetic segregation in successive generations. A genotype called *MiMe* in rice (Mitosis rather than Meiosis) was created by identifying crucial Meiosis-related genomes, and haploid plants, have also been developed using technology with CRISPR. The clonal progeny-maintained genome-wide paternal heterozygosity when *MiMe* was paired with haploidy in hybrid rice, showing the viability of asexual reproduction by seed propagation in crops [15].

A while back, CRISPR technology has been used to tranquilize *Solanum pimpinellifolium*, a wild tomato that is surprisingly stress-tolerant but low in fruit production. In one study 6 loci that are critical for production and yield were tested, and increase in fruit size, fruit number and fruit lycopene accumulation were seen in the engineered lines [16].

Low-gluten, transgene-free wheat is another example. It was derived from *alpha-gliadin* genes by multiplex knockout. Gluten proteins are present in wheat grains. Among these, the alpha-gliadin family is the primary protein group associated with coeliac disease and non-coeliac gluten growth.

High immunogenic wheat varieties have not been reached by conventional mutagenesis and plant breeding for the individuals with coeliac [17]. The technology of CRISPR/Cas9 was used to decrease the volume of alpha-gliadins reliably and effectively. It may be used to produce food products with minimal gluten and serving as a source to incorporate this characteristic into exclusive crops such as wheat.

In case of tomato, the use of NHEJ in CRISPR was used for the deletion of the autoinhibitory domain by the insertion of a stop codon, *GAD2* and *GAD3* genes encoding a central enzyme in *γ-aminobutyric acid* biosynthesis, leading to an increased level of *γ-aminobutyric acid GABA* (non-proteinogenic amino acid that has hypotensive impact) in tomatoes. These tomatoes may be helpful for the treatment of patients with moderate, elevated, or high average blood pressure [18].

Orphan plants, such as sweet potatoes, groundnuts, cassava, bananas, etc. Quinoa and quinoa are essential crops of high nutritional value locally with good nutritive adaptations and qualities. Nonetheless, considering their broad capacity for improving the protection of food and nutrition, the undesirable characteristics (such as low yield, sprawling growth and fruit drop) avoid larger planting of orphan crops. CRISPR technology provides a powerful tool for accelerating the domestication of orphan crops, which is inexpensive, fast, accurate, and capable of editing multiple sites and modifying gene regulation. It has recently been used to target genes that regulate the architecture of plants, flower production and fruit production. Groundcherry, a semi-domesticated orphan crop, and *modified* plants displayed enhanced domestication traits [9, 19].

4.1.3 DISADVANTAGE OF CRISPR IN PLANT BIOTECHNOLOGY

Even though CRISPR/Cas9 system is sturdy and straightforward, it has the restriction of only being able to recognize sequences upstream of the proper 5′-NGG-3′ PAMs. This limits its capacity to locate additional potential targets. Other CRISPR/Cas9 variations were required to circumvent this limitation. The CRISPR/Cpf1 technology is most commonly utilized to overcome such problems. In contrast to Cas9's blunt endings, Cpf1 generates sticky ends with 4–5 nucleotide overhangs by recognizing T-rich PAM [20]. CRISPR/Cas9 and CRISPR/Cpf1 engineering systems can both be used with DNA-free GE and/or base editing. Crop improvement has made significant use of the CRISPR/Cpf technology. Rice and tobacco were mutated using *Francisella novicida* Cpf1 (FnCpf1) engineered mutagenesis [21] (Figure 4.2).

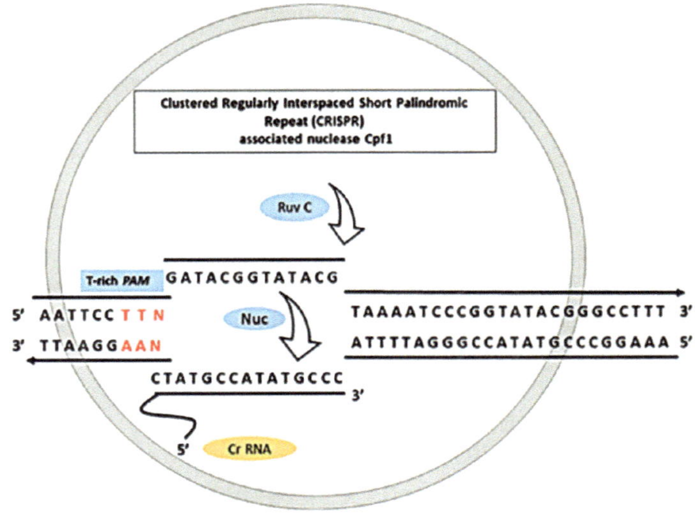

FIGURE 4.2 Plant genome editing using CRISPR/Cpf1 is depicted schematically. The Nuc and RuvC domains of the Cpf1 nuclease target complementary and non-complementary DNA, respectively, to create a sticky end DSB.

Source: Adapted from Ref. [22]

4.2 MICROBES

Food matrices are the ideal host for a variety of microbiota due to the presence of balanced substrates. This microbiota consists of beneficial and detrimental microbes resulting in both positive and negative outcomes. The expedient microbes play a fundamental role in food processing and preservation processes such as enhancing nutritional content, inhibiting pathogens, administering health benefits by adding valuable live microbes to the host gut system, increasing organoleptic characteristics of food, etc. [23]. Concurrently, the advantageous bacteria are used as probiotics and as primary starter cultures in the fermentation process to produce cheese, yogurt, bread, etc. Notwithstanding, the detrimental microbes result in life threatening foodborne diseases and food spoilage [24]. The reports estimated that more than a quarter of food spoilage around the globe is due to microbial contamination hence creating a financial burden on the world's economy. Though fermentation practices have been utilized for a millennium, still the fermentation failures have been documented due to the bacteriophage-mediated spoilage compromising the product quality, composition, and cultural purity. Hence, the dire need to ensure food safety, food security, and food longevity could only be possible after managing this microbiota throughout the food supply chain. This forced the food technology to struggle for unwittingly managing both advantageous and disadvantageous microbiota concomitantly to find a promising technology to overcome this emerging threat [25].

Recently, advances in microbiology have upgraded the industrial abilities to unravel the role of microbiota in palatable and healthy food products. The potential of the "CRISPR craze" has revolutionized the field of food biotechnology. This groundbreaking tool has many applications in the food supply chain such as strain typing, GE, vaccination, phage resistance, and anti-microbial activity [25] (Figure 4.3).

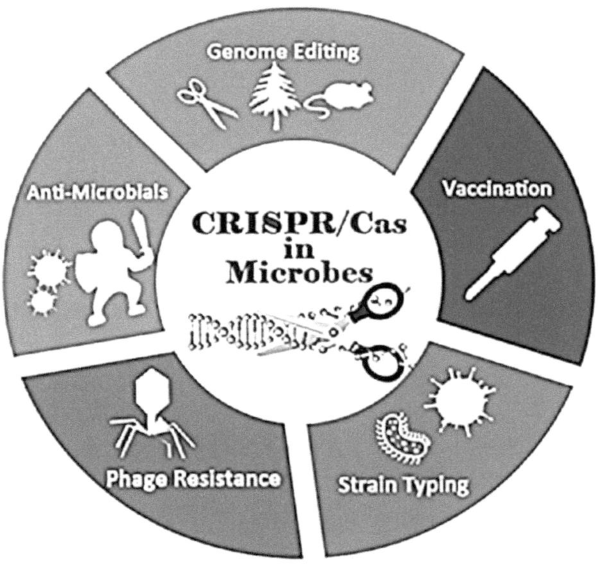

FIGURE 4.3 Applications of CRISPR/Cas in microbes (Figure 4.1 shows the application of CRISPR/Cas in microbes such as genome editing, vaccination, strain typing, phage resistance, and anti-microbials which are explained in subsections in detail).

4.2.1 STRAIN TYPING

The peculiar feature of the CRISPR/Cas system is the acquisition of novel spacers in the CRISPR array with time as it encounters the threats of invasive mobile elements. These unique sequences acquired in the hyper variable region provide a valuable basis for tracing the evolutionary trajectory, path, and origin of the isolates [26]. By exploring these novel spacer sequences from the arrays of the host genome, we can get geographical, environmental, and historical insights into host-viral interplay. This marvelous aspect of CRISPR/Cas makes it a dynamic approach for exploring strain divergence, convergence, microbial evolution, strain identification, ecology, and studying genotypes in complex environmental samples at the population level [27]. This tremendous technology could be used to identify and locate the detrimental microbes across encountered the food supply chain [28]. Using CRISPR/Cas strain typing, the spoilage microorganism can easily be tracked down in the starter culture industry as many fermentation microbes also contain a high content of CRISPR/Cas making this a supreme candidate for monitoring the strain identification of pathogenic outbreaks [28].

With the evolution of CRISPR-based strain typing, many studies have been reported using this technology for typing of probiotics and starter cultures, e.g., *streptococci* and *lactobacilli* [27, 29]. The genotyping of *S. Thermophilus* revealed that the existence of CRISPR/Cas in variable strains within the species provides profitable insights into the origins of many strains [29]. Because of the existence of natural CRISPR Cas1 system in the divergent strains within the species, the sequencing of the single locus for genotyping of even clonal strains has been eased to a lot of extent. The trailer end of the repeated spacer CRISPR arrays contains the periodically conserved ancestral spacer sequences while the leader end of the arrays consists of recently acquired spacer sequences. Even the identical isolates obtained from the same ancestral strains contain diverse spacer sequences as a result of this spacer acquisition ability due to foreign exposure. For instance, the CRISPR typing in *S. Thermophilus* revealed the presence of four divergent CRISPR systems, i.e., Universal and Variable CRISPR/Cas systems including two systems that actively acquired the spacers with time [29, 30]. This disclosed that how the strains that have been isolated within a few months have acquired different spacers on exposure. These active CRISPR loci can attain the novel spacers on daily basis [31]. Hence, the CRISPR/Cas technology can be largely implemented for strain typing of dairy cultures obtained from industrial and public sources.

CRISPR/Cas based strain typing has been used to aid food security and safety by tracking the food-borne pathogens and spoilage microbes. For instance, the strain typing of *Salmonella* has provided valuable insights into the microevolution and epidemiology of these strains in outbreaks [29]. *S. thermophilus* which is routinely used to manufacture yogurt and cheese in industries is also subjected to CRISPR typing for studying the ecology of various strains of *S. thermophilus*. The role of CRISPR Cas in developing variations in the genome of the probiotic bacteria, i.e., *Lactobacillus casei* and *Lactobacillus paracasei* was exposed using CRISPR based typing approach [32]. The sublineage of *Lactobacillus rhamnosus*, used in dairy products, and closely associated strains were traced using this approach [33]. CRISPR Cas genotyping is also implemented in strain typing of *Enterococcus faecalis*, a starter microbe utilized in the fermented meats [34]. The valuable insights into evolutionary relationships were discovered by scanning the spacer sequences of *Bifidobacterium* and *Lactobacillus gasseri*, used as probiotics [35]. The spoilage microbe, *Lactobacillus buchneri*, used in cucumber fermentations has also been successfully typed using this innovative tool [36]. Also. To analyze the source of transmission of *Salmonella enterica* serovar *Enteritidis* from poultry and pigs to human beings, 329 isolates of *Salmonella enterica*, 18 isolates of pig, 3 isolates of goose, 6 isolates of duck, and 38 isolates of humans were assessed in a study in China. During the major outbreaks of salmonellosis (food-borne illness) caused *by Salmonella enterica* serovars *Typhimurium* and *Heidelberg*, CRISPR typing was used to monitor the strains in the population. The genotyping of *Helicobacter pylori* showed low DI (Discrimination Index) with traditional typing techniques, but CRISPR-virulence typing made this job easy for studying the epidemiology of this species [37]. Many other detrimental microbes have undergone the CRISPR based genotyping including *Corynebacterium diphtheria, Vibrio parahaemolyticus, Staphylococcus aureus, Legionella pneumophila, Clostridium difficile, E. coli, Campylobacter jejuni* [38].

4.2.2 STRAIN TYPING'S USING CRISPR/CAS APPROACH

4.2.2.1 SPOLIGOTYPING

Spacer oligonucleotide typing or spoligotyping was the first method in which CRISPR was used for subtyping. This technique was first implemented in *M. tuberculosis* [39]. According to this technique, the CRISPR arrays were amplified with PCR using DR sequence-specific primers. The amplified PCR products were hybridized to membrane-bearing probes of required spacer sequences. This technique was very valuable for MTBC (*M. tuberculosis* complex) subtyping [40]. As the CRISPR technology made progress, reports revealed that many spacer sequences are lost with time hence strain to strain variability was unable to be detected by using this procedure as it can detect only pre-defined spacer sequences. To overcome the shortcomings of this process, the microbead-spoligotyping was developed [41]. This next-generation approach was implemented in typing of *Salmonella* spp., *Corynbacterium diptheriae*, and *Legionella pneumophila*. When used in typing *Legionella* it was found that not all the strains have CRISPR/Cas system hence this approach was unable to be used in routine genotyping [38(c), 42].

4.2.2.2 SEQUENCE BASED CRISPR TYPING

With advances in molecular biology, Sequence-based CRISPR typing was developed. *Streptococcus* was the first species of bacteria on which this approach was implemented [43]. The method revealed that PCR products obtained after amplification of spacer sequences were interrelated with the RFLP types of two insertion spacer sequences, i.e., IS1548 and IS1562. After successful implementation of CRISPR typing on *Streptococcus*, *Y. pestis* biovars were subjected to the CRISPR bases typing analysis. *Y. pestis* is the causative agent of plaque disease and the analysis revealed that it contains three arrays of CRISPR (Ypa, Ypb, and Ypc) [44]. It was reported from the study of strains representing these arrays that CRISPR based typing separated the isolates having identical MLVA types and these isolates had variable geographical origins [45]. The *Salmonella* spp. has been subjected to several CRISPR typing experiments and the sequencing results of about 1,000 CRISPR arrays of *Salmonella* disclosed that this species doesn't acquire the spacer sequences in a rapid manner [41(b), 46]. Correspondingly, *Campylobacter jejuni*, the leading causative agent of gastroenteritis was also subjected to sequence-based CRISPR typing in several studies. A clear harmony of multiple isolates with epidemiological data was obtained from three *C. jejuni* outbreaks using sequence-based CRISPR typing [46(d)].

4.2.2.3 CRISPR-MVLST APPROACH

A high-resolution typing approach than multi-virulence locus sequence typing as MVLST is CRISPR-MVLST as a combination of both approaches. None of the typing approaches were able to give detailed discrimination between strains of clonal serotypes of *Salmonella Enteritidis and Salmonella Heidelberg* but the combinatorial approach of CRISPR-MVLST and PFGE has been reported to give a high discrimination index [46(d)]. The efficiency of this combinatorial approach is proved by the study in which 141 *S. Enteritidis* serotypes were screened of which PFGE described 22 sequence types with DI (Discrimination Index) of 0.79 and CRISPR MVLST describes 13 sequences with DI of 0.71 while the combinatorial approach (CRISPR-MVLST and PFGE) defined 45 sequence types and a higher DI as 0.92 [46(b)].

Similarly, other alternative approaches using CRISPR as a typing tool include Real-Time PCR based CRISPR Typing and using CRISPR locus size for subtyping. In the former approach, specific pathogenic spacers acquired in a strain can be identified using specifically designed fluorescent probes for that spacer using Real-time PCR approach. While in the latter one, the dissimilarities in the number of CRISPR array spacers in the isolates are used in screening as the difference in the number of spacers results in a variable length of CRISPR arrays [47].

4.2.3 GENOME EDITING (GE)

Manipulation of the microbiota in the food supply chain is the key factor in controlling food spoilage and enhancing the nutritional content. The concept of fermentation has been introduced for decades even before the advent of food Science Fields. This well-known process has been utilized to process and preserve food such as yogurt, cheese, beer, wine, salami, sausages, pickles, sauerkraut, and so on. A complex set of microbes is used as a starter culture and encompasses a variety of microbiota in the way to get the desired outcome as a food product. These microbial communities not only valuable for the preservation of food but also defines the shelf-life, flavor, texture, and security of food. The CRISPR/Cas technology provides a promising solution to the food supply chain by ensuring all these qualities in food products [1].

The incredible property of CRISPR/Cas to be used as an antibacterial tool has a remarkable potential to be used widely to limit food spoilage around the globe. This technology can be manipulated to kill the spoilage microorganisms or target those microbes which repress these spoilage microorganisms, for instance, the lactic acid bacteria. The spoilage microbes can be eradicated from the population by utilizing extracellular vesicle and phage particles to introduce CRISPR/Cas in the genome of bacteria. This CRISPR/Cas system will specifically target the gene of interest (GOI) required to eliminate the spoilage microbes. Consequently, the shelf life of the food products will be enhanced because of the successful removal of spoilage microbes. Such as the bacteriophage therapy was used to limit the population of *Pseudomonas*, the spoilage bacteria [1]. This methodology can be made specific by several manipulations such as the target sequence used to design CRISPR nucleases, phages or extracellular vesicles used as a host to engineer CRISPR, and the universal endogenous CRISPR/Cas occurring in the target species. On the contrary, the adaptive immunity of the advantaged microbes such as microbes utilized in starter culture can be enhanced to suppress the population of spoiled bacteria. For instance, the robustness of milk fermenting starter culture was enhanced by boosting the adaptive immunity of *Streptococcus thermophiles* via the CRISPR/Cas system [1]. Similarly, *L. acidophilus* which is a probiotic bacteria can be edited using CRISPR/Cas for delivering the vaccine. Additionally, the GE attribute of CRISPR/Cas can be exploited to confiscate a variety of undesirable or detrimental characters in these microbes including toxic genes, antibiotic resistance genes, virulence factors, etc. [48].

CRISPR/Cas has the potential to transform the coming probiotic era into the CRISPR biotic era for improving the food industries. The taste, aroma, color, texture, flavor, and nutritional content of food can be modified by altering the metabolic pathways using CRISPR/Cas technique. For instance, the aroma and taste of beer were enhanced by manipulating *Saccharomyces cerevisiae*. Likewise, for improving the biofilm formation and gut colonization, the EPS, i.e., Exopolysaccharide level could be enhanced using CRISPR/Cas [49]. Many industrially valuable microbes have been manipulated by CRISPR/Cas to produce secondary metabolites, biochemical, and biofuels for example, *Streptomyces, Lactobacillus, Pseudomonas, Staphylococcus, Bacillus, Clostridium, Mycobacterium, E. coli* [50]. The color pigment of the food products can be increased by engineering the genes encoding for color pigments in the genome of the target host using the CRISPR/Cas system [51]. Filamentous fungi are very complex creatures to be manipulated by genetic engineering because of the low frequency of gene targeting. But astonishingly, CRISPR/Cas has made this tedious task possible with increased frequency [52]. This has been used to manipulate the red color-producing filamentous fungi as *Talaromyces atroroseus*. Other filamentous fungi in which CRISPR/Cas has been engineered are *Magnaporthe oryzae, Penicillium chrysogenum* [53], *Neurospora crassa* [54], *Trichoderma resssei, Aspergillus nidulans*, etc. [52] (Figure 4.4 and Table 4.1).

4.2.4 VACCINATION

The exposure of bacteria to phage infection typically results in the death of the majority of bacterial species but a minor population of them develops sequence-directed immunity after acquiring the spacer DNA sequences [106]. The potential of CRISPR/Cas technology to acquire spacer DNA makes bacteria vaccinated against these foreign DNA particles [106, 138]. CRISPR/Cas averts the uptake of foreign DNA by cleaving it thus inhibiting

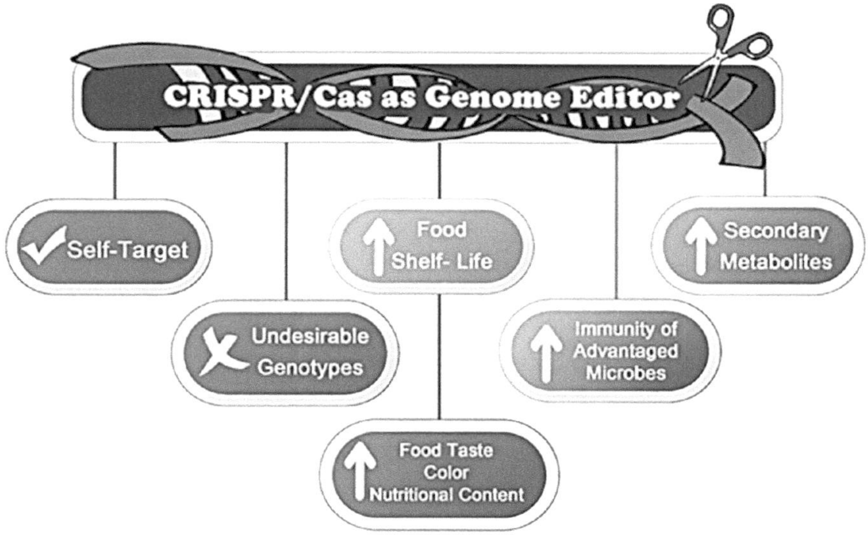

FIGURE 4.4 Applications of CRISPR/Cas as genome editor.

the host cells to develop antibiotic resistance enacted by these plasmids [139]. The studies revealed that theoretically, all the plasmids have in-borne "Malthusian fitness burden" inherited to the host cells. This fitness burden is required as an energy source for the host cell to aid in replication of the foreign DNA or plasmid [140]. Though this CRISPR/Cas mediated plasmid vaccination or immunization is a universal event occurring in bacteria, it can be programmed artificially for developing immunity in bacteria. Hence this approach can be utilized to vaccinate industrial microbes harboring CRISPR against plasmids and bacteriophages, etc. [25].

As the plasmids harboring antibiotic resistance genes can enact this attribute to other bacteria spreading the antibiotic resistance, this transferrable antibiotic resistance has become a major concern in the food industry. This Horizontal Gene Transfer event frequently occurs via conjugation, transformation, or phage transduction [138]. For example, MRSA and VRSA (Methicillin-resistant *staphylococcus* and vancomycin-resistant *staphylococcus*) have developed this resistance attribute because of HGT [141]. Not only CRISPR/Cas has the potential to develop phage resistance, but CRISPR/Cas can be manipulated to target the plasmids harboring antibiotic resistance genes. For instance, a successful event of transferring heterologous immunity in *S. thermophilus* has been reported [142]. *S. thermophilus* has been infected

TABLE 4.1 Microbes Edited Using CRISPR/Cas

Species	Function	Technique Used	Achievement	References
		Bacteria		
Bacillus subtilis	Industrially valuable enzymes low-molecular-weight chemicals production.	Recombination	Successfully optimized for genome editing.	[55]
Bacillus subtilis	Hyaluronic acid (HA) production.	CRISPRi	HA titer increased.	[56]
Bacillus subtilis	N-acetylglucosamine production.	Xylose-induced CRISPRi.	103.1 g/L NAG achieved in fermentation.	[57]
Bacillus subtilis	L-valine and β-cyclodextrin glycosyltransferase production.	Cas9-mediated genome editing.	>14-fold increase of valine titer. >2.5-fold increase of β-cyclodextrin glycosyltransferase titer.	[58]
Bacillus smithii	Involved in metabolism of C5 and C6 sugar.	Cas9-mediated genome editing.	Successfully optimized for genome editing.	[59]
Clostridium ljungdahlii	Butyric acid production.	CRISPR/Cas9	Synthesis gas (CO_2/CO).	[60]
Clostridium ljungdahlii	Ethanol production from synthesis gas.	Cas9-mediated genome editing.	Successfully optimized for genome editing.	[61]
–	3-Hydroxybutyrate (3HB) production.	CRISPRi	2.3-fold increased production of 3HB up to 21 mM titer.	[62]
Clostridium tyrobutyricum	n-Butanol production.	Type I-B CRISPR-Cas/Cas9-mediated genome editing.	Enhanced production of butanol.	[63]
Clostridium pasteurianum	Involved in conversion of waste glycerol to butanol.	Cas9-mediated genome editing.	Successfully optimized for genome editing.	[64]
Clostridium autoethanogenum	Fermentation of CO, CO_2, and H_2 to yield ethanol, biofuel, and 2,3-butanediol.	Recombination	Improved gene deletion efficiency to 50%.	[65]

TABLE 4.1 *(Continued)*

Species	Function	Technique Used	Achievement	References
Clostridium beijerinckii	Production of acetone, butanol, and ethanol.	Recombination and CRISPRi.	Successfully optimized for genome editing.	[66]
Clostridium beijerinckii	–	dCas9-mediated CRISPRi. nCas9-mediated genome editing, dCas9-mediated CRISPRi.	Successfully optimized for genome editing.	[67]
Clostridium beijerinckii	–	Knock out	Formation of twin-clostridial consortium of *C. beijerinckii* and *C. cellulovorans.*	[68]
Clostridium cellulolyticum	Conversion of lignocellulosic biomass to important Biochemicals.	nCas9-mediated genome editing.	Enhanced RNA repression. 27% decrease of unwanted Acetate titer.	[69]
Clostridium acetobutylicum	Isopropanol butanol-ethanol production.	Knock out in ctfAB, adc, adh.	Gene insertion up to 3.6 kb achieved.	[70]
Clostridium acetobutylicum	–	dCas9-mediated CRISPRi.	Relief of catabolic repression.	[71]
C. saccharoperbutylacetonicum	Butanol production.	Cas9-mediated genome editing.	19.0 g/L butanol titer.	[72]
C. difficile	Pathogenic bacteria causing nonsocomic diarrhea.	Cas12a-mediated genome editing.	Successful deletion of fur, tetM, and ermB1/2.	[73]
C. cellulovorans	Production of solvents (acetone, butanol, and ethanol).	dCas9-mediated CRISPRi.	Formation of twin-clostridial consortium of *C. beijerinckii* and *C. cellulovorans*.	[68]
Corynebacterium glutamicum	Glutamate production	Cas9 nickase (D10A).	3-fold increased production.	[74]
Corynebacterium glutamicum	Amino acid production.	CRISPRi	Successful genome editing.	[75]

TABLE 4.1 *(Continued)*

Species	Function	Technique Used	Achievement	References
Corynebacterium glutamicum	–	Cas12a-mediated genome editing.	Successful genome editing.	[76]
Corynebacterium glutamicum	–	Cas9-mediated genome editing.	Successful genome editing.	[74]
Corynebacterium glutamicum	L-lysine, L-glutamate, and homo-butyrate production.	dCas9-mediated CRISPRi.	0.52 ± 0.02 g/L butyrate production. Increased 1.39-fold L-lysine. Successful repression of Pgi, Pck, Pyk increased the lysine and glutamate production.	[77]
Corynebacterium glutamicum	L-proline production	CRISPR Cpf1 mediated genome editing.	Successful genome editing.	[76]
Escherichia coli	5-Aminolevulinic acid production.	CRISPR-Cas12a	Achieved up to 1.55 g/L titer.	[78]
Escherichia coli	Succinate production.	Orthogonal CRISPR, CRISPRi.	178% increased production up to 2.5 g/L titer.	[79]
Escherichia coli	Naringenin 7-sulfate production.	CRISPRi	2.83-fold increase of bioconversion rate.	[80]
Escherichia coli	3HP production	iCREATE	60-fold increase up to 30 g/L titer.	[81]
Escherichia coli	1-Propanol production.	PS-Brick assembly and CRISPR/Cas9.	1.35 g/L titer achieved.	[82]
Escherichia coli	Uridine production.	CRISPR/Cas9	70.3 g/L titer achieved.	[83]
Escherichia coli	Octanoic acid production.	CRISPR/Cas9	Increased by 61% increase up to 442 mg/L titer.	[84]
Escherichia coli	Itaconic acid production	CRISPR/Cas9	3.06 g/L titer achieved.	[85]
Escherichia coli	Isopentenol production	CRISPRi	98% increased production.	[86]
Escherichia coli	Uridine production	CRISPR/Cas9	5.6 g/L titer achieved.	[87]

TABLE 4.1 *(Continued)*

Species	Function	Technique Used	Achievement	References
Escherichia coli	Lycopene, 4-hydroxybutyrate, malate, butanol, naringenin, malonyl-CoA, mevalonate production.	dCas9-mediated CRISPRi.	8-fold increase of lycopene production, 2.3-fold increase of malate production, 2.83-fold increase in naringenin.	[88]
Escherichia coli	4HB hydroxy-butyrate production.	CRISPRi	4HB increased to 18.4 mol%.	[89]
Escherichia coli	Adipic acid, β-carotene, and isopropanol production.	CRISPR/Cas9	7.1 g/L isopropanol titer. 76.9% increase of uridine production. 93.1% succinyl COA production.	[83, 90]
Escherichia coli	Medium chain fatty acids production.	–	3.8 g/L MCFA obtained.	[91]
Francisella novicida	Pathogenic bacteria.	CRISPR/Cas9	Successful repression of bacterial lipoproteins.	[92]
Halomonas sp.	3-Hydroxybutyrate/3-hydroxy valerate (PHBV) and PHB production.	CRISPR/Cas9	16-fold increased 3HV with a fraction at 11.81 mol%.	[89]
Halomonas bluephagenesis	Poly(3-hydroxybutyrate co-3-hydroxyvalerate) (PHBV) production	CRISPR/Cas9	6.3 g/L cell dry weight (CDW), 65% PHBV in CDW and 25 mol% 3HV in PHBV.	[93]
Klebsiella pneumoniae	n-Butanol production	CRISPRi	154% increase butanol production.	[94]
Klebsiella pneumoniae	3HP production	CRISPRi	37% increased 3HP production titer by successful reduction of lactic acid synthesis 36.7 g/L titer of 3HP achieved.	[95]
Klebsiella pneumoniae	Pathogenic bacteria	Cas9-mediated genome editing.	Successful deletion of nine genes of phiKpS2.	[96]

TABLE 4.1 *(Continued)*

Species	Function	Technique Used	Achievement	References
K. pneumoniae	1,3-propanediol, 2,3-butanediol and 3-hydroxy propionic acid (3-HP) production.	dCas9-mediated CRISPRi.	Reduced the lactic acid to 1 g/L.	[95]
Legionella pneumophila	Pathogenic bacteria	CRISPR/Cas	Successful genetic manipulation.	[97]
Lactobacillus reuteri	Probiotic strain biotherapeutics production.	Recombination approach (CRISP/Cas9 +ssDNA recombineering.	Successful application of codon saturation mutagenesis.	[98]
L. casei	–	Cas9-mediated genome editing.	Increased efficiency of single gene deletion and insertion.	[99]
M. tuberculosis	Pathogenic bacteria.	dCas9-mediated CRISPRi	Studied essential gene function. 20- to 100-fold knockdown efficiency achieved. Successful gene silencing.	[100]
Neisseria meningitidis	Pathogenic bacteria.	CRISPR/Cas	Successful genome manipulation.	[101]
S. aureus	Pathogenic bacteria.	DNA base editing.	Base editing system developed.	[102]
S. aureus	Pathogenic bacteria.	Cas9-mediated CRISPR.	Efficient genome editing system developed.	[103]
S. aureus	Pathogenic bacteria.	dCas9-mediated CRISPRi.	dCas9-mediated CRISPRi.	[104]
P. aeruginosa	–	dCas9-mediated CRISPRi.	Robust gene depletion system developed.	[105]
Streptococcus thermophilus	Probiotic strain. Fermentation strains.	CRISPR/Cas	Engineered immunity.	[106]
Streptomyces albus	Secondary metabolites production.	Cas9-mediated genome editing.	Activation of silent BGCs.	[107]
Streptomyces albus	Secondary metabolites production.	Recombination	70–100% efficiency of chromosomal deletions.	[107(a)]

TABLE 4.1 *(Continued)*

Species	Function	Technique Used	Achievement	References
Streptomyces coelicolor	Chloramphenicol. Anti-tumor compound. YM-216391 production.	CRISPR/Cas	2–23 folds increase in productivity of BGCs in chloramphenicol.	[108]
Streptomyces coelicolor	Secondary metabolites production.	Recombination and CRISPRi/Cas9-mediated genome editing.	Successful genetic manipulation.	[107(a), 109]
Streptomyces lividans	Secondary metabolites production.	Cas9-mediated genome editing.	Activation of silent BGCs.	[107]
Streptomyces lividans	Secondary metabolites production.	Recombination	70–100% efficiency of chromosomal deletions.	[107(a)]
Streptomyces viridochromogenes	Natural antibiotic (streptazolin) and herbicide (bialaphos) production.	Cas9-mediated genome editing.	Activation of silent BGCs.	[107]
S. hygroscopicus	Antibiotics and biochemicals production.	Cas12a-mediated genome editing.	Successful genetic manipulation.	[110]
S. venezuelae	Secondary metabolites production.	Cas9-mediated genome editing.	Activation of silent BGCs.	[107(b)]
S. rimosus	Producer of oxytetracycline and tetracycline.	Cas9-mediated genome editing.	36.8% increased yield of oxytetracycline.	[111]
Streptomyces pristinaespiralis	Pristinamycin I (PI)	CRISPR/Cas	2.4-fold higher titer of PI.	[112]
Streptomyces viridochromogenes	Secondary metabolites production.	Recombination	70–100% efficiency of chromosomal deletions.	[107(a)]
Tatumella citrea	Vitamin C precursor (2-keto-D-gluconic acid) production.	Recombination	100% efficiency in multi-gene editing.	[113]
Y. pestis	Pathogenic bacteria.	Cas12a-mediated genome editing.	Successful genome editing.	[114]
		Yeast		
Pichia pastoris	Common production strain.	Multiplex mutagenesis and recombination.	100% gene disruption efficiency obtained.	[115]

TABLE 4.1 *(Continued)*

Species	Function	Technique Used	Achievement	References
Aspergillus aculeatus	Enzyme production.	Mutagenesis	Efficient genome editing system developed.	[116]
Aspergillus brasiliensis	Enzyme production	Mutagenesis	Efficient genome editing system developed.	[116]
Aspergillus carbonarius	Enzyme production	Mutagenesis	Efficient genome editing system developed.	[116]
Aspergillus luchuensis	Enzyme production	Mutagenesis	Efficient genome editing system developed.	[116]
Aspergillus nidulans	Enzyme production	Mutagenesis	Efficient genome editing system developed.	[116]
Aspergillus niger	Enzyme production	Mutagenesis	Efficient genome editing system developed.	[116]
Aspergillus oryzae	Enzyme production	Mutagenesis	Efficient genome editing system developed.	[117]
Candida albicans	Common production strain. Involved in catabolism of phenol and formaldehyde.	Recombination and multiplexed recombination.	Efficient genome editing system developed.	[118]
Kluyveromyces lactis	Common production strain.	Multiplexed recombination.	64% increased efficiency of gene knockouts and point mutations.	[119]
Myceliophthora thermophile	Thermophilic strain. Cellulases production.	Mutagenesis, recombination, multiplexed recombination.	Lignocellulose and extracellular proteins titer increased 5 to 13 folds.	[120]
Neurospora crassa	Lignocellulosic enzymes production.	CRISPR/Cas9	Increased expression of cellulases.	[54]
Penicillium chrysogenum	b-lactam antibiotics production.	CRISPR/Cas9	Efficient genome editing system developed.	[121]

TABLE 4.1 *(Continued)*

Species	Function	Technique Used	Achievement	References
Saccharomyces cerevisiae	β-amyrin production.	CRISPRi	156.7 g/L titer achieved.	[122]
Saccharomyces cerevisiae	–	CRISPRa	Successful development of CRISPR/Cas-based gene activation library, improved thermo-tolerance.	[123]
Saccharomyces cerevisiae	–	CRISPR-mediated genome shuffling.	Improved thermo-tolerance.	[124]
Saccharomyces cerevisiae	Cis, cis-muconic acid production.	CRISPRi and *in vivo* assembly.	5–21% increase in titer.	[125]
Saccharomyces cerevisiae	2,3-Butanediol (BDO) production.	CRISPRi, construction of tRNA-sgRNA operons using LEGO.	2-fold increase of BDO titer.	[126]
Saccharomyces cerevisiae	3HP production	CRISPRa	Increased by 15–36%.	[127]
Trichoderma reesei	Common production strain.	Mutagenesis and recombination.	Efficient genome editing system developed.	[128]
Ustilage maydis	Itaconic acid production.	CRISPR/Cas9	10.2-fold increased IA, yield achieved at 19.4 g/L, yield further elevated to 53.5 g/L.	[123]
Ustilage maydis	Biochemical production, causative agent of corn smut	Mutagenesis	Concentration of acid up to 47 ± 1 g L^{-1} obtained.	[129]
Yarrowia lipolytica	Biochemical production.	Multiplexed mutagenesis and recombination.	Double and triple gene editing efficiency of 36.7 or 19.3% achieved.	[130]
Algae				
Anabaena	Biotechnological uses for genetic manipulation.	Cas12a-mediated genome editing.	Successful point mutations, knock-ins/outs achieved.	[131]
S. elongatus	Succinate production.	Cas9-mediated genome editing.	11-fold increase of succinate titer.	[132]

TABLE 4.1 *(Continued)*

Species	Function	Technique Used	Achievement	References
Synechocystis sp.	Biotechnological uses for genetic manipulation.	CRISPR/Cas	Fatty alcohols.	[133]
Synechocystis sp.	Biotechnological uses for genetic manipulation.	Cas12a-mediated genome editing.	Successful point mutations, knock-ins/outs achieved.	[131]
Synechocystis sp.	Biotechnological uses for genetic manipulation.	Cas9-mediated genome editing.	Efficient genome editing system developed.	[134]
Synechocystis sp. PCC *6803*	N butanol production.	Inducible CRISPRi.	5-fold increase of carbon partitioning to n-butanol.	[135]
Synechicystis sp.	Fatty alcohols production.	CRISPRi	3-fold Increase of octadocanol at 10.3 mg/g DCW.	[133]
Synechococcus sp.	Lactate production.	CRISPRi	2-fold more lactate titer achieved.	[136]
Synechococcus sp.	Succinate production.	CRISPRi	12.5-fold increase in succinate titer, 0.58–0.63 mg/L titer achieved.	[137]
Synechococcus sp.	Biotechnological uses for genetic manipulation.	Cas12a-mediated genome editing.	Successful point mutations, knock-ins/outs achieved.	[131]

with plasmids harboring antibiotic genes. The result was the development of immunity in some strains of *S. Thermophiles* which have acquired the spacer sequences for antibiotic resistance genes. This technique can be used to vaccinate CRISPR-free fermenting microbes against antibiotic resistance, phage infection, plasmids, and uptake of undesirable elements [143].

CRISPR/Cas takes various routes to ensure the limited uptake of undesirable elements. For instance, *Staphylococcus epidermidis* was reported to embrace a specific spacer sequence similar to sequence of nickase gene occurring in conjugative plasmids of practically all staphylococcal strains. CRIPSRi direct targets DNA by inserting a self-splicing intron sequence in nickase and averting the conjugation event as an alternative route. The studies show resilient affirmation that CRISPR/Cas systems are transferrable among divergent species in Archea and Bacteria through HGT event [138].

4.2.5 ANTIMICROBIALS

CRISPR/Cas as an antimicrobial toolbox serves solution to a variety of issues encountered by the fermentation industries such as cultural contamination in mixed cultures, modulation of the mixed cultures phages infection, and eradication of multiple undesirable strains instantaneously. The concept of using mixed cultures to reduce the risk of bacteriophage infection has certain positive outcomes but also provides us with some shortcomings such as contamination and characterization of mixed cultures and exertion in isolating the undesirable strains [144]. Management and isolation of strains from the mixed culture populations is a hectic task because of the comparable physiological attributes of various strains. CRISPR/Cas holds the potential to isolate strains differing in even single bases with high specificity thus sparing the desired strains and eradicating the target strains. Also, it can target multiple strains simultaneously revealing its multiplexed nature. For instance, two 99% identical *E. coli* strains were successfully separated using CRISPR anti-microbials [145]. This technology can be manipulated to explore the natural and artificial microbial populations. When complete eradication of undesirable strains is unmanageable, the composition of mixed cultures can be altered using CRISPR/Cas antimicrobials. This is done by managing the dosage of antimicrobials through an optimal blend of target and non-target plasmids which will be transformed later on into the mixed culture. For instance, this strategy was fruitfully experimented in modulating the *E. coli* mixed cultures. Hence, CRISPR/Cas programmable antimicrobials are effective toolboxes for selectively eradicating any targeted strains from required mixed cultures [25]. Also, this approach was implemented on the strains that lacked native CRISPR/Cas machinery by utilizing phagemids. Phagemids were developed due to the exertion of introducing a 160 kD complex of protein-RNA through membranes of gram-positive and gram-negative bacteria [146]. Phagemids are particularly designed plasmids with phage packing signals. These signals contain sequences that code for Cas9 nucleases, crRNA, and tracrRNA. The results of this study divulged that CRISPR/Cas antimicrobial highly distinguished two strains differing in a single base pair. Hence CRISPR/Cas can be programmed to target multiple strains at once by manipulating multiple spacers in a single phagemid. The resulting various crRNAs will target the multiple strains leaving closely associated strains safe. The extensive killing of the target pathogen can be done by elevating the count of phagemids [147].

CRISPR/Cas antimicrobials can be programmed to target virulence genes and antibiotic resistance genes from a population. The host cells can be infected with specially programmed phagemid against the targeted pathogen and at last CRISPR antimicrobials will eradicate the targeted pathogen by cleaving their genome thus saving the food products from spoilage microbes as experimentally proved in *S. aureus* and *Enterobacteriaceae* [148]. Correspondingly, it can be manipulated to curtail the spread of antibiotic resistance. A study used specifically programmed Cas9 to delete the resistance genes in pathogenic bacteria thus reverting the AMR (anti-microbial resistance) genes to antibiotic sensitive genes. CRISPR/Cas microbials seem to be a promising alternative to antibiotics in near future [147].

Double-stranded DNA (dsDNA) breaks can be lethal for prokaryotic organisms due to a lack of robust DNA damage repair pathways. This lethality virtue of CRISPR/Cas can be used in self-targeting undesirable bacterial populations [149]. This lethal attribute was exposed in a study in 2013 indicating that self-targeting via CRISPR/Cas can cause irreversible $\sim$105 reductions in viable cells [150]. Type-I CRISPRi inflicts the most lethal damage to DNA. This system uses an exonuclease Cas3 enzyme which causes deletions up to 40 kb distance in the genome. This type of damage elicited to the genome is usually repaired by low fidelity end-joining repair mechanism at a very low frequency [150]. Type-II system uses Cas9 enzyme to inflict dsDNA breaks in the genome which are usually repaired by high fidelity homologous recombination (HR) pathway [151]. Although, DNA repaired by this high-fidelity pathway is still subjected to target by CRISPR/Cas. Type-I system as a self-targeting approach seems more promising in targeting required pathogens as the death is not affected by chromosomal location and expression level. But this system desires an accurate designing of spacers that can differentiate closely resembling strains. This self-targeting CRISPR/Cas approach is highly promising in targeting spoilage microbes and sparing the advantageous microbiota of food matrices [149].

4.2.6 PHAGE RESISTANCE

Phage infection is the leading cause of failed fermentation in food industries. When the lytic phages reproduce, they result in host devastation creating fermentation an arduous process. They occur in raw foods and strict sanitization measures like pasteurization, ionization, radiation, and thermal energy treatment are not enough to eradicate them at once [152]. Cultural rotations, sanitization, modulation of process, ventilation systems, starter cultures, are all the substitutes to circumvent phage exposure. But these measures still didn't promise a complete solution. It is of no wonder to know that there is the high occurrence of CRISPR/Cas in fermenting microbes due to high preponderance of phage infection in the fermentation process. Here the most promising solution can be contributed by the immunization attribute of the native CRISPR/Cas system to enhance the phage resistance. Utilizing this native spacer acquisition event of the CRISPR/Cas vaccination system, we can immunize the required fermenting strains artificially by exposing them with iterative rounds of multiple phages attack and selecting native CRISPR mutants thus developing immunity against multiple industrially significant phage infections [27]. It must not be forgotten to use the genetically diverse and industrially relevant phages in the immunization process. This would help us to establish a natural genetic tagging system by the distinctive spacer combination generated through exposing the strains to multiple rounds of phage infections. The iterative acquisition of novel spacers provides us with CRISPR variants that have a broad spectrum of phage resistance. For achieving a high lifespan of industrial microbes, CRISPR variants that have been obtained after iterative rounds of phage exposures can be pooled in the cultural rotation process. Although several studies divulged that ancestral spacer sequences are infrequently lost with time, but it is experimentally evidenced that newly acquired spacers remain intact for a long period [27].

An example is the successful demonstration of this theory in *S. thermophiles*. *S. thermophiles* was immunized by challenging with four various lytic phages in four iterative rounds to obtain CRISPR variants. Additionally, at the end of each iterative round, the phage insensitive variants were confirmed to have acquired the required spacer sequences. Hence, the CRISPR/Cas vaccination is a novel experimental framework to enhance phage resistance in fermenting microbes [153].

4.3 LIVESTOCKS

Large-animal genetic modification can be used for a range of uses, including food processing, animal welfare and pest control, and the advancement of animal biomedicine and basic science. The biggest concern about modern technologies is whether it should be used in the first place. In the case of livestock, the global background of food demand and supply, as well as current questions regarding environmental sustainability, are both factors to consider [154, 155].

One of the greatest obstacles facing humanity's future is the increasing need for food to feed the world's developing population. As per the FAO, the UN, and the World Bank, this intake trend is being driven up not only by global population growth, which is expected to reach well over 9 billion by 2050, as well as by animal protein consumption being higher per capita than vegetable carbohydrates. Poverty levels have fallen significantly in recent years, and residents now have more access to nutritious food [156, 157]. The increased demand has already put strain on livestock supply, and this inclination is expected to continue in approaching years [158].

Traditional methods of increasing growth, such as bettering animal welfare, diet, biology, fertility, and management, will play an important role. This, however, would not be sufficient to guarantee the required output transition. In this case, preconceived ideas of novel and revolutionary technologies are needed, and CRISPR is seen as a valuable tool to aid in the transformation of the world's livestock. How will the future of farm animals influence a novel's productivity, disease tolerance, or even the disappearance of pests across history? We are on the verge of achieving these goals thanks to CRISPR [158, 159].

In 2011, Sir John Beddington led a group of experts who looked at the complicated relationship between the world's food, oil, and water needs. In the context of climate change, he referred to the same rise in demand for these goods as a "healthy wind." Global population development, market demand, improvements in domestic and global governance, climate change, demand for vital resources (e.g., clean water), and climate change are six main stresses on global food supply structures that are now unable to feed the population, according to the FORESIGHT report consumer behavior [159]. Our species faces a huge obstacle in producing more food and consuming less resources. We continue to address the possible future effects of transgenic technologies and genome editing (GE), as well as the impact of genomics, transgenesis, and genome planning on topics impacting animal and animal production, health and well-being, and our food production capability [159].

4.3.1 IMPACT OF GENETIC DEVELOPMENT ON ANIMAL PRODUCTION

Despite the fact that many domesticated animals have been bred for thousands of years, the underlying management processes in effect have resulted in significant increases in productivity. Animal farming has become quicker, cheaper, cleaner, and more sustainable as a result of genetic advancements, with less environmental effects. For example, from the 1960s to 2005, selected breeds resulted in larger pipeline sizes of 50% in pigs, an increase of 37% thin pork, and a double lean pork per kilogram; in chickens, 2 kg weight days decreased from 100 to 40 days, the breast meat ratio soared up from 12 to 20%, the rate of fodder conversion was split in halves, eggs per year increased by 30% and eggs per ton diet increased by 80%; and finally, in cattle, milk production increased by 67%. This steady rise in food production has shown tremendous success over the past few decades, even though the benefits have been immeasurably seen in developed lands [160].

4.3.2 EXAMPLES OF GENETIC MODIFICATION

Domesticated animals have been genetically modified (GM) for decades to improve food production quality, animal health and well-being, while also reducing environmental traits. Industry and academia are welcoming CRISPR technology because it promises new scientific areas, accelerated product growth, and solutions to global problems. CRISPR has been found to be crucial in plants, animals, yeast, fungi, and bacteria in studies. It is no longer a question that when, CRISPR products will be produced. Other items, such as brown-colored apples and salmon that grows quickly, have already been launched in some markets [161]. Other products, such as pork and CRISPR-modified crops like maize and beans, have started commercial production research and will be ready in two years. CRISPR's technology has allowed agriculture to manufacture next-generation goods even quicker than the government's regulatory strategy can be applied, thanks to its numerous advantages [162, 163].

4.3.3 GENOME PLANNING IN AGRICULTURAL BREEDING

The advancement of flexible genetic engineering technology has changed the livestock industry in recent years. Given the ease at which vast organisms can be changed, it's not shocking that far more elite phenotypes have been created in the last few years than in the previous three decades. Lactation quality, meat yield, disease tolerance, and bioreactors are all potential agricultural benefits of these livestock that cannot be readily obtained by traditional breeding methods [164, 163].

4.3.4 ENHANCED PERFORMANCE

One of the striking examples of knockout is of gene myostatin (*MSTN*). *MSTN* restraints the growth of skeletal muscle as it is the negative regulator family hormone. *MSTN* Knockout results in enhanced bone formation it can be useful in producing meat under certain conditions. The natural version of the *MSTN-KO* phenotype is famed from cattle breed Belgian Blue and Piedmontese. With the help of DNA nucleases, the *MSTN-KO* phenotype was successful in cattle, pigs, sheep, and goats [165]. The animals looked dissimilar from the wild people with noticeable muscle differences, which was also noted during muscle comparisons. Continuity of operation is possible, because for example by identifying certain SNPs related to key features or breeding livestock that produce interest in the sex you want (male offspring in the beef cattle, the female chickens, and the pigs). An imperative use of genetic engineering is the preservation of a preservative animals which will operate the production of gametes from the germ cells findings to someone else. First for example such a model of animal pigs that carry knocks out of *NANOS2* Pigs with this mutation lack the virus line. These *NANOS2*-null male has the capability to provide an appropriate habitation to treat viral cells from a genetic male, too so it expands its inheritance power [166].

4.3.5 MAKING MILK BETTER

β-Lactoglobulin (BLG) is one of the chief whey proteins as a key allergen in milk from ruminants (goats, cows but usually not in human milk), and results in allergies that can be mild or life threatening. The two of the most extensively used approaches to lessen the allergenic effects of *BLG* include heat processing and enzymatic hydrolysis. But these techniques are very costly and can upset the nutritional value of milk by generating unpredicted products. Genetic modification can be directly used to lower *BLG* levels in ruminants. Reports have demonstrated that using ZFN and TALENs have been successful in modifying BLG in the cattle. It was reported that in the TALEN system, cows with *BLG* mutations lacked mature *BLG*. Furthermore, *BLG* is also a reliable source of mammary gland bioreactors, and the *lactoferrin* gene *(hLF)* was extracted using TALEN in goats [164, 166]. Phenotyping in goats revealed high levels of *hLF* and the lack of *BLG* in the milk. Using a similar strategy, Luo et al. have obtained high levels of production of human serum album in cow's milk. By means of these precise genetic engineering procedures, majority pharmaceutical proteins are anticipated to be produced in animal milk sooner or later [163].

In a study conducted by Cui et al., around 10 valuable goats were found in possession of mono-allelic conversion of *blg* type. After that, the second *blg* falls asleep was directed at those animals by knocking on *hlf* gene (human lactoferrin), a glycoprotein involved in intolerant immune responses and iron adsorption in the intestinal tract. After breastfeeding, milk analysis demonstrated that mono-allelic mutations have a decreased level of *blg* in milk, and the biallelic-converted milk was entirely free beta-lactoglobulin and contains moderate quantities of human lactoferrin [165].

4.3.6 MEAT PRODUCTION, TEXTURE, AND QUALITY

Myostatin (*MSTN*) is a protein that majorly plays its role as a negative regulator in controlling the growth of muscles and is secreted in muscle tissue. The mutations of *MSTN* resulted in double muscle behavior, first reported in cattle and then in sheep, dogs, and humans. This makes *MSTN* a striking factor in genome planning for growing lean meat in livestock. In goats, *MSTN* locus was also used to insert the fat-1 gene, which converts n6 polyunsaturated fatty acid (*PUFA*) into n–3 *PUFA*. Hence it was experimentally proved that the GM goats had a better muscle growth and improved healthier meat [167].

4.3.7 PRODUCTION OF NON-ALLERGEN ANIMAL-DERIVED PRODUCTS

Genes such as ovalbumin and ovomucoid were extracted utilizing CRISPR/Cas approach to remove two chief allergenic components from egg white. This can make the eggs chew for many users who can eat chicken eggs [168].

4.3.8 RESISTANCE TO DISEASE

Foot-and-mouth disease virus (FMDV) is an economically debilitating infection upsetting the swine industry throughout the globe. Transgenic pigs were clearly exposed to intermediate FMDV-specific RNAs derived from small hairpin RNAs (shRNAs). These piglets showed no clinical signs of infection compared to the wild-type pigs, providing an additional genetic model of genetic engineering, and diseases [169] (Table 4.2).

TABLE 4.2 Examples of Genome Edited Livestock for Different Purposes

Traits	Species	Editing Machinery	Technique	Genome Edit Target	References
Increased muscle growth (double-muscle phenotype).	Cow, sheep, goat, channel catfish, pig.	ZFN/ TALENS/ CRISPR/ Cas9	SCNT/ CMI	*Myostatin* (*GDF8*)	[170]
Hornlessness (polled)	Cow	TALEN	SCNT	*Pc Polled*	[171]
Boretaint (hormone release during sexual maturity leading to undesired meat taste).	Pig	CRISPR/ Cas9	SCNT	*KISS1R*	[171]
Sterility	Salmon	CRISPR/ Cas9	CMI	Dead end protein.	[172]
PRRSV porcine reproductive and respiratory syndrome virus.	Pig	CRISPR/ Cas9	CMI/SCNT	*Nanos 2*	[171]
ASFV African swine fever virus.	Pig	CRISPR/ Cas9	SCNT	*RELA 2*	[173]
Mannheimia (Pasteurella) haemolytica resilience.	Cow	–	–	*CD18*	[174]
Bovine tuberculosis resilience.	Cow	TALEN/ CRISPR/ Cas9	CMI/SCNT	*NRAMP1*	[175]
Xenotransplantation (removal of endogenous retroviruses).	Pig	CRISPR/ Cas9	SCNT	Porcine endogenous retrovirus genes.	[161(a)]

Abbreviations: CMI: Cytoplasmic microinjection; SCNT: Somatic cell nuclear transfer; TALENs: Transcription activator like endonucleases; and ZFNs: Zinc-finger nucleases.

Source: Adapted from: Tait-Burkard et al. [169(b)].

4.3.9 ANIMAL WELFARE

Physical removal of horns from cattle to guard animals and producers from accidental harm is expensive and aching for animals. Genetic analysis has identified a distinct predisposition to horn-lined cows (e.g., 'pooled'), a trait common to cattle but rare in dairy breeds. Thus, the candidate's allele was 'tested' and concentrated in dairy cows using TALEN-mediated genome mutation and reproductive integration [176].

4.3.10 BIOREACTORS

Human products can be manufactured using livestock as a bioreactor. Blood-based serum albumin (HSA) results in various life-threatening illnesses, e.g., liver failure and severe panic disorder, and is much needed. Owing to the shortage of human blood transfusions and the hazards of infection accompanying human blood, alternative methods for HAS production have long been sought. Victory was accomplished when CRISPR/Cas9 was used to knock on a person's cDNA album in the Albin endus native locus, which resulted in transgenic pigs with a human album in their blood [177].

In summary, genome-classified animals can be presented more proficiently and precisely, and advanced planning tools offer great potential in accelerating genetic development in pets. To meet these potentials, many genomic markers that regulate genetic variation or genetics that are the most important drugs in livestock will need to be revealed, simplifying powerful genetic engineering technologies. As a matter of fact, the nome-nuclease-mediated genome planning technology will equip researchers with new effective tools to upgrade farm animal breeding, leading them to a new '*in vitro* breeding' or 'artificial breeding environment.

4.4 CONCLUSIONS

We live in a CRISPR universe, and the future has arrived. CRISPR is impacting the global food supply at any stage, from starter cultures to crop and livestock improvement, from farm to plate. CRISPR is only one example of how microbes form our environment on a daily basis. Innovations in microbial and also in metagenomics are showing just how much promise still remains. This is particularly the case in the food processing industry, where microbes and yeast strains are being used extensively in fermentation processes and the production of a wide range of foods and liquor. This opens up opportunities to use many CRISPR-based innovations in farm microbiomes (e.g., feed microbiome, livestock microbiome, and soil microbiome), as well as production facilities (e.g., brewing containers, supply chains, health and safety monitoring stations, manufacturing environments) and the customer (e.g., oral and gut microbiomes), although microbiota became farm to table, by CRISPR. The effect of soil microorganisms on crop yield is now being studied. Ultimately, the microbiome(s) of the human gut have a major effect on its wellbeing, which is also influenced by food. It is clear that the capabilities of CRISPR in the food supply chain are far from being depleted, and that CRISPR-based technologies for the manufacture of safe and environmentally efficient food products have a long way to go.

KEYWORDS

- **antimicrobials**
- **genome editing**
- **livestocks**
- **microbes**
- **phage resistance**
- **plant biotechnology**
- **spoligotyping**
- **vaccination**

REFERENCES

1. Barrangou, R., & Notebaart, R. A., (2019). CRISPR-directed microbiome manipulation across the food supply chain. *Trends in Microbiology, 27*(6), 489–496.
2. Tilman, D., Balzer, C., Hill, J., & Befort, B. L., (2011). Global food demand and the sustainable intensification of agriculture. *Proceedings of the National Academy of Sciences, 108*(50), 20260–20264.
3. Chen, K., Wang, Y., Zhang, R., Zhang, H., & Gao, C., (2019). CRISPR/Cas genome editing and precision plant breeding in agriculture. *Annual Review of Plant Biology, 70*, 667–697.
4. Ray, D. K., Mueller, N. D., West, P. C., & Foley, J. A., (2013). Yield trends are insufficient to double global crop production by 2050. *PloS One, 8*(6), e66428.
5. Puchta, H., Dujon, B., & Hohn, B., (1993). Homologous recombination in plant cells is enhanced by *in vivo* induction of double strand breaks into DNA by a site-specific endonuclease. *Nucleic Acids Research, 21*(22), 5034–5040.

6. Wright, D. A., Townsend, J. A., Winfrey, Jr. R. J., Irwin, P. A., Rajagopal, J., Lonosky, P. M., Hall, B. D., et al., (2005). High-frequency homologous recombination in plants mediated by zinc-finger nucleases. *The Plant Journal, 44*(4), 693–705.
7. Shahzad, R., Jamil, S., Ahmad, S., Nisar, A., Amina, Z., Saleem, S., Iqbal, M. Z., et al., (2021). Harnessing the potential of plant transcription factors in developing climate resilient crops to improve global food security: Current and future perspectives. *Saudi Journal of Biological Sciences.*
8. Zhu, H., Li, C., & Gao, C., (2020). Applications of CRISPR–Cas in agriculture and plant biotechnology. *Nature Reviews Molecular Cell Biology, 21*(11), 661–677.
9. Zhang, Y., Pribil, M., Palmgren, M., & Gao, C., (2020). A CRISPR way for accelerating improvement of food crops. *Nature Food, 1*(4), 200–205.
10. Zeng, Y., Wen, J., Zhao, W., Wang, Q., & Huang, W., (2020). Rational improvement of rice yield and cold tolerance by editing the three genes OsPIN5b, GS3, and OsMYB30 with the CRISPR–Cas9 system. *Frontiers in Plant Science, 10*, 1663.
11. Du, B., Chen, R., Guo, J., & He, G., (2020). Current understanding of the genomic, genetic, and molecular control of insect resistance in rice. *Molecular Breeding, 40*(2), 1–25.
12. Cai, C., Wang, W., Ye, S., Zhang, Z., Ding, W., Xiang, M., Wu, C., & Zhu, Q., (2020). Overexpression of a novel *Arabidopsis* gene SUPA leads to various morphological and abiotic stress tolerance alternations in *Arabidopsis* and poplar. *Frontiers in Plant Science, 11*, 1766.
13. Jiang, N., Yan, J., Liang, Y., Shi, Y., He, Z., Wu, Y., Zeng, Q., et al., (2020). Resistance genes and their interactions with bacterial blight/leaf streak pathogens (*Xanthomonas oryzae*) in rice (*Oryza sativa* L.)—An updated review. *Rice, 13*(1), 1–12.
14. Xie, J., Guo, G., Wang, Y., Hu, T., Wang, L., Li, J., Qiu, D., et al., (2020). A rare single nucleotide variant in Pm5e confers powdery mildew resistance in common wheat. *New Phytologist, 228*(3), 1011–1026.
15. Khanday, I., & Sundaresan, V., (2021). Plant zygote development: Recent insights and applications to clonal seeds. *Current Opinion in Plant Biology, 59*, 101993.
16. Gasparini, K., Dos, R. M. J., Peres, L. E. P., & Zsögön, A., (2021). De novo domestication of wild species to create crops with increased resilience and nutritional value. *Current Opinion in Plant Biology, 60*, 102006.
17. Bai, M., Yuan, J., Kuang, H., Gong, P., Li, S., Zhang, Z., Liu, B., et al., (2020). Generation of a multiplex mutagenesis population via pooled CRISPR-Cas9 in soya bean. *Plant Biotechnology Journal, 18*(3), 721–731.
18. Gramazio, P., Takayama, M., & Ezura, H., (2020). Challenges and prospects of new plant breeding techniques for GABA improvement in crops: Tomato as an example. *Frontiers in Plant Science, 11.*
19. Kwon, C. T., Heo, J., Lemmon, Z. H., Capua, Y., Hutton, S. F., Van, E. J., Park, S. J., & Lippman, Z. B., (2020). Rapid customization of *Solanaceae* fruit crops for urban agriculture. *Nature Biotechnology, 38*(2), 182–188.
20. Zetsche, B., Gootenberg, J. S., Abudayyeh, O. O., Slaymaker, I. M., Makarova, K. S., Essletzbichler, P., Volz, S. E., et al., (2015). Cpf1 is a single RNA-guided endonuclease of a class 2 CRISPR-Cas system. *Cell, 163*(3), 759–771.
21. Endo, A., Masafumi, M., Kaya, H., & Toki, S., (2016). Efficient targeted mutagenesis of rice and tobacco genomes using Cpf1 from *Francisella novicida. Scientific Reports, 6*(1), 1–9.

22. Zhao, Y., Yang, X., Zhou, G., & Zhang, T., (2020). Engineering plant virus resistance: From RNA silencing to genome editing strategies. *Plant Biotechnology Journal, 18*(2), 328–336.
23. Papadimitriou, K., Pot, B., & Tsakalidou, E., (2015). How microbes adapt to a diversity of food niches. *Current Opinion in Food Science, 2*, 29–35.
24. Douglas, G. L., & Klaenhammer, T. R., (2010). Genomic evolution of domesticated microorganisms. *Annual Review of Food Science and Technology, 1*, 397–414.
25. Stout, E., Klaenhammer, T., & Barrangou, R., (2017). CRISPR-Cas technologies and applications in food bacteria. *Annual Review of Food Science and Technology, 8*, 413–437.
26. Arslan, Z., Hermanns, V., Wurm, R., Wagner, R., & Pul, Ü., (2014). Detection and characterization of spacer integration intermediates in type IE CRISPR–Cas system. *Nucleic Acids Research, 42*(12), 7884–7893.
27. Barrangou, R., & Horvath, P., (2012). CRISPR: New horizons in phage resistance and strain identification. *Annual Review of Food Science and Technology, 3*, 143–162.
28. Barrangou, R., & Dudley, E. G., (2016). CRISPR-based typing and next-generation tracking technologies. *Annual Review of Food Science and Technology, 7*, 395–411.
29. Horvath, P., Romero, D. A., Coûté-Monvoisin, A. C., Richards, M., Deveau, H., Moineau, S., Boyaval, P., et al., (2008). Diversity, activity, and evolution of CRISPR loci in *Streptococcus thermophilus*. *Journal of Bacteriology, 190*(4), 1401–1412.
30. (a) Barrangou, R., (2013). CRISPR-Cas systems and RNA-guided interference. *Wiley Interdisciplinary Reviews: RNA, 4*(3), 267–278; (b) Horvath, P., & Barrangou, R., (2010). CRISPR/Cas, the immune system of bacteria and archaea. *Science, 327*(5962), 167–170.
31. Paez-Espino, D., Morovic, W., Sun, C. L., Thomas, B. C., Ueda, K. I., Stahl, B., Barrangou, R., & Banfield, J. F., (2013). Strong bias in the bacterial CRISPR elements that confer immunity to phage. *Nature Communications, 4*(1), 1–7.
32. (a) Broadbent, J. R., Neeno-Eckwall, E. C., Stahl, B., Tandee, K., Cai, H., Morovic, W., Horvath, P., et al., (2012). Analysis of the *Lactobacillus casei* supragenome and its influence in species evolution and lifestyle adaptation. *BMC Genomics, 13*(1), 1–18; (b) Smokvina, T., Wels, M., Polka, J., Chervaux, C., Brisse, S., Boekhorst, J., Van, H. V. J. E., & Siezen, R. J., (2013). *Lactobacillus paracasei* comparative genomics: Towards species pan-genome definition and exploitation of diversity. *PloS One, 8*(7), e68731.
33. Douillard, F. P., Ribbera, A., Kant, R., Pietilä, T. E., Järvinen, H. M., Messing, M., Randazzo, C. L., et al., (2013). Comparative genomic and functional analysis of 100 *Lactobacillus rhamnosus* strains and their comparison with strain GG. *PLoS Genet., 9*(8), e1003683.
34. Hullahalli, K., Rodrigues, M., Schmidt, B. D., Li, X., Bhardwaj, P., & Palmer, K. L., (2015). Comparative analysis of the orphan CRISPR2 locus in 242 Enterococcus faecalis strains. *PloS One, 10*(9), e0138890.
35. (a) Briner, A. E., Lugli, G. A., Milani, C., Duranti, S., Turroni, F., Gueimonde, M., Margolles, A., et al., (2015). Occurrence and diversity of CRISPR-Cas systems in the genus *Bifidobacterium*. *PloS One, 10*(7), e0133661; (b) Sanozky-Dawes, R., Selle, K., O'Flaherty, S., Klaenhammer, T., & Barrangou, R., (2015). Occurrence and activity of a type II CRISPR-Cas system in *Lactobacillus gasseri*. *Microbiology, 161*(9), 1752–1761.
36. Briner, A. E., & Barrangou, R., (2014). *Lactobacillus buchneri* genotyping on the basis of clustered regularly interspaced short palindromic repeat (CRISPR) locus diversity. *Applied and Environmental Microbiology, 80*(3), 994–1001.

37. Zhao, X., Yu, Z., & Xu, Z., (2018). Study the features of 57 confirmed CRISPR loci in 38 strains of *Staphylococcus aureus*. *Frontiers in Microbiology, 9*, 1591.
38. (a) Kovanen, S., Kivistö, R., Rossi, M., & Hänninen, M. L., (2014). A combination of MLST and CRISPR typing reveals dominant *Campylobacter jejuni* types in organically farmed laying hens. *Journal of Applied Microbiology, 117*(1), 249–257; (b) Hargreaves, K. R., Flores, C. O., Lawley, T. D., & Clokie, M. R., (2014). Abundant and diverse clustered regularly interspaced short palindromic repeat spacers in *Clostridium difficile* strains and prophages target multiple phage types within this pathogen. *MBio, 5*(5); (c) Mokrousov, I., Vyazovaya, A., Kolodkina, V., Limeschenko, E., Titov, L., & Narvskaya, O., (2009). Novel macroarray-based method of *Corynebacterium diphtheriae* genotyping: Evaluation in a field study in Belarus. *European Journal of Clinical Microbiology & Infectious Diseases, 28*(6), 701–703; (d) Jiang, Y., Yin, S., Dudley, E. G., & Cutter, C. N., (2015). Diversity of CRISPR loci and virulence genes in pathogenic *Escherichia coli* isolates from various sources. *International Journal of Food Microbiology, 204*, 41–46; (e) D'Auria, G., Jiménez-Hernández, N., Peris-Bondia, F., Moya, A., & Latorre, A., (2010). Legionella pneumophila pangenome reveals strain-specific virulence factors. *BMC Genomics, 11*(1), 1–13; (f) Kinnevey, P. M., Shore, A. C., Brennan, G. I., Sullivan, D. J., Ehricht, R., Monecke, S., Slickers, P., & Coleman, D. C., (2013). Emergence of sequence type 779 methicillin-resistant *Staphylococcus aureus* harboring a novel pseudo staphylococcal cassette chromosome mec (SCCmec)-SCC-SCCCRISPR composite element in Irish hospitals. *Antimicrobial Agents and Chemotherapy, 57*(1), 524–531; (g) Sun, H., Li, Y., Shi, X., Lin, Y., Qiu, Y., Zhang, J., Liu, Y., et al., (2015). Association of CRISPR/Cas evolution with *Vibrio parahaemolyticus* virulence factors and genotypes. *Foodborne Pathogens and Disease, 12*(1), 68–73.
39. Groenen, P. M., Bunschoten, A. E., Soolingen, D. V., & Errtbden, J. D. V., (1993). Nature of DNA polymorphism in the direct repeat cluster of *Mycobacterium tuberculosis*; application for strain differentiation by a novel typing method. *Molecular Microbiology, 10*(5), 1057–1065.
40. (a) Kamerbeek, J., Schouls, L., Kolk, A., Van, A. M., Van, S. D., Kuijper, S., Bunschoten, A., et al., (1997). Simultaneous detection and strain differentiation of *Mycobacterium tuberculosis* for diagnosis and epidemiology. *Journal of Clinical Microbiology, 35*(4), 907–914; (b) Streicher, E., Victor, T., Van, D. S. G., Sola, C., Rastogi, N., Van, H. P., & Warren, R., (2007). Spoligotype signatures in the *Mycobacterium tuberculosis* complex. *Journal of Clinical Microbiology, 45*(1), 237–240.
41. (a) Cowan, L. S., Diem, L., Brake, M. C., & Crawford, J. T., (2004). Transfer of a *Mycobacterium tuberculosis* genotyping method, Spoligotyping, from a reverse line-blot hybridization, membrane-based assay to the Luminex multianalyte profiling system. *Journal of Clinical Microbiology, 42*(1), 474–477; (b) Fabre, L., Zhang, J., Guigon, G., Le Hello, S., Guibert, V., Accou-Demartin, M., De Romans, S., et al., (2012). CRISPR typing and subtyping for improved laboratory surveillance of *Salmonella* infections. *PloS One, 7*(5), e36995.
42. (a) Mokrousov, I., Narvskaya, O., Limeschenko, E., & Vyazovaya, A., (2005). Efficient discrimination within a *Corynebacterium diphtheriae* epidemic clonal group by a novel macroarray-based method. *Journal of Clinical Microbiology, 43*(4), 1662–1668; (b) Mokrousov, I., Limeschenko, E., Vyazovaya, A., & Narvskaya, O., (2007). *Corynebacterium diphtheriae* spoligotyping based on combined use of two CRISPR loci. *Biotechnology Journal: Healthcare Nutrition Technology, 2*(7), 901–906; (c)

Ginevra, C., Jacotin, N., Diancourt, L., Guigon, G., Arquilliere, R., Meugnier, H., Descours, G., et al., (2012). Legionella pneumophila sequence type 1/Paris pulsotype subtyping by spoligotyping. *Journal of Clinical Microbiology, 50*(3), 696–701.
43. Hoe, N., Nakashima, K., Grigsby, D., Pan, X., Dou, S. J., Naidich, S., Garcia, M., et al., (1999). Rapid molecular genetic subtyping of serotype M1 group A *Streptococcus strains*. *Emerging Infectious Diseases, 5*(2), 254.
44. Pourcel, C., Salvignol, G., & Vergnaud, G., (2005). CRISPR elements in yersinia pestis acquire new repeats by preferential uptake of bacteriophage DNA, and provide additional tools for evolutionary studies. *Microbiology, 151*(3), 653–663.
45. Cui, Y., Li, Y., Gorgé, O., Platonov, M. E., Yan, Y., Guo, Z., Pourcel, C., et al., (2008). Insight into microevolution of yersinia pestis by clustered regularly interspaced short palindromic repeats. *PloS One, 3*(7), e2652.
46. (a) Liu, F., Barrangou, R., Gerner-Smidt, P., Ribot, E. M., Knabel, S. J., & Dudley, E. G., (2011). Novel virulence gene and clustered regularly interspaced short palindromic repeat (CRISPR) multilocus sequence typing scheme for subtyping of the major serovars of *Salmonella enterica* subsp. enterica. *Applied and Environmental Microbiology, 77*(6), 1946–1956; (b) Shariat, N., DiMarzio, M. J., Yin, S., Dettinger, L., Sandt, C. H., Lute, J. R., Barrangou, R., & Dudley, E. G., (2013). The combination of CRISPR-MVLST and PFGE provides increased discriminatory power for differentiating human clinical isolates of *Salmonella enterica* subsp. enterica serovar enteritidis. *Food Microbiology, 34*(1), 164–173; (c) Shariat, N., Kirchner, M. K., Sandt, C. H., Trees, E., Barrangou, R., & Dudley, E. G., (2013). Subtyping of *Salmonella enterica* serovar Newport outbreak isolates by CRISPR-MVLST and determination of the relationship between CRISPR-MVLST and PFGE results. *Journal of Clinical Microbiology, 51*(7), 2328–2336; (d) Liu, F., Kariyawasam, S., Jayarao, B. M., Barrangou, R., Gerner-Smidt, P., Ribot, E. M., Knabel, S. J., & Dudley, E. G., (2011). Subtyping *Salmonella enterica* serovar enteritidis isolates from different sources by using sequence typing based on virulence genes and clustered regularly interspaced short palindromic repeats (CRISPRs). *Applied and Environmental Microbiology, 77*(13), 4520–4526.
47. Shariat, N., & Dudley, E. G., (2014). CRISPRs: Molecular signatures used for pathogen subtyping. *Applied and Environmental Microbiology, 80*(2), 430–439.
48. Pijkeren, J. P. V., & Barrangou, R., (2018). Genome editing of food-grade lactobacilli to develop therapeutic probiotics. *Bugs as Drugs: Therapeutic Microbes for the Prevention and Treatment of Disease,* 389–408.
49. Dank, A., Smid, E. J., & Notebaart, R. A., (2018). CRISPR-Cas genome engineering of esterase activity in *Saccharomyces cerevisiae* steers aroma formation. *BMC Research Notes, 11*(1), 1–6.
50. Cho, S., Shin, J., & Cho, B. K., (2018). Applications of CRISPR/Cas system to bacterial metabolic engineering. *International Journal of Molecular Sciences, 19*(4), 1089.
51. Sen, T., Barrow, C. J., & Deshmukh, S. K., (2019). Microbial pigments in the food industry—Challenges and the way forward. *Frontiers in Nutrition, 6*, 7.
52. Zhang, C., Meng, X., Wei, X., & Lu, L., (2016). Highly efficient CRISPR mutagenesis by microhomology-mediated end joining in *Aspergillus fumigatus*. *Fungal Genetics and Biology, 86*, 47–57.
53. Arazoe, T., Miyoshi, K., Yamato, T., Ogawa, T., Ohsato, S., Arie, T., & Kuwata, S., (2015). Tailor-made CRISPR/Cas system for highly efficient targeted gene replacement in the rice blast fungus. *Biotechnology and Bioengineering, 112*(12), 2543–2549.

54. Matsu-Ura, T., Baek, M., Kwon, J., & Hong, C., (2015). Efficient gene editing in *Neurospora Crassa* with CRISPR technology. *Fungal Biology and Biotechnology, 2*(1), 1–7.
55. Altenbuchner, J., (2016). Editing of the *Bacillus subtilis* genome by the CRISPR-Cas9 system. *Applied and Environmental Microbiology, 82*(17), 5421–5427.
56. Westbrook, A. W., Ren, X., Oh, J., Moo-Young, M., & Chou, C. P., (2018). Metabolic engineering to enhance heterologous production of hyaluronic acid in *Bacillus subtilis*. *Metabolic Engineering, 47*, 401–413.
57. Wu, Y., Chen, T., Liu, Y., Lv, X., Li, J., Du, G., Ledesma-Amaro, R., & Liu, L., (2018). CRISPRi allows optimal temporal control of N-acetylglucosamine bioproduction by a dynamic coordination of glucose and xylose metabolism in *Bacillus subtilis*. *Metabolic Engineering, 49*, 232–241.
58. (a) Westbrook, A. W., Moo-Young, M., & Chou, C. P., (2016). Development of a CRISPR-Cas9 tool kit for comprehensive engineering of *Bacillus subtilis*. *Applied and Environmental Microbiology, 82*(16), 4876–4895; (b) Westbrook, A. W., Ren, X., Moo-Young, M., & Chou, C. P., (2018). Metabolic engineering of *Bacillus subtilis* for l-valine overproduction. *Biotechnology and Bioengineering, 115*(11), 2778–2792; (c) Zhang, K., Duan, X., & Wu, J., (2016). Multigene disruption in undomesticated *Bacillus subtilis* ATCC 6051a using the CRISPR/Cas9 system. *Scientific Reports, 6*(1), 1–11; (d) Zhang, K., Su, L., & Wu, J., (2018). Enhanced extracellular pullulanase production in *Bacillus subtilis* using protease-deficient strains and optimal feeding. *Applied Microbiology and Biotechnology, 102*(12), 5089–5103.
59. Mougiakos, I., Bosma, E. F., Weenink, K., Vossen, E., Goijvaerts, K., Van, D. O. J., & Van, K. R., (2017). Efficient genome editing of a facultative thermophile using mesophilic spCas9. *ACS Synthetic Biology, 6*(5), 849–861.
60. Huang, H., Chai, C., Yang, S., Jiang, W., & Gu, Y., (2019). Phage serine integrase-mediated genome engineering for efficient expression of chemical biosynthetic pathway in gas-fermenting *Clostridium ljungdahlii*. *Metabolic Engineering, 52*, 293–302.
61. Huang, H., Chai, C., Li, N., Rowe, P., Minton, N. P., Yang, S., Jiang, W., & Gu, Y., (2016). CRISPR/Cas9-based efficient genome editing in *Clostridium ljungdahlii*, an autotrophic gas-fermenting bacterium. *ACS Synthetic Biology, 5*(12), 1355–1361.
62. Woolston, B. M., Emerson, D. F., Currie, D. H., & Stephanopoulos, G., (2018). Rediverting carbon flux in *Clostridium ljungdahlii* using CRISPR interference (CRISPRi). *Metabolic Engineering, 48*, 243–253.
63. Zhang, J., Zong, W., Hong, W., Zhang, Z. T., & Wang, Y., (2018). Exploiting endogenous CRISPR-Cas system for multiplex genome editing in *Clostridium tyrobutyricum* and engineer the strain for high-level butanol production. *Metabolic Engineering, 47*, 49–59.
64. Pyne, M. E., Bruder, M. R., Moo-Young, M., Chung, D. A., & Chou, C. P., (2016). Harnessing heterologous and endogenous CRISPR-Cas machineries for efficient markerless genome editing in *Clostridium*. *Scientific Reports, 6*(1), 1–15.
65. Nagaraju, S., Davies, N. K., Walker, D. J. F., Köpke, M., & Simpson, S. D., (2016). Genome editing of *Clostridium autoethanogenum* using CRISPR/Cas9. *Biotechnology for Biofuels, 9*(1), 1–8.
66. (a) Wang, Y., Zhang, Z. T., Seo, S. O., Choi, K., Lu, T., Jin, Y. S., & Blaschek, H. P., (2015). Markerless chromosomal gene deletion in *Clostridium beijerinckii* using CRISPR/Cas9 system. *Journal of Biotechnology, 200*, 1–5; (b) Wang, Y., Zhang, Z. T., Seo, S. O., Lynn, P., Lu, T., Jin, Y. S., & Blaschek, H. P., (2016). Bacterial genome

editing with CRISPR-Cas9: Deletion, integration, single nucleotide modification, and desirable "clean" mutant selection in *Clostridium beijerinckii* as an example. *ACS Synthetic Biology, 5*(7), 721–732.
67. Wang, Y., Zhang, Z. T., Seo, S. O., Lynn, P., Lu, T., Jin, Y. S., & Blaschek, H. P., (2016). Gene transcription repression in *Clostridium beijerinckii* using CRISPR-dCas9. *Biotechnology and Bioengineering, 113*(12), 2739–2743.
68. Wen, Z., Minton, N. P., Zhang, Y., Li, Q., Liu, J., Jiang, Y., & Yang, S., (2017). Enhanced solvent production by metabolic engineering of a twin-clostridial consortium. *Metabolic Engineering, 39*, 38–48.
69. (a) Xu, T., Li, Y., Shi, Z., Hemme, C. L., Li, Y., Zhu, Y., Van, N. J. D., et al., (2015). Efficient genome editing in *Clostridium cellulolyticum* via CRISPR-Cas9 nickase. *Applied and Environmental Microbiology, 81*(13), 4423–4431; (b) Xu, T., Li, Y., He, Z., Van, N. J. D., & Zhou, J., (2017). Cas9 nickase-assisted RNA repression enables stable and efficient manipulation of essential metabolic genes in *Clostridium cellulolyticum. Frontiers in Microbiology, 8*, 1744.
70. Wasels, F., Jean-Marie, J., Collas, F., López-Contreras, A. M., & Ferreira, N. L., (2017). A two-plasmid inducible CRISPR/Cas9 genome editing tool for *Clostridium acetobutylicum. Journal of Microbiological Methods, 140*, 5–11.
71. Bruder, M. R., Pyne, M. E., Moo-Young, M., Chung, D. A., & Chou, C. P., (2016). Extending CRISPR-Cas9 technology from genome editing to transcriptional engineering in the genus *Clostridium. Applied and Environmental Microbiology, 82*(20), 6109–6119.
72. Wang, S., Dong, S., Wang, P., Tao, Y., & Wang, Y., (2017). Genome editing in *Clostridium saccharoperbutylacetonicum* N1-4 with the CRISPR-Cas9 system. *Applied and Environmental Microbiology, 83*(10).
73. Hong, W., Zhang, J., Cui, G., Wang, L., & Wang, Y., (2018). Multiplexed CRISPR-Cpf1-mediated genome editing in *Clostridium difficile* toward the understanding of pathogenesis of C. difficile infection. *ACS Synthetic Biology, 7*(6), 1588–1600.
74. Wang, Y., Liu, Y., Liu, J., Guo, Y., Fan, L., Ni, X., Zheng, X., et al., (2018). MACBETH: Multiplex automated *Corynebacterium glutamicum* base editing method. *Metabolic Engineering, 47*, 200–210.
75. Cleto, S., Jensen, J. V., Wendisch, V. F., & Lu, T. K., (2016). *Corynebacterium glutamicum* metabolic engineering with CRISPR interference (CRISPRi). *ACS Synthetic Biology, 5*(5), 375–385.
76. Jiang, Y., Qian, F., Yang, J., Liu, Y., Dong, F., Xu, C., Sun, B., et al., (2017). CRISPR-Cpf1 assisted genome editing of *Corynebacterium glutamicum. Nature Communications, 8*(1), 1–11.
77. (a) Liu, J., Wang, Y., Lu, Y., Zheng, P., Sun, J., & Ma, Y., (2017). Development of a CRISPR/Cas9 genome editing toolbox for *Corynebacterium glutamicum. Microbial Cell Factories, 16*(1), 1–17; (b) Peng, F., Wang, X., Sun, Y., Dong, G., Yang, Y., Liu, X., & Bai, Z., (2017). Efficient gene editing in *Corynebacterium glutamicum* using the CRISPR/Cas9 system. *Microbial Cell Factories, 16*(1), 1–13; (c) Cho, J. S., Choi, K. R., Prabowo, C. P. S., Shin, J. H., Yang, D., Jang, J., & Lee, S. Y., (2017). CRISPR/Cas9-coupled recombineering for metabolic engineering of *Corynebacterium glutamicum. Metabolic Engineering, 42*, 157–167; (d) Wang, B., Hu, Q., Zhang, Y., Shi, R., Chai, X., Liu, Z., Shang, X., et al., (2018). A RecET-assisted CRISPR–Cas9 genome editing in *Corynebacterium glutamicum. Microbial Cell Factories, 17*(1), 1–16.

78. Ao, X., Yao, Y., Li, T., Yang, T. T., Dong, X., Zheng, Z. T., Chen, G. Q., et al., (2018). A multiplex genome editing method for *Escherichia coli* based on CRISPR-Cas12a. *Frontiers in Microbiology, 9*, 2307.
79. Sung, L. Y., Wu, M. Y., Lin, M. W., Hsu, M. N., Truong, V. A., Shen, C. C., Tu, Y., et al., (2019). Combining orthogonal CRISPR and CRISPRi systems for genome engineering and metabolic pathway modulation in *Escherichia coli. Biotechnology and Bioengineering, 116*(5), 1066–1079.
80. Webb, H. E., Brichta-Harhay, D. M., Brashears, M. M., Nightingale, K. K., Arthur, T. M., Bosilevac, J. M., Kalchayanand, N., et al., (2017). *Salmonella* in peripheral lymph nodes of healthy cattle at slaughter. *Frontiers in Microbiology, 8*, 2214.
81. Swartz, J. R., (2018). Expanding biological applications using cell-free metabolic engineering: An overview. *Metabolic Engineering, 50*, 156–172.
82. Liu, S., Xiao, H., Zhang, F., Lu, Z., Zhang, Y., Deng, A., Li, Z., Yang, C., & Wen, T., (2019). A seamless and iterative DNA assembly method named PS-Brick and its assisted metabolic engineering for threonine and 1-propanol production. *Biotechnology for Biofuels, 12*(1), 1–20.
83. Wu, H., Li, Y., Ma, Q., Li, Q., Jia, Z., Yang, B., Xu, Q., et al., (2018). Metabolic engineering of *Escherichia coli* for high-yield uridine production. *Metabolic Engineering, 49*, 248–256.
84. Tan, Z., Yoon, J. M., Chowdhury, A., Burdick, K., Jarboe, L. R., Maranas, C. D., & Shanks, J. V., (2018). Engineering of *E. coli* inherent fatty acid biosynthesis capacity to increase octanoic acid production. *Biotechnology for Biofuels, 11*(1), 1–15.
85. Yang, Z., Wang, H., Wang, Y., Ren, Y., & Wei, D., (2018). Manufacturing multienzymatic complex reactors *in vivo* by self-assembly to improve the biosynthesis of itaconic acid in *Escherichia coli. ACS Synthetic Biology, 7*(5), 1244–1250.
86. Tian, T., Kang, J. W., Kang, A., & Lee, T. S., (2019). Redirecting metabolic flux via combinatorial multiplex CRISPRi-mediated repression for isopentenol production in *Escherichia coli. ACS Synthetic Biology, 8*(2), 391–402.
87. Li, Y., Yan, F., Wu, H., Li, G., Han, Y., Ma, Q., Fan, X., et al., (2019). Multiple-step chromosomal integration of divided segments from a large DNA fragment via CRISPR/Cas9 in *Escherichia coli. Journal of Industrial Microbiology and Biotechnology, 46*(1), 81–90.
88. (a) Kim, S. K., Han, G. H., Seong, W., Kim, H., Kim, S. W., Lee, D. H., & Lee, S. G., (2016). CRISPR interference-guided balancing of a biosynthetic mevalonate pathway increases terpenoid production. *Metabolic Engineering, 38*, 228–240; (b) Lv, L., Ren, Y. L., Chen, J. C., Wu, Q., & Chen, G. Q., (2015). Application of CRISPRi for prokaryotic metabolic engineering involving multiple genes, a case study: Controllable P (3HB-co-4HB) biosynthesis. *Metabolic Engineering, 29*, 160–168; (c) Gao, C., Wang, S., Hu, G., Guo, L., Chen, X., Xu, P., & Liu, L., (2018). Engineering *Escherichia coli* for malate production by integrating modular pathway characterization with CRISPRi-guided multiplexed metabolic tuning. *Biotechnology and Bioengineering, 115*(3), 661–672; (d) Kim, S. K., Seong, W., Han, G. H., Lee, D. H., & Lee, S. G., (2017). CRISPR interference-guided multiplex repression of endogenous competing pathway genes for redirecting metabolic flux in *Escherichia coli. Microbial Cell Factories, 16*(1), 1–15; (e) Chu, L. L., Dhakal, D., Shin, H. J., Jung, H. J., Yamaguchi, T., & Sohng, J. K., (2018). Metabolic engineering of *Escherichia coli* for enhanced production of naringenin 7-sulfate and its biological activities. *Frontiers in Microbiology, 9*, 1671;

(f) Wu, J., Du, G., Chen, J., & Zhou, J., (2015). Enhancing flavonoid production by systematically tuning the central metabolic pathways based on a CRISPR interference system in *Escherichia coli*. *Scientific Reports, 5*(1), 1–14; (g) Li, S., Jendresen, C. B., Grünberger, A., Ronda, C., Jensen, S. I., Noack, S., & Nielsen, A. T., (2016). Enhanced protein and biochemical production using CRISPRi-based growth switches. *Metabolic Engineering, 38*, 274–284; (h) Nielsen, A. A., & Voigt, C. A., (2014). Multi-input CRISPR/C as genetic circuits that interface host regulatory networks. *Molecular Systems Biology, 10*(11), 763; (i) Fontana, J., Voje, W. E., Zalatan, J. G., & Carothers, J. M., (2018). Prospects for engineering dynamic CRISPR–Cas transcriptional circuits to improve bioproduction. *Journal of Industrial Microbiology and Biotechnology, 45*(7), 481–490.
89. Qin, Q., Ling, C., Zhao, Y., Yang, T., Yin, J., Guo, Y., & Chen, G. Q., (2018). CRISPR/Cas9 editing genome of extremophile *Halomonas* spp. *Metabolic Engineering, 47*, 219–229.
90. (a) Jiang, W., Bikard, D., Cox, D., Zhang, F., & Marraffini, L. A., (2013). RNA-guided editing of bacterial genomes using CRISPR-Cas systems. *Nature Biotechnology, 31*(3), 233–239; (b) Li, Y., Lin, Z., Huang, C., Zhang, Y., Wang, Z., Tang, Y. J., Chen, T., & Zhao, X., (2015). Metabolic engineering of *Escherichia coli* using CRISPR–Cas9 meditated genome editing. *Metabolic Engineering, 31*, 13–21; (c) Liang, L., Liu, R., Garst, A. D., Lee, T., Beckham, G. T., & Gill, R. T., (2017). CRISPR EnAbled trackable genome engineering for isopropanol production in *Escherichia coli*. *Metabolic Engineering, 41*, 1–10; (d) Zhao, M., Huang, D., Zhang, X., Koffas, M. A., Zhou, J., & Deng, Y., (2018). Metabolic engineering of *Escherichia coli* for producing adipic acid through the reverse adipate-degradation pathway. *Metabolic Engineering, 47*, 254–262.
91. Wu, J., Zhang, X., Xia, X., & Dong, M., (2017). A systematic optimization of medium chain fatty acid biosynthesis via the reverse beta-oxidation cycle in *Escherichia coli*. *Metabolic Engineering, 41*, 115–124.
92. Sampson, T. R., Saroj, S. D., Llewellyn, A. C., Tzeng, Y. L., & Weiss, D. S., (2013). A CRISPR/Cas system mediates bacterial innate immune evasion and virulence. *Nature, 497*(7448), 254–257.
93. Chen, Y., Chen, X. Y., Du, H. T., Zhang, X., Ma, Y. M., Chen, J. C., Ye, J. W., et al., (2019). Chromosome engineering of the TCA cycle in *Halomonas* bluephagenesis for production of copolymers of 3-hydroxybutyrate and 3-hydroxyvalerate (PHBV). *Metabolic Engineering, 54*, 69–82.
94. Wang, M., Liu, L., Fan, L., & Tan, T., (2017). CRISPRi based system for enhancing 1-butanol production in engineered *Klebsiella pneumoniae*. *Process Biochemistry, 56*, 139–146.
95. Wang, J., Zhao, P., Li, Y., Xu, L., & Tian, P., (2018). Engineering CRISPR interference system in *Klebsiella pneumoniae* for attenuating lactic acid synthesis. *Microbial Cell Factories, 17*(1), 1–12.
96. Shen, J., Zhou, J., Chen, G. Q., & Xiu, Z. L., (2018). Efficient genome engineering of a virulent *Klebsiella bacteriophage* using CRISPR-Cas9. *Journal of Virology, 92*(17).
97. Gunderson, F. F., & Cianciotto, N. P., (2013). The CRISPR-associated gene Cas2 of *Legionella pneumophila* is required for intracellular infection of amoebae. *MBio, 4*(2).
98. Oh, J. H., & Van, P. J. P., (2014). CRISPR–Cas9-assisted recombineering in *Lactobacillus reuteri*. *Nucleic Acids Research, 42*(17), e131–e131.

99. Song, X., Huang, H., Xiong, Z., Ai, L., & Yang, S., (2017). CRISPR-Cas9D10A nickase-assisted genome editing in *Lactobacillus casei*. *Applied and Environmental Microbiology, 83*(22).
100. (a) Choudhary, E., Thakur, P., Pareek, M., & Agarwal, N., (2015). Gene silencing by CRISPR interference in mycobacteria. *Nature Communications, 6*(1), 1–11; (b) Rock, J. M., Hopkins, F. F., Chavez, A., Diallo, M., Chase, M. R., Gerrick, E. R., Pritchard, J. R., et al., (2017). Programmable transcriptional repression in mycobacteria using an orthogonal CRISPR interference platform. *Nature Microbiology, 2*(4), 1–9; (c) Singh, A. K., Carette, X., Potluri, L. P., Sharp, J. D., Xu, R., Prisic, S., & Husson, R. N., (2016). Investigating essential gene function in *Mycobacterium tuberculosis* using an efficient CRISPR interference system. *Nucleic Acids Research, 44*(18), e143-e143.
101. McLeod, S. M., Kimsey, H. H., Davis, B. M., & Waldor, M. K., (2005). CTXφ and *Vibrio cholerae*: Exploring a newly recognized type of phage–host cell relationship. *Molecular Microbiology, 57*(2), 347–356.
102. Gu, T., Zhao, S., Pi, Y., Chen, W., Chen, C., Liu, Q., Li, M., et al., (2018). Highly efficient base editing in *Staphylococcus aureus* using an engineered CRISPR RNA-guided cytidine deaminase. *Chemical Science, 9*(12), 3248–3253.
103. Chen, W., Zhang, Y., Yeo, W. S., Bae, T., & Ji, Q., (2017). Rapid and efficient genome editing in *Staphylococcus aureus* by using an engineered CRISPR/Cas9 system. *Journal of the American Chemical Society, 139*(10), 3790–3795.
104. Dong, X., Jin, Y., Ming, D., Li, B., Dong, H., Wang, L., Wang, T., & Wang, D., (2017). CRISPR/dCas9-mediated inhibition of gene expression in *Staphylococcus aureus*. *Journal of Microbiological Methods, 139*, 79–86.
105. Tan, S. Z., Reisch, C. R., & Prather, K. L., (2018). A robust CRISPR interference gene repression system in *Pseudomonas*. *Journal of Bacteriology, 200*(7).
106. Barrangou, R., Fremaux, C., Deveau, H., Richards, M., Boyaval, P., Moineau, S., Romero, D. A., & Horvath, P., (2007). CRISPR provides acquired resistance against viruses in prokaryotes. *Science, 315*(5819), 1709–1712.
107. (a) Cobb, R. E., Wang, Y., & Zhao, H., (2015). High-efficiency multiplex genome editing of *Streptomyces* species using an engineered CRISPR/Cas system. *ACS Synthetic Biology, 4*(6), 723–728; (b) Zhang, M. M., Wong, F. T., Wang, Y., Luo, S., Lim, Y. H., Heng, E., Yeo, W. L., et al., (2017). CRISPR–Cas9 strategy for activation of silent *Streptomyces* biosynthetic gene clusters. *Nature Chemical Biology, 13*(6), 607–609.
108. Li, L., Zheng, G., Chen, J., Ge, M., Jiang, W., & Lu, Y., (2017). Multiplexed site-specific genome engineering for overproducing bioactive secondary metabolites in actinomycetes. *Metabolic Engineering, 40*, 80–92.
109. Tong, Y., Charusanti, P., Zhang, L., Weber, T., & Lee, S. Y., (2015). CRISPR-Cas9 based engineering of actinomycetal genomes. *ACS Synthetic Biology, 4*(9), 1020–1029.
110. Li, L., Wei, K., Zheng, G., Liu, X., Chen, S., Jiang, W., & Lu, Y., (2018). CRISPR-Cpf1-assisted multiplex genome editing and transcriptional repression in *Streptomyces*. *Applied and Environmental Microbiology, 84*(18).
111. Jia, H., Zhang, L., Wang, T., Han, J., Tang, H., & Zhang, L., (2017). Development of a CRISPR/Cas9-mediated gene-editing tool in *Streptomyces rimosus*. *Microbiology, 163*(8), 1148–1155.
112. Meng, J., Feng, R., Zheng, G., Ge, M., Mast, Y., Wohlleben, W., Gao, J., et al., (2017). Improvement of pristinamycin I (PI) production in *Streptomyces pristinaespiralis* by metabolic engineering approaches. *Synthetic and Systems Biotechnology, 2*(2), 130–136.

113. Jiang, Y., Chen, B., Duan, C., Sun, B., Yang, J., & Yang, S., (2015). Multigene editing in the *Escherichia coli* genome via the CRISPR-Cas9 system. *Applied and Environmental Microbiology, 81*(7), 2506–2514.
114. Yan, M. Y., Yan, H. Q., Ren, G. X., Zhao, J. P., Guo, X. P., & Sun, Y. C., (2017). CRISPR-Cas12a-assisted recombineering in bacteria. *Applied and Environmental Microbiology, 83*(17).
115. Weninger, A., Hatzl, A. M., Schmid, C., Vogl, T., & Glieder, A., (2016). Combinatorial optimization of CRISPR/Cas9 expression enables precision genome engineering in the methylotrophic yeast *Pichia pastoris*. *Journal of Biotechnology, 235*, 139–149.
116. Nødvig, C. S., Nielsen, J. B., Kogle, M. E., & Mortensen, U. H., (2015). A CRISPR-Cas9 system for genetic engineering of filamentous fungi. *PloS One, 10*(7), e0133085.
117. Katayama, T., Tanaka, Y., Okabe, T., Nakamura, H., Fujii, W., Kitamoto, K., & Maruyama, J. I., (2016). Development of a genome editing technique using the CRISPR/Cas9 system in the industrial filamentous fungus *Aspergillus oryzae*. *Biotechnology Letters, 38*(4), 637–642.
118. (a) Min, K., Ichikawa, Y., Woolford, C. A., & Mitchell, A. P., (2016). *Candida albicans* gene deletion with a transient CRISPR-Cas9 system. *MSphere, 1*(3); (b) Tsai, S. C., Tsai, L. D., & Li, Y. K., (2005). An isolated *Candida albicans* TL3 capable of degrading phenol at large concentration. *Bioscience, Biotechnology, and Biochemistry, 69*(12), 2358–2367; (c) Vyas, V. K., Barrasa, M. I., & Fink, G. R., (2015). A *Candida albicans* CRISPR system permits genetic engineering of essential genes and gene families. *Science Advances, 1*(3), e1500248.
119. Horwitz, A. A., Walter, J. M., Schubert, M. G., Kung, S. H., Hawkins, K., Platt, D. M., Hernday, A. D., et al., (2015). Efficient multiplexed integration of synergistic alleles and metabolic pathways in yeasts via CRISPR-Cas. *Cell Systems, 1*(1), 88–96.
120. Liu, Q., Gao, R., Li, J., Lin, L., Zhao, J., Sun, W., & Tian, C., (2017). Development of a genome-editing CRISPR/Cas9 system in thermophilic fungal *Myceliophthora* species and its application to hyper-cellulase production strain engineering. *Biotechnology for Biofuels, 10*(1), 1–14.
121. Pohl, C., Kiel, J., Driessen, A., Bovenberg, R., & Nygard, Y., (2016). CRISPR/Cas9 based genome editing of *Penicillium chrysogenum*. *ACS Synthetic Biology, 5*(7), 754–764.
122. Ni, J., Zhang, G., Qin, L., Li, J., & Li, C., (2019). Simultaneously down-regulation of multiplex branch pathways using CRISPRi and fermentation optimization for enhancing β-amyrin production in *Saccharomyces cerevisiae*. *Synthetic and Systems Biotechnology, 4*(2), 79–85.
123. Schwab, C., Tveit, A. T., Schleper, C., & Urich, T., (2014). Gene expression of *Lactobacilli* in murine forestomach biofilms. *Microbial Biotechnology, 7*(4), 347–359.
124. Mitsui, R., Yamada, R., & Ogino, H., (2019). Improved stress tolerance of *Saccharomyces cerevisiae* by CRISPR-Cas-mediated genome evolution. *Applied Biochemistry and Biotechnology, 189*(3), 810–821.
125. Kildegaard, K. R., Tramontin, L. R. R., Chekina, K., Li, M., Goedecke, T. J., Kristensen, M., & Borodina, I., (2019). CRISPR/Cas9-RNA interference system for combinatorial metabolic engineering of *Saccharomyces cerevisiae*. *Yeast, 36*(5), 237–247.
126. Deaner, M., Holzman, A., & Alper, H. S., (2018). Modular ligation extension of guide RNA operons (LEGO) for multiplexed dCas9 regulation of metabolic pathways in *Saccharomyces cerevisiae*. *Biotechnology Journal, 13*(9), 1700582.

127. Ferreira, R., Skrekas, C., Hedin, A., Sánchez, B. J., Siewers, V., Nielsen, J., & David, F., (2019). Model-assisted fine-tuning of central carbon metabolism in yeast through dCas9-based regulation. *ACS Synthetic Biology, 8*(11), 2457–2463.
128. Liu, R., Chen, L., Jiang, Y., Zhou, Z., & Zou, G., (2015). Efficient genome editing in filamentous fungus *Trichoderma reesei* using the CRISPR/Cas9 system. *Cell Discovery, 1*(1), 1–11.
129. (a) Geiser, E., Wiebach, V., Wierckx, N., & Blank, L. M., (2014). Prospecting the biodiversity of the fungal family *Ustilaginaceae* for the production of value-added chemicals. *Fungal Biology and Biotechnology, 1*(1), 1–10; (b) Schuster, M., Schweizer, G., Reissmann, S., & Kahmann, R., (2016). Genome editing in *Ustilago maydis* using the CRISPR–Cas system. *Fungal Genetics and Biology, 89*, 3–9.
130. (a) Gao, S., Tong, Y., Wen, Z., Zhu, L., Ge, M., Chen, D., Jiang, Y., & Yang, S., (2016). Multiplex gene editing of the *Yarrowia lipolytica* genome using the CRISPR-Cas9 system. *Journal of Industrial Microbiology and Biotechnology, 43*(8), 1085–1093; (b) Schwartz, C., Shabbir-Hussain, M., Frogue, K., Blenner, M., & Wheeldon, I., (2017). Standardized markerless gene integration for pathway engineering in *Yarrowia lipolytica*. *ACS Synthetic Biology, 6*(3), 402–409.
131. Ungerer, J., & Pakrasi, H. B., (2016). Cpf1 is a versatile tool for CRISPR genome editing across diverse species of cyanobacteria. *Scientific Reports, 6*(1), 1–9.
132. (a) Wendt, K. E., Ungerer, J., Cobb, R. E., Zhao, H., & Pakrasi, H. B., (2016). CRISPR/Cas9 mediated targeted mutagenesis of the fast-growing cyanobacterium *Synechococcus elongatus* UTEX 2973. *Microbial Cell Factories, 15*(1), 1–8; (b) Li, H., Shen, C. R., Huang, C. H., Sung, L. Y., Wu, M. Y., & Hu, Y. C., (2016). CRISPR-Cas9 for the genome engineering of cyanobacteria and succinate production. *Metabolic Engineering, 38*, 293–302.
133. Vickers, N. J., (2017). Animal communication: When I'm calling you, will you answer too? *Current Biology, 27*(14), R713–R715.
134. Xiao, Y., Wang, S., Rommelfanger, S., Balassy, A., Barba-Ostria, C., Gu, P., Galazka, J. M., & Zhang, F., (2018). Developing a Cas9-based tool to engineer native plasmids in *Synechocystis* sp. PCC 6803. *Biotechnology and Bioengineering, 115*(9), 2305–2314.
135. Shabestary, K., Anfelt, J., Ljungqvist, E., Jahn, M., Yao, L., & Hudson, E. P., (2018). Targeted repression of essential genes to arrest growth and increase carbon partitioning and biofuel titers in cyanobacteria. *ACS Synthetic Biology, 7*(7), 1669–1675.
136. Gordon, G. C., Korosh, T. C., Cameron, J. C., Markley, A. L., Begemann, M. B., & Pfleger, B. F., (2016). CRISPR interference as a titratable, trans-acting regulatory tool for metabolic engineering in the cyanobacterium *Synechococcus* sp. strain PCC 7002. *Metabolic Engineering, 38*, 170–179.
137. Huang, C. H., Shen, C. R., Li, H., Sung, L. Y., Wu, M. Y., & Hu, Y. C., (2016). CRISPR interference (CRISPRi) for gene regulation and succinate production in cyanobacterium S. elongatus PCC 7942. *Microbial Cell Factories, 15*(1), 1–11.
138. Marraffini, L. A., & Sontheimer, E. J., (2008). CRISPR interference limits horizontal gene transfer in staphylococci by targeting DNA. *Science, 322*(5909), 1843–1845.
139. Brouns, S. J., Jore, M. M., Lundgren, M., Westra, E. R., Slijkhuis, R. J., Snijders, A. P., Dickman, M. J., et al., (2008). Small CRISPR RNAs guide antiviral defense in prokaryotes. *Science, 321*(5891), 960–964.

140. Levin, B. R., (2010). Nasty viruses, costly plasmids, population dynamics, and the conditions for establishing and maintaining CRISPR-mediated adaptive immunity in bacteria. *PLoS Genet., 6*(10), e1001171.
141. (a) Weigel, L. M., Clewell, D. B., Gill, S. R., Clark, N. C., McDougal, L. K., Flannagan, S. E., Kolonay, J. F., et al., (2003). Genetic analysis of a high-level vancomycin-resistant isolate of *Staphylococcus aureus*. *Science, 302*(5650), 1569–1571; (b) Furuya, E. Y., & Lowy, F. D., (2006). Antimicrobial-resistant bacteria in the community setting. *Nature Reviews Microbiology, 4*(1), 36–45.
142. De Melo, A. G., Levesque, S., & Moineau, S., (2018). Phages as friends and enemies in food processing. *Current Opinion in Biotechnology, 49*, 185–190.
143. Hynes, A. P., Labrie, S. J., & Moineau, S., (2016). Programming native CRISPR arrays for the generation of targeted immunity. *MBio, 7*(3).
144. Beisel, C. L., Gomaa, A. A., & Barrangou, R., (2014). A CRISPR design for next-generation antimicrobials. *Genome Biology, 15*(11), 1–4.
145. Gomaa, A. A., Klumpe, H. E., Luo, M. L., Selle, K., Barrangou, R., & Beisel, C. L., (2014). Programmable removal of bacterial strains by use of genome-targeting CRISPR-Cas systems. *MBio, 5*(1).
146. Citorik, R. J., Mimee, M., & Lu, T. K., (2014). Sequence-specific antimicrobials using efficiently delivered RNA-guided nucleases. *Nature Biotechnology, 32*(11), 1141–1145.
147. Shabbir, M. A. B., Shabbir, M. Z., Wu, Q., Mahmood, S., Sajid, A., Maan, M. K., Ahmed, S., et al., (2019). CRISPR-Cas system: Biological function in microbes and its use to treat antimicrobial resistant pathogens. *Annals of Clinical Microbiology and Antimicrobials, 18*(1), 1–9.
148. (a) Jinek, M., Chylinski, K., Fonfara, I., Hauer, M., Doudna, J. A., & Charpentier, E., (2012). A programmable dual-RNA–guided DNA endonuclease in adaptive bacterial immunity. *Science, 337*(6096), 816–821; (b) Van, D. O. J., Westra, E. R., Jackson, R. N., & Wiedenheft, B., (2014). Unravelling the structural and mechanistic basis of CRISPR–Cas systems. *Nature Reviews Microbiology, 12*(7), 479–492.
149. Selle, K., & Barrangou, R., (2015). CRISPR-Based technologies and the future of food science. *Journal of Food Science, 80*(11), R2367–R2372.
150. Vercoe, R. B., Chang, J. T., Dy, R. L., Taylor, C., Gristwood, T., Clulow, J. S., Richter, C., et al., (2013). Cytotoxic chromosomal targeting by CRISPR/Cas systems can reshape bacterial genomes and expel or remodel pathogenicity islands. *PLoS Genet., 9*(4), e1003454.
151. (a) Garneau, J. E., Dupuis, M. È., Villion, M., Romero, D. A., Barrangou, R., Boyaval, P., Fremaux, C., et al., (2010). The CRISPR/Cas bacterial immune system cleaves bacteriophage and plasmid DNA. *Nature, 468*(7320), 67–71; (b) Gasiunas, G., Barrangou, R., Horvath, P., & Siksnys, V., (2012). Cas9–crRNA ribonucleoprotein complex mediates specific DNA cleavage for adaptive immunity in bacteria. *Proceedings of the National Academy of Sciences, 109*(39), E2579–E2586.
152. Marcó, M. B., Moineau, S., & Quiberoni, A., (2012). Bacteriophages and dairy fermentations. *Bacteriophage, 2*(3), 149–158.
153. Barrangou, R., Coûté-Monvoisin, A. C., Stahl, B., Chavichvily, I., Damange, F., Romero, D. A., Boyaval, P., et al., (2013). Genomic impact of CRISPR immunization against bacteriophages. *Biochemical Society Transactions, 41*(6), 1383–1391.
154. McGonigle, P., & Ruggeri, B., (2014). Animal models of human disease: Challenges in enabling translation. *Biochemical Pharmacology, 87*(1), 162–171.

155. Bendixen, E., Danielsen, M., Larsen, K., & Bendixen, C., (2010). Advances in porcine genomics and proteomics—A toolbox for developing the pig as a model organism for molecular biomedical research. *Briefings in Functional Genomics, 9*(3), 208–219.
156. Komor, A. C., Badran, A. H., & Liu, D. R., (2017). CRISPR-based technologies for the manipulation of eukaryotic genomes. *Cell, 168*(1, 2), 20–36.
157. Anzalone, A. V., Koblan, L. W., & Liu, D. R., (2020). Genome editing with CRISPR–Cas nucleases, base editors, transposases, and prime editors. *Nature Biotechnology, 38*(7), 824–844.
158. Cong, Á., Ran, F., Cox, D., Lin, S., Barretto, R., Habib, N., Hsu, P., Wu, X., et al., (2013). Multiplex genome engineering using CRISPR/Cas systems. *Science, 339*(6121), 819–823.
159. Huang, J., Wang, Y., & Zhao, J., (2018). CRISPR editing in biological and biomedical investigation. *Journal of Cellular Physiology, 233*(5), 3875–3891.
160. Fussner, E., Djuric, U., Strauss, M., Hotta, A., Perez-Iratxeta, C., Lanner, F., Dilworth, F. J., et al., (2011). Constitutive heterochromatin reorganization during somatic cell reprogramming. *The EMBO Journal, 30*(9), 1778–1789.
161. (a) Yang, L., Güell, M., Niu, D., George, H., Lesha, E., Grishin, D., Aach, J., et al., (2015). Genome-wide inactivation of porcine endogenous retroviruses (PERVs). *Science, 350*(6264), 1101–1104; (b) Niu, D., Wei, H. J., Lin, L., George, H., Wang, T., Lee, I. H., Zhao, H. Y., et al., (2017). Inactivation of porcine endogenous retrovirus in pigs using CRISPR-Cas9. *Science, 357*(6357), 1303–1307.
162. (a) Darwin, C., & Bynum, W. F., (2009). *The Origin of Species by Means of Natural Selection: Or, the Preservation of Favored Races in the Struggle for Life*. AL Burt New York; (b) Wen, J., Mu, P., & Deng, Y., (2016). Mycotoxins: Cytotoxicity and biotransformation in animal cells. *Toxicology Research, 5*(2), 377–387.
163. Wei, J., Wagner, S., Maclean, P., Brophy, B., Cole, S., Smolenski, G., Carlson, D. F., et al., (2018). Cattle with a precise, zygote-mediated deletion safely eliminate the major milk allergen beta-lactoglobulin. *Scientific Reports, 8*(1), 1–13.
164. Yu, S., Luo, J., Song, Z., Ding, F., Dai, Y., & Li, N., (2011). Highly efficient modification of beta-lactoglobulin (BLG) gene via zinc-finger nucleases in cattle. *Cell Research, 21*(11), 1638–1640.
165. Cui, C., Song, Y., Liu, J., Ge, H., Li, Q., Huang, H., Hu, L., et al., (2015). Gene targeting by TALEN-induced homologous recombination in goats directs production of β-lactoglobulin-free, high-human lactoferrin milk. *Scientific Reports, 5*(1), 1–11.
166. Luo, Y., Wang, Y., Liu, J., Cui, C., Wu, Y., Lan, H., Chen, Q., et al., (2016). Generation of TALE nickase-mediated gene-targeted cows expressing human serum albumin in mammary glands. *Scientific Reports, 6*(1), 1–11.
167. Kambadur, R., Sharma, M., Smith, T. P., & Bass, J. J., (1997). Mutations in myostatin (GDF8) in double-muscled Belgian blue and piedmontese cattle. *Genome Research, 7*(9), 910–915.
168. Van, E. A. L., (2019). In application of genome editing in farm animals: Cattle. *Transgenic Research* (pp. 93–100). Springer.
169. (a) Whitworth, K. M., Rowland, R. R., Ewen, C. L., Trible, B. R., Kerrigan, M. A., Cino-Ozuna, A. G., Samuel, M. S., et al., (2016). Gene-edited pigs are protected from porcine reproductive and respiratory syndrome virus. *Nature Biotechnology, 34*(1), 20–22; (b) Burkard, C., Lillico, S. G., Reid, E., Jackson, B., Mileham, A. J., Ait-Ali, T., Whitelaw, C. B. A., & Archibald, A. L., (2017). Precision engineering for PRRSV resistance in

pigs: Macrophages from genome edited pigs lacking CD163 SRCR5 domain are fully resistant to both PRRSV genotypes while maintaining biological function. *PLoS Pathogens, 13*(2), e1006206.
170. (a) Wang, X., Niu, Y., Zhou, J., Zhu, H., Ma, B., Yu, H., Yan, H., et al., (2018). CRISPR/Cas9-mediated MSTN disruption and heritable mutagenesis in goats causes increased body mass. *Animal Genetics, 49*(1), 43–51; (b) Khalil, K., Elayat, M., Khalifa, E., Daghash, S., Elaswad, A., Miller, M., Abdelrahman, H., et al., (2017). Generation of myostatin gene-edited channel catfish (*Ictalurus punctatus*) via zygote injection of CRISPR/Cas9 system. *Scientific Reports, 7*(1), 1–12.
171. Große-Brinkhaus, C., Storck, L. C., Frieden, L., Neuhoff, C., Schellander, K., Looft, C., & Tholen, E., (2015). Genome-wide association analyses for boar taint components and testicular traits revealed regions having pleiotropic effects. *BMC Genetics, 16*(1), 1–16.
172. Wargelius, A., Leininger, S., Skaftnesmo, K. O., Kleppe, L., Andersson, E., Taranger, G. L., Schulz, R. W., & Edvardsen, R. B., (2016). Dnd knockout ablates germ cells and demonstrates germ cell independent sex differentiation in Atlantic salmon. *Scientific Reports, 6*(1), 1–8.
173. Van, G. H., Van, B. W., Van, D. J., Delputte, P. L., & Nauwynck, H. J., (2010). Identification of the CD163 protein domains involved in infection of the porcine reproductive and respiratory syndrome virus. *Journal of Virology, 84*(6), 3101–3105.
174. Shanthalingam, S., Tibary, A., Beever, J. E., Kasinathan, P., Brown, W. C., & Srikumaran, S., (2016). Precise gene editing paves the way for derivation of *Mannheimia haemolytica* leukotoxin-resistant cattle. *Proceedings of the National Academy of Sciences, 113*(46), 13186–13190.
175. Gao, Y., Wu, H., Wang, Y., Liu, X., Chen, L., Li, Q., Cui, C., et al., (2017). Single Cas9 nickase induced generation of NRAMP1 knockin cattle with reduced off-target effects. *Genome Biology, 18*(1), 1–15.
176. (a) Zhang, X., Li, Z., Yang, H., Liu, D., Cai, G., Li, G., Mo, J., et al., (2018). Novel transgenic pigs with enhanced growth and reduced environmental impact. *Elife, 7*, e34286; (b) Wells, K. D., Bardot, R., Whitworth, K. M., Trible, B. R., Fang, Y., Mileham, A., Kerrigan, M. A., et al., (2017). Replacement of porcine CD163 scavenger receptor cysteine-rich domain 5 with a CD163-like homolog confers resistance of pigs to genotype 1 but not genotype 2 porcine reproductive and respiratory syndrome virus. *Journal of Virology, 91*(2).
177. (a) Zhang, D., Hussain, A., Manghwar, H., Xie, K., Xie, S., Zhao, S., Larkin, R. M., et al., (2020). Genome editing with the CRISPR-Cas system: An art, ethics, and global regulatory perspective. *Plant Biotechnology Journal, 18*(8), 1651–1669; (b) Li, Q., Qin, Z., Wang, Q., Xu, T., Yang, Y., & He, Z., (2019). Applications of genome editing technology in animal disease modeling and gene therapy. *Computational and Structural Biotechnology Journal, 17*, 689–698.

CHAPTER 5

CRISPR-BASED MULTIPLEX GENOME EDITING IN CROP PLANTS TO END HUNGER FROM PLANET EARTH

MUBEEN FATIMA,[1,2] SIDQUA ZAFAR,[1] KABEER HANEEF,[3] MUHAMMAD ASHAR AYUB,[4] SYED MUHAMMAD HASSAN ASKRI,[5] WAQAR SHAFAQAT,[6] RABIA SALEEM,[7] and WAJID UMAR[8]

[1]*Department of Forensic Science, Lahore University of Biological & Applied Sciences (UBAS), Lahore, Punjab, Pakistan*

[2]*Center for Applied Molecular Biology and Forensic Sciences, University of Punjab, Lahore, Pakistan*

[3]*School of Life Science, Tsinghua University, Beijing, China*

[4]*Institute of Agro-Industry and Environment, The Faculty of Agriculture and Environment, The Islamia University of Bahawalpur, 63100, Punjab, Pakistan*

[5]*Zhejiang Key Laboratory of Crop Germplasm Resource, Department of Agronomy, College of Agriculture and Biotechnology, Zhejiang University, Hangzhou, 310058, China*

[6]*Department of Forestry, College of Forest Resources, Mississippi State University, Mississippi State 39762, US*

[7]*Department of Biological Sciences, Forman Christian College and University, Lahore, Pakistan*

[8]*Leibniz Institute of Agricultural Engineering and Bioeconomy, 14469 Potsdam, Germany*

CRISPR-Based Gene Editing: A Solution for Global Food Security. Shakeel Ahmad, Asif Ali Khan, and Muhammad Arshad Javed (Eds.)

ABSTRACT

Food security for ever-increasing global population has become a lead issue of modern-day and genetic engineering of food crops can be a possible answer of this problem. Recent interventions have made the directed selection of resistance/susceptible genes feasible in crop species with the intricate possibility of offering a huge number of potential targets for crop protection. However, the latest progress in cutting-edge technologies showing directed effects and intricate specificity remains elusive for a long time. CRISPR/Cas9 is RNA-guided newly emerged technology undergoing sophisticated implementation in eukaryotic crop plants to overcome hunger issues. Using RNA-guided meticulous interventions, it has become possible to modulate the targeted genome using CRISPR/Cas9 technology to overcome the surplus issue of foods. Genetic knockout through CRISPR/Cas9 is now a feasible option showing unprecedented efficiency with the possibility of multiplexing several targets and even genome-wide mutagenesis screening. Availability of Cas9 enzymes from bacterial species has enabled to achieve better efficacy through specified interventions. In this review, we will summarize the CRISPR/Cas9 system, recent progress, and improvements ongoing to trigger yield production in crop plants. Furthermore, we discuss the overview of sgRNAs mediated interventions and specified selection issues of CRISPR/Cas9 to meet surplus issues. Additionally, we will discuss the need for advancements required as future dimensions in this CRISPR/Cas9 technique.

5.1 INTRODUCTION

Climate change and the growing population are two main concerns of the modern world. The total world population is expected to increase up to 9.7 billion by 2050. This huge growing population will be a challenge for the future food and nutritional security of the world [1]. Moreover, the undergoing cycles of climate change will impact the planet Earth negatively leading to the destruction of natural ecosystems. This destruction will impose environmental stresses that will affect agricultural productivity [2]. It is estimated that up to 70% rise in food production will be needed to keep up with the speed of population growth. So, there is an urgent need to develop such crop varieties that give rise to higher yields as well as the ability to cope up with biotic and abiotic stresses [1]. Classical plant breeding played an important role in high-yielding variety development but now it is unable to face future challenges because variety development is a resource-intensive and time-consuming process [1, 3]. Thus, a replaceable approach is required to develop new desirable varieties in the shortest period of time [1]. Genome editing (GE) is one of the new plant breeding technologies (NPBT) that will be able to commit sustainability to global food security [4].

GE is a kind of genetic engineering in which DNA can be inserted, deleted, or replaced in the living organism's genome that is depicted in Figure 5.1. A double-strand break in the DNA of an organism at a specific locus is needed to be generated for the initiation of GE [5]. Different nucleases that are also known as "molecular scissors" have been engineered by researchers for the precision of Double-stranded breaks [6]. Previously, scientists made efforts to understand and redesign different protein templates like TALEN and zinc finger nucleases (ZFNs) [7]. These tools played an important role in different GE applications including mutation and correction involving monogenic diseases [8]. DNA binding domains of both ZFN and TALEN contain tandem repeats which can be engineered for recognition of specific DNA sequences [9]. DSBs are generated by guiding the resulting chimeric nucleases to the target sequence. For each target site, new ZFN and TALENS chimeric proteins are required. So, this is the major problem in the use of both for GE because it is not an easy task to engineer the new protein [9].

Therefore, a new versatile system CRISPR-Cas was developed that has paved a way for a revolution in life sciences [8]. In the CRISPR-Cas system, Cas endonuclease is directed towards the target DNA by a short singe guided RNA (sgRNA). For each new target site, a new sgRNA is required. So, the gene-editing process is simplified by this method [9]. On the basis of morphological, phylogenetic, and structural characteristics of Cas proteins, CRISPR-Cas systems are classified into class 1 and class 2 that have different mechanisms of guided RNA (gRNA) biogenesis and target interference [10]. Class 1 includes type I, III, and IV while Class 2 includes type II, V, and V of Cas proteins. CRISPR-associated nucleases belonging to Class 2 are highly adapted for use of nucleic acid manipulation. However, CRISPR-Cas9 system that belongs to type II, is best characterized and extensively used for plant genome targeting [11].

CRISPR-Cas9 system was first established by three independent groups for use in Wheat *(Triticum aestivum)*, Rice (*Oryza sativa*), Arabidopsis thaliana, and Tobacco (*Nicotiana benthamiana*) in 2013 [12]. A synthetic sgRNA and Cas9 endonuclease are present in the CRISPR/Cas9 molecular immunity system. Cas9 endonucleases are guided by sgRNA to the target sequence that is complementary to its 20 nucleotides, leading to the PAM (Protospacer associated motif) which is required for the activity of Cas9 [13]. Several genomic loci can be targeted at the same time by using multiple sgRNA [14]. Due to the establishment of CRISPR-Cas9, plant breeders were able to control the precise introduction of targeted sequence variations, which proved to be a game-changer for rapid improvement in agricultural crops [15, 16].

Due to various improvements in CRISPR-Cas systems, such as nucleotide substitution tools for base editing and CRISPR/CPf1, GE has become a cost-effective, easy to use, and well adopted targeted genetic manipulation tool that has been used for different crops (Figure 5.1). Traits like biotic and abiotic stress resistance, yield, and quality have been modified by GE. This approach also played a role in the enhancement of hybrid-breeding techniques [17, 18]. Due to CRISPR-Cas9 technological advancements,

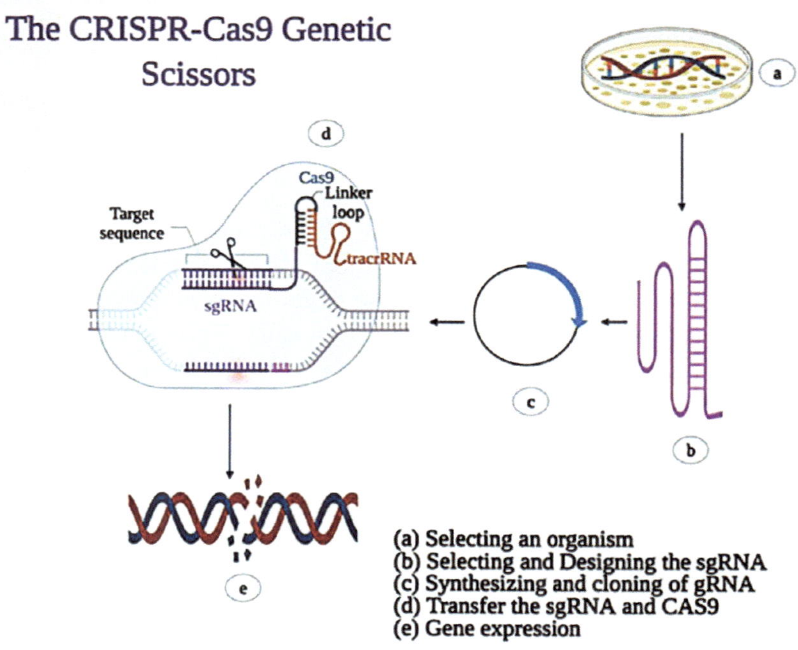

FIGURE 5.1 Workflow strategy for clustered regularly interspaced short palindromic repeats (CRISPR)-associated protein 9 (Cas9) to mediate genome editing: Single-guide RNA (sgRNA) as well as Cas9 enzyme make up the CRISPR/Cas9 complex. The pattern Cas9 enzyme is directed by sgRNA to generate double-strand DNA breaks in the target DNA. Reassortment and non-homologous final stage (NHEJ) processes are part of the cell's DNA repair machinery. Specifically, sgRNA sequences are designed according to compatibility of target genome to modulate genome sequence. In brief, specified selection of sgRNA is carried first which is subsequently cloned and transferred into particular host cells. Proper testing of desirable trait could be carried through western blot analysis, respectively to assess the expression profile.

traits can be modified within a single generation by the addition of desirable traits or elimination of unwanted traits in commercial varieties [16]. It is a source of efficient and rapid development of new high-yielding varieties. For Example, a single allele from one variety can be precisely transferred to another variety belonging to the same species, e.g., Transfer of disease-resistant allele from non-commercial to a commercial variety [19]. Rapid advancement in genome engineering of crops has been observed in the last five years [19]. Thus, enhancement of global food security and sustainable agriculture is possible by CRISPR-Cas.

5.2 SGRNAS-MEDIATED INTERVENTIONS TO TRIGGER PLANT GENOME EDITING (GE)

CRISPR/Cas systems are highly diversified. In natural systems, nuclease Cas9 and two short RNAs, CRISPR RNA (crRNA) and a small trans-activating CRISPR RNA (tracrRNA) form a complex feasible to cut a viral DNA or a foreign Plasmid entering the bacterial cell [20]. The crRNA has a target sequence of 20 nucleotides that guide Cas9 to a particular locus and to the sequence where Cas9 binds. The tracrRNA develops pre-crRNA into a mature one by connecting crRNA with Cas9 [21–23]. The CRISPR/Cas is transitioned into a genome engineering tool from a biological phenomenon by the work of Jinek et al. [24]. They changed the 20 nucleotides that were responsible for targeting DNA to a specific locus. They combined the specificity of crRNA and structural stability of tracrRNA into a single chimeric guide RNA (sgRNA). For this purpose, they fuse the 5′-end of tracrRNA and the 3′-end of crRNA with a GAAA tetraloop. Moreover, the parts of crRNA and tracrRNA that were not used in guiding DNA were cut. As a result, a more efficient and less complex system consisting of 2 components only came into being. The length of sgRNA is 20–22 nucleotides long, so it is very easy to design and synthesize them as oligonucleotides [25]. The DNA sequence along with 5′-N (20–22) NGG can be targeted by a Cas9 nuclease by bringing change in the 20–22 guided sequence [20]. The sgRNA guides the Cas9 endonuclease that cleaves a particular DNA sequence with the help of conserved RuvC and HNH nuclease domains [21]. The dependency of the cleavage is basically on PAM (protospacer adjacent motif) 5′-NGG-3′ that is a short sequence motif present next to the target sequence [20, 26, 27]. Hence, (DBS) double-stranded breaks are produced at 3 to 4 base pairs upstream to the PAM sequence within the target sequence [21].

The double-stranded breaks produced within the target sequence can be repaired by two mechanisms:

i. Homology directed repair (HDR) pathway; and
ii. Non-homologous end-joining (NHEJ) pathway.

If the repair template is absent, then the ligation of DSBs is mediated by the NHEJ pathway that is resulted in insertion or deletion (Indel) mutations at the target site. As an end result, gene knockout takes place as frameshift mutations and premature stop codons generated by indel. If the repair template is present in the form of double-stranded DNA (dsDNA) having homology arms flanking the insertion sequence or it is a single-stranded DNA oligonucleotide (ssODN), then the target sequences can be accurately corrected as planned through the HDR pathway [28–30]. But the efficiency of the knockout is much higher than the knock-in (KI) because the HDR is only effective in dividing cells. However, cell state, cell type, repair template, and genomic locus are the other factors that play an important role in the efficiency of HDR [21].

5.3 OVERVIEW OF DIRECTED CUSTOM-DESIGNED NUCLEASES

An effective method for genetic engineering is the use of engineered nucleases for targeting DNA at specific locations in the genome. Double-stranded breaks are induced specifically to the target site by these nucleases and repair is achieved through the natural process of non-homologous end joining (NHEJ) repair pathway and homologous recombination (HR) [31]. For catalyzing double-stranded breaks at specific sites genome engineered nucleases were designed which stimulate the occurrence of designed modification at the break site [32]. For knock out of the gene of interest (GOI), the NHEJ recombination pathway is useful because it results in deletion or insertion mutations, and it is also useful for site-directed mutagenesis. A repair template is not used for its occurrence. Repair template is used in homology-directed repair (HDR) [28]. The resulting insertions or deletions are precisely controlled by this method. It is less efficient as compared to the NHEJ pathway. Targeted gene replacement is achieved through HR [34].

For GE, four types of engineered endonucleases are suitable for inducing double-stranded breaks in Plants. Engineered mega nucleases (EMNS), ZFNs, transcription activator-like effector nucleases (TALENS), and (CRISPR: clustered regularly interspaced short palindromic repeats)/Cas CRISPR associated 9 are the four major endonucleases [31]. The general principle for operating all four nucleases is the same. These engineered nucleases consist of a DNA binding domain responsible for site-specificity and an endonuclease domain which works as an enzyme for double-stranded breaks [32]. DNA binding domains are customized to recognize effectively any sequence due to the versatility of these nucleases. Genome structure and function are affected by combining numerous effector domains to these DNA binding modules [35]. Modification of a plant gene through genome engineering mainly includes four steps: developing and designing of engineered nucleases, construct delivery into a plant through transformation, Nuclease expression induction, and plant screening for desired DNA sequence [32].

Homing endonucleases/meganucleases (MNs) are rare cutters. In eukaryotes, they have just a few natural recognition sites. For targeting of transgene and native sequence of plants and in other organisms. Rare cutters are very successful because of their high specificity and precise cleavage of target sites [36]. Mega nucleases (MGs) are naturally occurring gene targeting proteins that are encoded by mobile introns. Homodimers of two identical subunits of 160–200 AA in size are formed by these endonucleases. Remarkable specificity is provided by its DNA target site which is 20–30 bp [32].

ZFP are the proteins that were first used for genome engineering in the form of designed zinc-finger nucleases [31]. In the human genome frequently, encoded protein is ZFP. ZF domain consists of 30 AA in conserved ββα configuration [37]. Its molecular structure makes an attractive framework for designing custom-designed DNA binding proteins [35]. It is an artificial restriction enzyme in which the DNA cleavage domain of restriction enzyme *fokI* is fused with synthetic zinc finger binding domain ZFP. In this endonuclease, 3 bp sequence is recognized by 3–5 finger modules of the DNA binding domain [36]. They are powerful tools for gene targeting. The binding of nonspecific cleavage domain of *FOKI* endonuclease with zinc finger protein (ZFP) provides the biologists the way for site-specificity and permanent modification of plants and other genomes by HR pathway [38]. Authentic production of ZFPs which specifically recognize the target site within the genome is the base of designer ZFNs that perform DNA cleavage at predetermined sites. The best framework for the development of custom designed ZFNs with new sequence specifications is best offered by Cys_2His_2 ZFPs [39].

TALES are the naturally occurring proteins that are obtained from plant pathogenic bacteria genus Xanthomonas. Its DNA binding domain consists of a 33–35 AA repeat domain from which each is capable of recognizing the single base pair. The specificity of TALE is determined by repeat variable di residues (RVDs) two hypervariable amino acids [40]. Adjacent DNA sequences are recognized by modular TALE repeats like ZFNs. Re-engineering of linkage between repeats has not occurred in TALES contrast to ZFP [41]. Ligation independent cloning techniques, high throughput solid phase assembly, and golden gate molecular cloning have been developed for rapid assembly of custom-designed TALE arrays because TALES arrays face challenges due to identical repeat sequences [35]. The target sequence is recognized by engineered TALEs which are composed of RVD domains in TALENS.

The CRISPR Cas system has recently engineered an efficient alternative of ZFNs and TALENs for targeted genome alterations. This is an immune system response of bacteria that provides immunity in bacteriophages through RNA-guided DNA cleavage against foreign DNA [42]. In the CRISPR Cas genomic locus, a small segment of foreign DNA is integrated and transcribed to produce CRISPR RNA (crRNA). Annealing of these CrRNA is done for transactivating of CrRNA which performs silencing of invading DNA or sequence-specific cleavage by Cas proteins. After redesigning CrRNA CRISPR/Cas system can be used to effectively cleave any DNA sequence [35].

5.4 SELECTION OF TARGETED CRISPR SPECIFIC INTERVENTIONS TO ACHIEVE EFFICACY TO OVERCOME HUNGER

Advances in GE due to the availability of different crops genome sequences make it possible to target any desired trait. More precise gene targeting is possible through ZFNs and TALENs, but these techniques are complicated, time-consuming, and expensive. To reduce off-target cleavage and for improving target specificity CRISPR/Cas9 editing tool is used due to its simple design [42]. The application of this technique increases the yield by developing the non-GMO crops having desired traits. The provision of food to the growing population is a critical challenge today. An increase of 60–100% in global food production is required to provide food security to the growing population [43]. Almost for GE of 20 crop species CRISPR/Cas9 method has been used [44]. This technique has been used for Biotic and abiotic stress tolerance mechanisms and for yield improvement.

In wheat, corn, tomato, and rice use of CRISPR tools in agriculture has been shown. Resistance against powdery mildew fungus in wheat has been provided by using SDN-1 in wheat. Maize yield is improved during drought stress by applying SDN-3 to the Zar8 gene promotor [45]. Increase in tomato yields is achieved by using SDN-1. A rice variety has been developed by using this GE technique which produces 25 to 30% more grain [46].

5.5 CASE STUDIES

The growth and development of plants are majorly affected by abiotic stresses like salinity, drought, heavy metals, and temperature that results in huge yield losses. There are various new gene-editing techniques that can be used to employ targeted gene editing for desirable traits and resilient plant varieties can be developed. CRISPR/Cas9 is one of the gene-editing technology that has been used to knock out the non-desirable genes that were making plants more susceptible to abiotic stresses [47].

The main target of GE in cereal crops is to increase the production of food to provide food security to huge growing populations. Different traits that have been targeted by using CRISPR/Cas9 include resistance against insects and viral, fungal, and bacterial pathogens in cereal crops. Nutritional quality improvement and tolerance against herbicides were also main priorities in the case of cereals [48].

5.5.1 RICE

Food quality and food crop production are two main factors that contribute food security. As rice is a major crop that is consumed by a large population, it is important to develop high quality and high-yielding rice varieties so that 10 billion people can be fed by 2050. Conventional breeding has certain limitations to achieve this target, but the rice breeding has been revolutionizing by GE techniques like CRISPR/Cas9 [49].

Most trials in the field for the GE (Genome edited) plants exist for Rice and they took place in China. In one of the oldest studies, CRISPR/Cas9 mediated knockouts were used to assess four important regulators of yield-related traits. In this study, dense erect panicles increased grain numbers and size only occurred in two generations enormously stimulating breeding improvements of the elite lines. In other studies, eight different genes related to yield were knocked out in rice by CRISPR/Cas9 multiplex transformations and different phenotypes of multiple and single mutants were examined in field trials [50]. Moreover, CRISPR/Cas9 technology has been potentially employed to remove the mitogen-activated protein kinase 3 (slmapk3) gene, to get better understanding of the concerned regulatory mechanism. Tomato plants genetically knock-out with aforementioned gene potentially expressed severe environmental stress and harsh conditions in contrast to wild-type tomato plants. Another corresponding study have elaborated how CRISPR/Cas9 based technology have been employed potentially to efficiently generated point mutation in particularly rice plants, resulting in herbicide resistance plants. Correspondingly, CRISPR/Cas9 applications have been employed recently to generate overexpressing the maize auxin-interlinked genes involved in mediating organ size.

There are many healthy constituents that are present in Rice bran oil and one of them is Oleic acid. The conversion of oleic acid into linoleic acid in plants is catalyzed by fatty acid desaturase 2 enzyme (*FAD2*) and the mutants of *FAD* show altered linoleic and oleic acid content in different crops. The genome of rice (*Oryza sativa* L.) contains 3 functional genes of *FAD2* and *OsFAD2-1* among of all has the highest expression in rice seeds. *OsFAD2* gene was disrupted by Abe et al. [51] through CRISPR/Cas9-mediated targeted mutagenesis to obtain high oleic acid and low linoleic acid rice bran oil. They successfully produced a homozygous *OsFAD2* knockout rice plant. Oleic acid content was increased by more than twice present in the wild type, and the content of the linoleic acid was decreased in FAD2-1 mutant seeds of brown rice to an unpredictable level [51].

Large amounts of Cadmium Cd in Rice grains are a major problem for the people who consume rice as their staple food. It is a major challenge for the conventional breeders to develop new cultivars with low Cadmium (Cd). Tang et al. [52] reported the development of a novel Indica rice line having low Cadmium content and no transgenes. They carried out the knockout of *OsNramp5* (metal transporter gene) through the Cas9 system. Hydroponic cultures demonstrated that *OsNramp5* mutants contain decreased Cd content in their shoots and roots. While according to paddy field trials it was demonstrated that the Cd content in *OsNramp5* mutants was less than 0.05 mg/KG as compared to the wild type indica rice. Moreover, there were effects on the yield of *OsNramp5* mutants [52].

Yin et al. [48] used the CRISPR/Cas9 system to knock out an early developmental *EPFL9* gene (Epidermal patterning factor like-9) that is a positive regulator of stomatal development in Arabidopsis. It is also an orthologous gene in rice. The mutants present in the germline showing edits were carried out up to the T2 generation where homozygous mutants were obtained. The density of stomata on the abaxial surface of the homozygous mutant plants was reduced by more than eightfold [48].

Cereals that contain high amounts of Amylose and resistant starch (RS) shows potential health advantages. The physical properties and the fine structure of starch are majorly defined by *SBE* (starch branching enzyme). Starch branching is a major challenge in commercial lines. Sun et al. [69] created targeted mutagenesis in rice in *SBEI* and *SBEIIb* in rice by using the CRISPR/Cas9 system. The mutant plants only carrying frameshift mutations were recuperated in T1 generation following segregation. *SBEI* mutants show no difference from the wild type while in *SBEIIb* mutants' higher proportion of long chains were observed in debranched amylopectin. It resulted in a significant increase in Amylose content (AC) and RS, as high as 25.0 and 9.8% respectively. So, the following study demonstrated that rice having high AC and starch resistance can be produced by editing *SBEIIb* through the CRISPR/Cas9 system [54].

A single dominant waxy gene controls AC in rice. Zhang et al. [55] introduced a loss of function mutation in the waxy gene through the CRISPR/Cas9 system in two Japonica rice varieties. Their results showed that the AC is reduced by a mutation in the waxy gene and the rice is converted into the glutinous ones without affecting even another single desirable trait. This method offered an effective way to improve glutinosity in elite rice cultivars [55].

The small non-coding RNAs that play a significant role in stress responses and plant development is called microRNA (miRNA). Due to less availability of Knock out gene tools, the loss of function Studies of miRNA has been challenging traditionally. Zhou et al. [56] targeted single miRNA genes (*OsMIR528* and *OsMIR408*) and miRNA gene families (*miR820a/b/c* and *miR815a/b/c*) in rice through CRISPR/Cas9. The frequency for the targeted mutagenesis lies between 48% and 89% for each target site among regenerated T_0 lines. Biallelic mutations were generated by three independents gRNAs in the case of miRNA528 in all confirmed mutant lines. When two guide RNA were used for targeting in T_0 rice lines then the miRNA genes were deleted readily at frequency up to 60%. This study demonstrated that CRISPR/Cas9 is an efficient tool for knocking out miRNA genes in plants [56].

The *dep1* gene (dense and erect panicle gene 1) has been extensively used in the rice breeding for high-yielding Japonica cultivars in China. But the round and short grain shape is the outcome with the desired results as well that is not attractive for the potential global rice market. Mao et al. [57] created a series of NILs (Near Isogenic lines) (YF47*dep1-gs3*, YF47*dep1-gw8*, YF47*dep1-tgw6*, YF47d*ep1-gl7*, and YF47*dep1-qgl3*) in Yanfeng 47 (YF47*dep1*) background using CRISPR/Cas9 technology. The grain length to width ratio was significantly increased in all NILs in comparison to YF47*dep1* while YF47*dep1-gs3* was the exception. All of the NILs also showed reduced chalkiness degree as well. YF47*dep1-gw8* among all NILs was the one that showed greatest degree of length to width ration and reduced chalkiness. The yield of YF47*dep1-tgw6* and

YF47*dep1-gs3* was significantly greater than the YF47*dep1* while the yield of YF47*dep1-qgl3* was the lowest [57].

5.5.2 MAIZE

ARGOS8 in maize is a negative regulator for ethylene responses. According to a previous study, transgenic plants overexpress *ARGOS8* that has reduced ethylene sensitivity while grain yield has been improved under drought stress circumstances. Shi et al. [58] generated new variants of *ARGOS8* by using CRISPR/Cas9 technology. In this study, they have replaced the native promoter of gene ARGOS8 with GOS2 (a native maize promoter). This promoter confers a moderate level of expression, and it was inserted at the 5′ end of the concerned gene. *ARGOS8* variants resulted in an increased level of expression as compared to the native allele. These variants also have increased grain yield in flowering stress conditions by 5 bushels per acre and no yield loss in well-watered circumstances [58].

Male sterility provides a practical receding tool to tackle hybrid vigor for the production of hybrid seeds. The generation of new male sterile lines is a prerequisite for the development of hybrid seed production technology. Chen et al. [59] targeted the MS8 gene in maize through a CRISPR/Cas9 vector. As a result, eight independent T_0 transgenic lines were produced but the *MS8* gene was not mutated in these T_0 transgenic lines. However, in F1 and F2 progeny mutations in *MS8* were detected. F2 generation of male sterile lines was screened for transgene-free *MS8* male-sterile plants [59].

Gibberellin oxidase is also known as the green revolution gene harboring significant contribution in semi-dwarf phenotype that lodging resistance and product. In their study, Zhang et al. [60] edited the *GA20ox3* gene in maize by using CRISPR/Cas9 technology and generated semi-dwarf varieties. Dwarf phenotypes can be recovered by the application of exogenous gibberellin. It indicated that mutants are deficient in gibberellin. *GA20* Oxidase was disrupted which resulted in elevated *GA53* and *GA12* contents while the contents of various other *GA* precursors (*GA24*, *GA15*, *GA20*, *GAg*, and *GA44*) in mutants were decreased. The accumulation of *GA1* and *GA2* were also decreased which contributes to the semi-dwarf phenotype. Transgene-free dwarf maize plants were selected from F2 generation that can be used in any breeding program further [60].

5.5.3 WHEAT

Genome editing (GE) of the complex genome like wheat had been a challenge for molecular biology researchers. The CRISPR/Cas9 system is an efficient system for GE in diploid plants, but it is still a challenge for complex genomes. In their study, Kim et al. [70] have applied the CRISPR/Cas9 system for editing wheat ethylene-responsive factor 3 (*TaERF3*) and wheat dehydration responsive element-binding protein 2 (*TaDREB2*). T7 endonuclease assay, restriction enzyme digestion assay, and sequencing were used to analyze the effectiveness of mutagenesis in the protoplast of wheat. According to the overall results of this study, it is suggested that CRISPR/Cas9 system can be utilized for GE in wheat protoplast for crop improvement [61].

As we discussed earlier, food with high contents of RS and amylose are beneficial for humans consuming them. Triticum aestivum is a major staple food around the world but its new varieties have a low content of RS. Correspondingly, Li et al. [62] have generated wheat having high ACs by targeting in a spring wheat cv Bobwhite and a modern winter wheat cv Zhengmai 7698 (ZM) through CRISPR/Cas9 technology. They generated transgene-free mutant lines with either partial or triple-null *TaSBEIIa* alleles in Bobwhite and ZM, respectively. After the analysis of structure, properties, and starch composition it was revealed that the effect of the partial or triple-null allele was dependent on dosage. Triple-null lines exhibited more profound effects on the fine structure of amylopectin, starch composition, and nutritional and physicochemical properties. The flour of triple-null lines had significantly increased RS, protein, soluble pentosane, and ACs that are highly beneficial for human health. So collectively researchers were successful in modification of structure, properties, and starch content with the help of targeted mutagenesis by using the CRISPR/Cas9 system [62].

5.6 PROS AND CONS

CRISPR Cas is a very precise, faster, simpler, and less expensive tool when compared to other genetic engineering tools. Off-target mutation is one of the main issues of genetically modified (GM) organisms. But the rate of off-target mutation in CRISPR Cas is very low as compared to other methods [63]. A study of transgenic soybean was conducted, in which transgenes were inserted into Agrobacterium. The results showed that the genomic variation was very low as compared to the genomic variations caused by radiation. Moreover, 88% of scientists of the American

Association for The Advancement of Science (AAAS), believe that GM organisms are safe for consumption [64]. Moreover, If Cas9 protein is engineered with more precision, there are lesser chances of off-target mutations. CRISPR/Cas9 can be used to bring about multiple genetic changes at once [65].

CRISPR Cas also gained importance because of its ease of design and its gene knockout and KI efficiency is comparatively higher [64]. Despite all the ease, there are still certain problems associated with the CRISPR/Cas9 system, e.g., efficiency, positional effects, and delivery. It is still a problem to deliver a specific gene to a particular organ or a cell in a specific quantity at a proper time. Moreover, there is still a need to improve the effectiveness of gene-specific targeting. It is still a challenge to deliver Cas9 protein safely [66].

CRISPR/Cas systems also helped scientists to develop virus-resistant plant materials by mutating RNA and DNA based viruses. But this immunity against viruses were badly failed in eukaryotes. CRISPR/Cas systems are very limited against RNA and DNA viruses because viruses are rapidly replicating. To overcome these viruses, a CRISPR/Cas system is required that would be more effective, and efficient, e.g., CRISPR/Cas13. Due to a great potential against RNA viruses, CRISPR/Cas13 is a precise, robust, and scalable RNA editing system. So, in the future, CRISPR/Cas13 is a far better choice for RNA genome targeting as compared to CRISPR/Cas9 [67].

5.7 FUTURE DIMENSIONS

Nutritional characteristics, quality, and production of horticultural crops can be potentially enhanced by using novel breeding techniques like CRISPR/Cas9, etc. There is a high expectation that this technique will be proved to be a tool to sort out all the food deficit problems in the world. The main concern about gene-edited products is the acceptance by the producers and consumers [68]. In the CRISPR/Cas9 working system, a 20-nucleotides long guide RNA (sgRNA) is required that must be specific to the target DNA sequence. But with the achievement of more precision in the gene targeting system, there would be no need for a long guide RNA. But a smaller and simpler sgRNA can be used [64].

CRISPR/Cas is an efficient system for inserting and deleting desirable DNA sequences like genes, promoters, and transcriptional regulatory elements. The main limiting factor in crop growth and development is

Nitrogen. But the dependence of Cereal crops on Nitrogen fertilizers can be reduced by transferring the genetic elements of the Nod factor signaling pathway from legumes to Cereals by using CRISPR/Cas [53]. Recent advances in GE in plants show the potential of synthetic biology to fulfill urgent ecological and nutritional demands. But before the general commercial use of these engineered products, many associated technologies need to be mature. By successful, fast, and widespread implementation of synthetic biology, many game-changing effects can be brought out in nutrition and agriculture [33].

KEYWORDS

- **CRISPR/Cas**
- **gene-specific targeting**
- **horticultural crops**
- **nitrogen fertilizers**
- **nucleotides**
- **plant genome editing**
- **transcriptional regulatory elements**

REFERENCES

1. Tofazzal, I., (2019). CRISPR-Cas technology in modifying food crops. *CAB Reviews, 14*(050), 1–16.
2. Georges, F., & Ray, H., (2017). Genome editing of crops: A renewed opportunity for food security. *GM Crops & Food, 8*(1), 1–12.
3. Ma, X., Mau, M., & Sharbel, T. F., (2018). Genome editing for global food security. *Trends in Biotechnology, 36*(2), 123–127.
4. Zaidi, S. S. E. A., Vanderschuren, H., Qaim, M., Mahfouz, M. M., Kohli, A., Mansoor, S., & Tester, M., (2019). New plant breeding technologies for food security. *Science, 363*(6434), 1390–1391.
5. National Academies of Sciences, Engineering, and Medicine, (2017). *Human Genome Editing: Science, Ethics, and Governance*. National Academies Press.
6. Petersen, B., & Niemann, H., (2015). Molecular scissors and their application in genetically modified farm animals. *Transgenic Research, 24*(3), 381–396.
7. Saha, S. K., Saikot, F. K., Rahman, M. S., Jamal, M. A. H. M., Rahman, S. K., Islam, S. R., & Kim, K. H., (2019). Programmable molecular scissors: Applications of a new tool for genome editing in biotech. *Molecular Therapy-Nucleic Acids, 14*, 212–238.

8. Paul, A., Bharati, J., Punetha, M., Kumar, S., Mallesh, V. G., Chouhan, V. S., & Sarkar, M., (2019). Transcriptional regulation of thrombospondins and its functional validation through CRISPR/Cas9 mediated gene editing in corpus luteum of water buffalo (*Bubalus bubalis*). *Cell Physiol. Biochem: International Journal of Experimental Cellular Physiology, Biochemistry, and Pharmacology, 52*(3), 532–552.
9. Mao, Y., Zhang, H., Xu, N., Zhang, B., Gou, F., & Zhu, J. K., (2013). Application of the CRISPR–Cas system for efficient genome engineering in plants. *Molecular Plant, 6*(6), 2008–2011.
10. Hille, F., Richter, H., Wong, S. P., Bratovič, M., Ressel, S., & Charpentier, E., (2018). The biology of CRISPR-Cas: Backward and forward. *Cell, 172*(6), 1239–1259.
11. Gosavi, G., Yan, F., Ren, B., Kuang, Y., Yan, D., Zhou, X., & Zhou, H., (2020). Applications of CRISPR technology in studying plant-pathogen interactions: Overview and perspective. *Phytopathology Research, 2*(1), 1–9.
12. Arora, L., & Narula, A., (2017). Gene editing and crop improvement using CRISPR-Cas9 system. *Frontiers in Plant Science, 8*, 1932.
13. Karvelis, T., Gasiunas, G., Young, J., Bigelyte, G., Silanskas, A., Cigan, M., & Siksnys, V., (2015). Rapid characterization of CRISPR-Cas9 protospacer adjacent motif sequence elements. *Genome Biology, 16*(1), 1–13.
14. Ali, Z., Abulfaraj, A., Idris, A., Ali, S., Tashkandi, M., & Mahfouz, M. M., (2015). CRISPR/Cas9-mediated viral interference in plants. *Genome Biology, 16*(1), 1–11.
15. Ahmad, S., Tang, L., Shahzad, R., Mawia, A. M., Rao, G. S., Jamil, S., & Tang, S., (2021). CRISPR-based crop improvements: A way forward to achieve zero hunger. *Journal of Agricultural and Food Chemistry*.
16. Chen, K., Wang, Y., Zhang, R., Zhang, H., & Gao, C., (2019). CRISPR/Cas genome editing and precision plant breeding in agriculture. *Annual Review of Plant Biology, 70*, 667–697.
17. Ahmar, S., Saeed, S., Khan, M. H. U., Ullah, K. S., Mora-Poblete, F., Kamran, M., & Jung, K. H., (2020). A revolution toward gene-editing technology and its application to crop improvement. *International Journal of Molecular Sciences, 21*(16), 5665.
18. Kamburova, V. S., Nikitina, E. V., Shermatov, S. E., Buriev, Z. T., Kumpatla, S. P., Emani, C., & Abdurakhmonov, I. Y., (2017). Genome editing in plants: An overview of tools and applications. *International Journal of Agronomy, 2017*.
19. Hsu, P. D., Lander, E. S., & Zhang, F., (2014). Development and applications of CRISPR-Cas9 for genome engineering. *Cell, 157*(6), 1262–1278.
20. Wolter, F., & Puchta, H., (2018). Application of CRISPR/Cas to understand Cis-and trans-regulatory elements in plants. In: *Plant Transcription Factors* (pp. 23–40). Humana Press, New York, NY.
21. Deng, H., Huang, W., & Zhang, Z., (2019). Nanotechnology based CRISPR/Cas9 system delivery for genome editing: Progress and prospect. *Nano-Research*, 1–14.
22. Xu, T., Li, Y., Van, N. J. D., He, Z., & Zhou, J., (2014). Cas9-based tools for targeted genome editing and transcriptional control. *Applied and Environmental Microbiology, 80*(5), 1544–1552.
23. Jiang, F., & Doudna, J. A., (2017). CRISPR–Cas9 structures and mechanisms. *Annual Review of Biophysics, 46*, 505–529.
24. Jinek, M., Krzysztof, C., Ines, F., Hauer, M., Jennifer, D., & Emmanuelle, C., (2012). A programmable dual-RNA-guided DNA endonuclease in adaptive bacterial immunity. *Science, 337*, 816–821.

25. Schmidt, T., Schmid-Burgk, J. L., & Hornung, V., (2015). Synthesis of an arrayed sgRNA library targeting the human genome. *Scientific Reports, 5*(1), 1–10.
26. Bernard, G., Gagneul, D., Alves, D. S. H., Etienne, A., Hilbert, J. L., & Rambaud, C., (2019). Efficient genome editing using CRISPR/Cas9 technology in chicory. *International Journal of Molecular Sciences, 20*(5), 1155.
27. Liang, G., Zhang, H., Lou, D., & Yu, D., (2016). Selection of highly efficient sgRNAs for CRISPR/Cas9-based plant genome editing. *Scientific Reports, 6*(1), 1–8.
28. Lieber, M. R., (2010). The mechanism of double-strand DNA break repair by the nonhomologous DNA end-joining pathway. *Annual Review of Biochemistry, 79*, 181–211.
29. Tang, X. D., Gao, F., Liu, M. J., Fan, Q. L., Chen, D. K., & Ma, W. T., (2019). Methods for enhancing clustered regularly interspaced short palindromic repeats/Cas9-mediated homology-directed repair efficiency. *Frontiers in Genetics, 10*, 551.
30. Liu, M., Rehman, S., Tang, X., Gu, K., Fan, Q., Chen, D., & Ma, W., (2019). Methodologies for improving HDR efficiency. *Frontiers in Genetics, 9*, 691.
31. Osakabe, Y., & Osakabe, K., (2015). Genome editing with engineered nucleases in plants. *Plant and Cell Physiology, 56*(3), 389–400.
32. Curtin, S. J., Voytas, D. F., & Stupar, R. M., (2012). Genome engineering of crops with designer nucleases. *The Plant Genome, 5*(2), 42–50.
33. Roell, M. S., & Zurbriggen, M. D., (2020). The impact of synthetic biology for future agriculture and nutrition. *Current Opinion in Biotechnology, 61*, 102–109.
34. Costa, J. R., Bejcek, B. E., McGee, J. E., Fogel, A. I., Brimacombe, K. R., & Ketteler, R., (2017). Genome editing using engineered nucleases and their use in genomic screening. In: *Assay Guidance Manual [Internet]*. Eli Lilly & Company and the National Center for Advancing Translational Sciences.
35. Gaj, T., Gersbach, C. A., & Barbas, III. C. F., (2013). ZFN, TALEN, and CRISPR/Cas-based methods for genome engineering. *Trends in Biotechnology, 31*(7), 397–405.
36. Tzfira, T., Weinthal, D., Marton, I., Zeevi, V., Zuker, A., & Vainstein, A., (2012). Genome modifications in plant cells by custom-made restriction enzymes. *Plant Biotechnology Journal, 10*(4), 373–389.
37. Beerli, R. R., & Barbas, C. F., (2002). Engineering polydactyl zinc-finger transcription factors. *Nature Biotechnology, 20*(2), 135–141.
38. Wu, J., Kandavelou, K., & Chandrasegaran, S., (2007). Custom-designed zinc finger nucleases: What is next? *Cellular and Molecular Life Sciences, 64*(22), 2933–2944.
39. Zaidi, S. S. E. A., Vanderschuren, H., Qaim, M., Mahfouz, M. M., Kohli, A., Mansoor, S., & Tester, M., (2019). New plant breeding technologies for food security. *Science, 363*(6434), 1390–1391.
40. Deng, D., Yan, C., Pan, X., Mahfouz, M., Wang, J., Zhu, J. K., & Yan, N., (2012). Structural basis for sequence-specific recognition of DNA by TAL effectors. *Science, 335*(6069), 720–723.
41. Gaj, T., Gersbach, C. A., & Barbas, III. C. F., (2013). ZFN, TALEN, and CRISPR/Cas-based methods for genome engineering. *Trends in Biotechnology, 31*(7), 397–405.
42. Bhatta, B. P., & Malla, S., (2020). Improving horticultural crops via CRISPR/Cas9: Current successes and prospects. *Plants, 9*(10), 1360.
43. Jaganathan, D., Ramasamy, K., Sellamuthu, G., Jayabalan, S., & Venkataraman, G. (2018). CRISPR for crop improvement: an update review. *Frontiers in Plant Science, 9*, 364675.

44. Ricroch, A., Clairand, P., & Harwood, W., (2017). Use of CRISPR systems in plant genome editing: Toward new opportunities in agriculture. *Emerging Topics in Life Sciences, 1*(2), 169–182.
45. Ansari, W. A., Chandanshive, S. U., Bhatt, V., Nadaf, A. B., Vats, S., Katara, J. L., & Deshmukh, R., (2020). Genome editing in cereals: Approaches, applications, and challenges. *International Journal of Molecular Sciences, 21*(11), 4040.
46. Gao, C., (2018). The future of CRISPR technologies in agriculture. *Nature Reviews Molecular Cell Biology, 19*(5), 275, 276.
47. Ahmad, S., Sheng, Z., Jalal, R. S., Tabassum, J., Ahmed, F. K., Hu, S., & Tang, S., (2021). CRISPR–Cas technology towards improvement of abiotic stress tolerance in plants. In: *CRISPR and RNAi Systems* (pp. 755–772). Elsevier.
48. Yin, X., Biswal, A. K., Dionora, J., Perdigon, K. M., Balahadia, C. P., Mazumdar, S., & Bandyopadhyay, A., (2017). CRISPR-Cas9 and CRISPR-Cpf1 mediated targeting of a stomatal developmental gene EPFL9 in rice. *Plant Cell Reports, 36*(5), 745–757.
49. Tabassum, J., Ahmad, S., Hussain, B., Mawia, A. M., Zeb, A., & Ju, L., (2021). Applications and potential of genome-editing systems in rice improvement: Current and future perspectives. *Agronomy, 11*(7), 1359.
50. Metje-Sprink, J., Sprink, T., & Hartung, F., (2020). Genome-edited plants in the field. *Current Opinion in Biotechnology, 61*, 1–6.
51. Abe, K., Araki, E., Suzuki, Y., Toki, S., & Saika, H., (2018). Production of high oleic/low linoleic rice by genome editing. *Plant Physiology and Biochemistry, 131*, 58–62.
52. Tang, L., Mao, B., Li, Y., Lv, Q., Zhang, L., Chen, C., & Zhao, B., (2017). Knockout of OsNramp5 using the CRISPR/Cas9 system produces low Cd-accumulating indica rice without compromising yield. *Scientific Reports, 7*(1), 1–12.
53. Adhikari, P., & Poudel, M., (2020). CRISPR-Cas9 in agriculture: Approaches, applications, future perspectives, and associated challenges. *Malaysian Journal of Halal Research, 3*(1), 6–16.
54. Arora, L., & Narula, A., (2017). Gene editing and crop improvement using CRISPR-Cas9 system. *Frontiers in Plant Science, 8*, 1932.
55. Zhang, J., Zhang, H., Botella, J. R., & Zhu, J. K., (2017). Generation of new glutinous rice by CRISPR/Cas9-targeted mutagenesis of the waxy gene in elite rice varieties. *Journal of Integrative Plant Biology, 60*(5), 369–375.
56. Zhou, J., Deng, K., Cheng, Y., Zhong, Z., Tian, L., Tang, X., & Zhang, Y., (2017). CRISPR-Cas9 based genome editing reveals new insights into microRNA function and regulation in rice. *Frontiers in Plant Science, 8*, 1598.
57. Mao, T., Zhu, M., Sheng, Z., Shao, G., Jiao, G., Mawia, A.M., Ahmad, S., Xie, L., Tang, S., Wei, X., & Hu, S., (2021). Effects of grain shape genes editing on appearance quality of erect-panicle geng/japonica rice. *Rice, 14*(1), 74.
58. Shi, J., Gao, H., Wang, H., Lafitte, H. R., Archibald, R. L., Yang, M., & Habben, J. E., (2017). ARGOS 8 variants generated by CRISPR-Cas9 improve maize grain yield under field drought stress conditions. *Plant Biotechnology Journal, 15*(2), 207–216.
59. Chen, R., Xu, Q., Liu, Y., Zhang, J., Ren, D., Wang, G., & Liu, Y., (2018). Generation of transgene-free maize male sterile lines using the CRISPR/Cas9 system. *Frontiers in Plant Science, 9*, 1180.
60. Zhang, J., Zhang, X., Chen, R., Yang, L., Fan, K., Liu, Y., & Liu, Y., (2020). Generation of transgene-free semidwarf maize plants by gene editing of gibberellin-oxidase20-3 using CRISPR/Cas9. *Frontiers in Plant Science, 11*, 1048.

61. Tang, L., Mao, B., Li, Y., Lv, Q., Zhang, L., Chen, C., & Zhao, B., (2017). Knockout of OsNramp5 using the CRISPR/Cas9 system produces low Cd-accumulating indica rice without compromising yield. *Scientific Reports, 7*(1), 1–12.
62. Li, J., Jiao, G., Sun, Y., Chen, J., Zhong, Y., Yan, L., & Xia, L., (2020). Modification of starch composition, structure, and properties through editing of TaSBEIIa in both winter and spring wheat varieties by CRISPR/Cas9. *Plant Biotechnology Journal.*
63. Ayanoğlu, F. B., Elçin, A. E., & Elçin, Y. M., (2020). Bioethical issues in genome editing by CRISPR-Cas9 technology. *Turkish Journal of Biology, 44*(2), 110–120.
64. Eş, I., Gavahian, M., Marti-Quijal, F. J., Lorenzo, J. M., Khaneghah, A. M., Tsatsanis, C., & Barba, F. J., (2019). The application of the CRISPR-Cas9 genome editing machinery in food and agricultural science: Current status, future perspectives, and associated challenges. *Biotechnology Advances, 37*(3), 410–421.
65. Siddique, H. R., (2016). CRISPR/Cas9 technology: A revolutionary molecular scissors for genome editing and genetic research. *MOJ Cell Sci. Rep., 3*(1), 00046.
66. Chaudhary, K., Chattopadhyay, A., & Pratap, D., (2018). The evolution of CRISPR/Cas9 and their cousins: Hope or hype? *Biotechnology Letters, 40*(3), 465–477.
67. Ahmad, S., Wei, X., Sheng, Z., Hu, P., & Tang, S., (2020). CRISPR/Cas9 for development of disease resistance in plants: Recent progress, limitations, and future prospects. *Briefings in Functional Genomics, 19*(1), 26–39.
68. Bhatta, B. P., & Malla, S., (2020). Improving horticultural crops via CRISPR/Cas9: Current successes and prospects. *Plants, 9*(10), 1360.
69. Sun, Y., Jiao, G., Liu, Z., Zhang, X., Li, J., & Xia, L. (2017). Generation of high-amylose rice through CRISPR/Cas9-mediated targeted mutagenesis of starch branching enzymes. *Frontiers in plant science*, *8*, 249451.
70. Kim, D., Alptekin, B., & Budak, H. (2018). CRISPR/Cas9 genome editing in wheat. *Functional & Integrative Genomics*, *18*, 31–41.
71. Jaganathan, D., Ramasamy, K., Sellamuthu, G., Jayabalan, S., & Venkataraman, G. (2018). CRISPR for crop improvement: an update review. *Frontiers in Plant Science*, *9*, 364675.
72. Mao, T., Zhu, M., Sheng, Z., Shao, G., Jiao, G., Mawia, A.M., Ahmad, S., Xie, L., Tang, S., Wei, X., & Hu, S., (2021). Effects of grain shape genes editing on appearance quality of erect-panicle geng/japonica rice. *Rice*, *14*(1), 74.

CHAPTER 6

ROLE OF CRISPR-CAS SYSTEM IN THE DEVELOPMENT OF CLIMATE SMART CROPS

ZULQURNAIN KHAN, ZULFIQAR ALI, and ASIF ALI KHAN

Institute of Plant Breeding and Biotechnology, MNS University of Agriculture, Multan, Pakistan

ABSTRACT

After ensuring food security for several decades, the green revolution has gained its potential and reaching its biological boundaries which is reflected by the stagnation in yield. Meeting the growing food demands of the masses, due to increasing population, is the biggest challenge for plant scientists. CRISPR-edited crops hold tremendous potential for food security and global agriculture. CRISPR technology has become a new face of genome editing (GE) with rapidly increasing applications in academic research (functional genomics and reverse genetics) as well as applied research for developing crop varieties with new or improved traits. Climate change is an additional threat to crop growth, and productivity which is resulting in many biotic and abiotic stresses to the plants. After successful deployment of gene editing technology to modify simple traits, more ambitious ventures to combat food security challenges in the wake of growing world population with changing climate are in progress. The chapter reviews and outlines the uses of CRISPR/Cas technology as a new breeding technique in modifying crops beyond simple traits targeting to fruitage the desirable diversity which was lost during domestication and manipulation for complex traits. Special emphasis will be given on the new breeding techniques (NBTs) which may be employed for the development of climate smart crops.

CRISPR-Based Gene Editing: A Solution for Global Food Security. Shakeel Ahmad, Asif Ali Khan, and Muhammad Arshad Javed (Eds.)

6.1 INTRODUCTION

Agricultural production and cultivation practices are strongly influenced owing to climate change especially of major crops in different geographical areas of the world [1]. Rapidly fluctuating temperature and change in the frequency and the magnitude of extreme weather conditions are directly affecting agricultural production [2]. Pollination, flowering, and grain-filling of most grain crops is affected by water scarcity. In contrast, along with an impact on the yield and quality, abundant rainfalls may damage plants due to higher relative humidity, which increases susceptibility for most of the diseases [3]. Crop yield is also badly affected by drought which is resulted in growth stagnancy and early maturing of the crops [4]. The impact of heat waves has been analyzed in wheat [5], rice [6], maize [7], and soybean [8]. It has been reported that an increase of 1°C of seasonal temperatures resulted in decrease in yield ranging from 3.1% in soybean to 7.4% in maize [9]. Increase amount of atmospheric CO_2 has variable effects on crops. In a positive sense, it increases plant photosynthesis and growth. In contrast, it may negatively affect the nutritional quality and health of the crops [10].

Due to climate change, rising in temperature and sea levels have greater impact on the arable land which is the major threat to agriculture. Rising temperature accelerate the locusts swarms which adversely damage the crops of Asia, Middle East, and East Africa in 2020 [11]. As population increasing which reduce the cultivated land area. Climate change threats the crops productivity and yield, the only way to maintains the food security is the production of climate smart crops. Resource efficient, with high water and nutrient use efficiency, and pest-resistant crops will be helpful in reducing environmental impacts due to wise and judicious use of fertilizers and pesticides for best optimal production of the crops.

In the challenging climate scenario, miscellaneous strategies are needed for developing climate-resilient crops varieties and expediting the rate of genetic gains [12]. Understanding physiological, biochemical, and molecular mechanisms is very crucial for the adaptation of the plants in climate change conditions. The identification of traits involved in adaptation to variable environment is one of the major objectives of the new breeding techniques (NBTs). Basic prerequisites of next generation breeding are the availability of large populations and germplasm for breeding, sophisticated high-throughput technologies, tools for big data analysis, and advanced molecular breeding techniques. It will be resulted in shortening the time frame and increasing precision to get desired traits fixed in the new varieties having high genetic potential to perform under climate change conditions.

Moreover, advances in omics, tissue culture and transformation techniques facilitated advanced biotechnology methods such as cis-genesis and genome editing (GE) [13], which has paved the way to produce a variety of value-added products promoting sustainable agriculture to address future challenges.

In last few years, GE technologies have been evolved that used for the targeted or site-specific editing such as hybrid enzymes or clustered regularly interspaced short palindromic repeat (CRISPR)/CRISPR-associated protein (Cas) [14]. Homology-directed repairs (HDR) is used to introduce the site-specific DNA modification in a genome, researchers prefer to induce the DNA double stand breaks (DSB) at specific site in a genome and stimulating the nonhomologous end-joining (NHEJ). In case of the crop breeding, transgene-free improved varieties can be generated through GE. Approaches that based on the hybrid enzyme contain fused DNA binding domain and nucleases, i.e., zinc-finger nucleases (ZFNs) and transcription activator-like effector nucleases (TALENs). These DNA binding domain and nucleases have been used foe the targeted GE of many crops, e.g., soybean [15] and wheat [16]. Although, the downside of the hybrid enzymes is its high cost and complexity of the protein engineering that necessary for development. The advantages of developing engineered crops using GE techniques are many including minimum time duration (Figure 6.1). Genetic engineering already has this advantage over conventional breeding. But gene editing has further improved time frame for developing genetically modified (GM) crops.

6.2 DEVELOPING CLIMATE RESILIENT CROPS THROUGH CRISPR-CAS SYSTEM

Global climate change is posing increased abiotic stresses such as drought, heat stress, salinity, heavy metal toxicity, cold stress, etc., to the crop plants and impact on yield and quality of the farm produce [17]. Abiotic stresses have been becoming serious constraints to crop production for world agriculture in the changing climate [18]. Abiotic stress is governed by multiple genes as a complex quantitative trait which is not easy to fix [17(b), 19]. In this regard, CRISPR/Cas9 system relies on simple RNA/DNA hybrids that confer sequence specificity and can manipulate nearly any sequence in the genome to reveal its function [20]. Being highly efficient, easy to construct and capable of engineering multiple sites simultaneously in the genome of an organism, has derailed the scientific community from

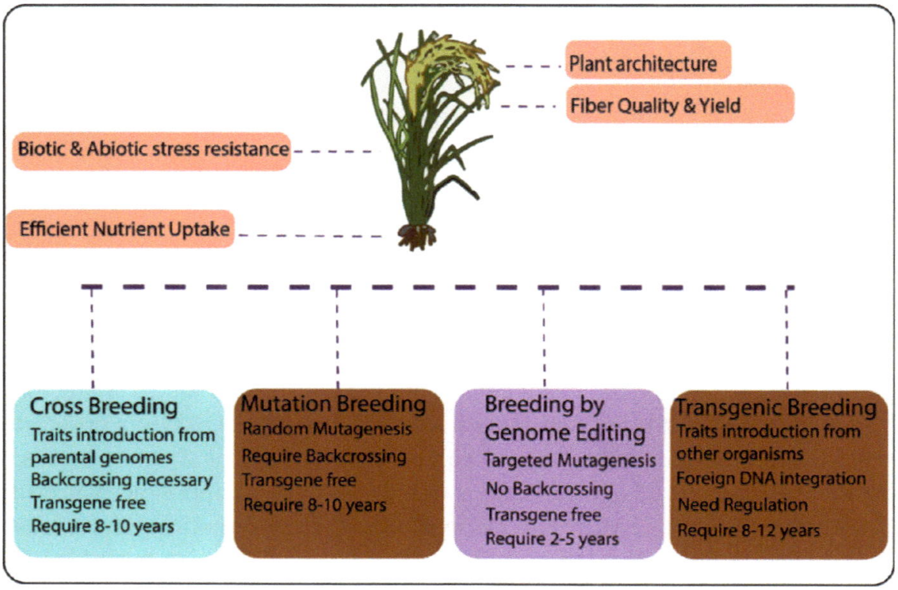

FIGURE 6.1 Comparison of breeding techniques for developing crop varieties.

preceding GE technologies [21]. The CRISPR-Cas GE shows high promise to address these environmental problems and develop resistant plants. For example, by knocking out *OsARM1*, *OsNramp5* and *OsHAK1* genes in rice, the breeders have developed rice strains tolerant to arsenic, cadmium, and radioactive cesium, respectively [22]. The CRISPR/Cas9-based editing of *OsPYL* abscisic acid (ABA) receptor gene family and pyl1/4/6 triple knockout resulted in rice with increased grain yield, greater high temperature tolerance, and reduced pre-harvest sprouting compared with wild type [23]. The *OsRR22* gene encodes a 696-amino acid B-type response regulator transcription factor that is involved in both cytokinin signal transduction and metabolism. Loss of its function has been reported to significantly increase the salt tolerance in rice [24]. Zhang et al. also demonstrated enhanced rice salinity tolerance via CRISPR-Cas9-targeted mutagenesis of the *OsRR22* gene [25]. Innovative transformation protocols of plants for abiotic stress tolerance are a challenging issue. Other genome modifications of plants using CRISPR-Cas technology for abiotic stress tolerance include herbicide tolerance in soybean [26], flax [27], in maize [28], rice [29], potato [30]; drought tolerance in rice [31]; salinity tolerance in cotton [32] and rice and tolerance to heavy metals such as arsenic [22(a)] and cadmium in rice [33].

Two new terminologies sensitivity genes (or *S* genes)' and 'tolerance genes (or *T* genes)' are introduced for defining the genetic factors which are responsible for the positive and negative regulation of plant response in relation with the abiotic stresses. However, the *S* genes and *R* genes are most establish terminologies and frequently well accepted in the area of biotic stresses.

6.2.1 STRUCTURAL GENES ARE TARGETED FOR ABIOTIC STRESS TOLERANCE

Structural genes are known as the important class of targets that can be used to enhance specific features of stress tolerance. Reactive oxygen species (ROS) are signaling molecules in plants that regulate gene expression [34], and also use to develop virus resistance in plants [35], ROS play a crucial role in plant because it can be used in symbiotic nitrogen fixation between the different plants and soil rhizobia [36]. Although, the excessive production of ROS, which is a typical plant response to abiotic and oxidative stresses, can cause a variety of growth abnormalities, including reduced photosynthesis rate, increased cell death, and even male sterility, resulting in lower crop yield [37]. Therefore, controlling ROS production and scavenging is critical for sustaining redox potential in cells. The large number of genes are used to encode antioxidant enzymes like glutathione reductases (GRs), catalases (CATs), glutathione S-transferases (GSTs), and few other peroxidases (PODs) are involved for searching and collecting the ROS molecules. These genes are known as *T* genes, and they help with abiotic stress tolerance. Until now, various *S* genes have been identified that cause excessive ROS production (also known as oxidative stress), decrease the activity of antioxidant, and increased programmed cell death (PCD), as well as genes that cause hormonal homeostasis disruption, making plants susceptible to abiotic stresses [38].

Molecular breeders have discovered many *T* genes linked to abiotic stress resistance, which they have introduced into plants to achieve tolerance. Under abiotic stress conditions, for example, papain-like cysteine proteases (PLCPs) are found to be enhanced in a variety of plant species, including sweet potato (SPCP2) and wheat (TaCP). Arabidopsis lines such as overexpressing SPCP2 and TaCP represent the improved verity of drought stress tolerance [145–147]. Melatonin biosynthetic genes are another example of this group. Melatonin is an antioxidative molecule that aids plants in collecting and searching the reactive oxygen and nitrogen species (ROS and RNS). Plants that overexpress melatonin biosynthesis genes have been shown to be resistant to abiotic stresses [148–150]. Furthermore, during the crop domestication programs several *T* gene alleles have been lost, but wild species are thought to still have them. This may be because breeders focus on selecting 'yield-contributing genes' during selection programs, ignoring genes that contribute to stress tolerance. So far, the *S* genes that negatively control biotic stresses have been understudied. By disrupting the pathways involved, knocking out *S* genes can contribute to stress tolerance. *Oryza sativa* stress-related RING finger protein 1 (*OsSRFP1*), for example, is an E3 ubiquitin ligase that acts as a negative regulator for various abiotic stresses by enhancing the level of H_2O_2 (an essential ROS species) and reducing the activity of antioxidant enzyme in plant tissues [39]. By disrupting H_2O_2 biosynthesis and positively controlling antioxidant activities, knocking down *OsSRFP1* enhanced plant tolerance to abiotic stresses. RNAi-mediated gene silencing has been used to monitor many sensitivity genes, including *OsDIS1* (*O. sativa* drought-induced SINA protein 1) and DST (drought and salt tolerance) [40]. While some *S* genes have been identified in various crop species, there have been few studies on how to engineer them to improve abiotic stress tolerance. However, some *S* genes that are responsible to cause susceptibility to biotic stresses (diseases and insect pests) have been targeted (mutated) in order to achieve biotic stress tolerance in various crops [41].

6.2.2 ENHANCING ABIOTIC STRESS TOLERANCE BY TARGETING REGULATORY GENES

Regulatory genes, encoding transcription factors (TFs), phosphatases, and kinases are another critical class of targets for modulating the expression of many downstream genes and activating several stress signals. In Arabidopsis, for example, the ANAC069 NAM-ATAF1/2 and CUC2 (NAC) TF gene acts as a down regulator of abiotic stresses (S gene). By binding to the core motif sequence C[A/G]CG[T/G] in the promoter region of multiple stress sensitive genes, ANAC069 controls their expression. This results in a decreased ability to scavenge ROS and a high degree of proline biosynthesis, resulting in increased salt sensitivity and osmotic stress [42]. T-DNA insertion knockdown mutants of ANAC069 showed improved resistance to salt and osmotic stress [42]. Likewise, increasing the ABA-induced stomatal closure by overexpressing *T* genes like AtMYB44 confers DST [151] and as well as the Overexpression of ZmWRKY106 also improves drought and heat tolerance in transgenic plants by controlling the expression of stress-related genes, lowering ROS levels, and increasing antioxidant enzyme activity [43]. miRNAs, which are small non-coding RNAs that regulate the expression of many genes at the post-transcriptional level by targeting particular mRNAs, are another significant target in this group. miRNAs regulate a variety of biological processes by silencing genes through complementary base pairing with mRNA transcripts. Few miRNAs have been identified as being involved in the regulation of the abiotic stress response, which involves the targeting of particular stress-responsive genes. E.g., osa-MIR393 regulates the expression of three important stress-responsive genes (*LOC Os02g06260*, *LOC Os05g41010*, and *LOC Os05g05800*) to negatively regulate salt and alkali stress tolerance [44]. As a result, silencing this gene can improve tolerance to these stresses. In Arabidopsis, another miRNA, miR399, controls phosphorus homeostasis by targeting the ubiquitin conjugating E2 enzyme (AtUBC24). As phosphorus is depleted, miR399 is upregulated, suppressing its target E2 enzyme. Inorganic phosphorus (Pi) is readily taken up by wild-type plants from their roots and transported back to shoots and other plant tissues. Plants overexpressing miR399, on the other hand, had inhibited Pi remobilization, which resulted in Pi toxicity due to excessive accumulation in the shoots [45]. While, using precise GE techniques to identify and target specific miRNAs involved in abiotic stress control, crops' abiotic stress tolerance could be improved.

6.3 CRISPR/CAS9-ASSOCIATED TARGETING OF ABIOTIC STRESS RESISTANCE GENE

The discovery of programmable nucleases that produce double-strand breaks can altered molecular biology by allowing for targeted GE [17(b)]. Zinc finger nucleases (ZFNs) are credited with being the first GE method to use programmable nucleases, resulting in an advancement in genome engineering [46]. TALENs, which are based on bacterial TALEs, soon followed, extending genome engineering's capabilities even further. These methods were easily adapted for genome engineering in over 40 different species [46]. Nevertheless, due to its obvious advantages over ZFNs and TALENs, the discovery of CRISPR/Cas9 drew a lot of interest from scientists all over the world.

While in case of mutagenesis, CRISPR/Cas9 can also be used for the activation (CRISPRa) and repression (CRISPR interference (CRISPRi)) gene expression by combining catalytically inactive Cas9 (dCas9) with a transcriptional activator or repressor [47]. As a result, it has the ability to take the place of traditional GM-based overexpression and gene silencing techniques. Since GE techniques allow for direct alteration or manipulation of a GOI, they have an advantage over traditional genetic engineering approaches in which transgene incorporation is random (in the case of overexpression), which may suppress or trigger other genes, and unintended silencing of other genes would be done by using the siRNA-mediated RNAi. In CRISPR, proper designing of gRNA will solve the off-targeting constraint. Though there have been some successful examples of CRISPRa/i in animal cells [48], as well as in plant systems [49]. CRISPR/Cas9-mediated GE has been widely used in a variety of plant plants, including wheat, rice, corn, and cotton. Even so, there are few studies on the targeting of abiotic stress resistance genes [17(b)]. This is due to the fact that previous research has concentrated on biotic stresses such as viruses and insect pests.

Recent studies indicate that targeting an *S* gene, *SlAGAMOUS-LIKE 6* (*SlAGL6*) in tomato is responsible of heat tolerance and with help of CRISPR/Cas9-mediated GE breeders increased fruit setting under heat stress [50]. In addition, targeting multiple genes in a single organism using CRISPR/Cas9 has also been introduced and successfully optimized for different crops such as rice [51], wheat [52], maize [53], and cotton [54]. By simultaneously targeting many *S* genes in an elite high-yielding yet vulnerable cultivar, CRISPR/Cas9 has the ability to produce crops that are resistant to multiple stresses. *T* genes may also be overexpressed using CRISPR-mediated gene amplification, as previously mentioned (CRISPRa). CRISPR/Cas9 implementations have recently been applied to genome-wide screening for selective trait enhancement thanks to the advent of a CRISPR/Cas base editing method [55]. Mutants of 'loss of function' or 'gain of function' may arise from precise base changes at key sites in a gene. This strategy is thought to have the potential to replace conventional plant breeding methods, which focus on identifying plant populations with enough genetic diversity to incorporate desirable traits into elite crop cultivars. CRISPR/Cas9 base editing can produce novel allelic variants in a population, and the novel alleles corresponding to a desired phenotype can then be classified by gRNA sequencing [56]. As a result, CRISPR/Cas9 base editing has a lot of promise for crop enhancement programs that want to create new traits quickly.

6.4 RECENT ADVANCES IN CRISPR/CAS SYSTEM

CRISPR-Cas9 GE uses inbuilt cellular repair systems that react to double strand breaks (DSBs). When Cas9 nuclease and a corresponding sgRNA are delivered together, DSBs are created at the locus of interest. Depending on the cell cycle phase and whether a homologous sequence is present or not, one of two pathways, non-homologous end joining (NHEJ), or homology directed repair (HDR), is then used to repair the break. Insertions and deletions (indels) are produced by the error prone NHEJ process, whereas precise genetic alterations can be inserted using HDR by using a synthetic donor. In numerous *in vitro* and *in vivo* models, this fundamental mechanism has been comprehensively investigated and used to carry out targeted single and multiplexed gene deletions as well as recombination [57].

The creation of off-target DSBs and the system's restriction to loci expressing the PAM sequence are two notable constraints. But there has been a lot of improvement in fixing these flaws. Researchers have investigated a variety of strategies to achieve targeted gene editing while reducing off target consequences. Controlling Cas9 activity both spatially and temporally is one tactic. Either sending Cas9 to a particular tissue or putting it downstream of a tissue-specific promoter can result in spatial regulation. Contrarily, the Tet-on method can be used to establish temporal regulation, in which Cas9 expression is positioned under a promoter with a tetracycline response element (TRE) such that Doxycycline can trigger Cas9 expression [58]. This technique, known as inducible CRISPR (iCRISPR), has been used to create human pluripotent stem cells (hPSCs) knockouts at various phases of development, enabling systematic genetic research in hPSCs. A promising application of iCRISPR *in vivo* is the robust gene disruption caused by doxycycline-induced Cas9 expression in different tissues and the modulation of target gene changes' frequency and size by inducible Cas9 nickase (Cas9n) [59]. In addition to iCRISPR, alternative systems have also been developed in which Cas9 has been engineered to respond to other small molecules such as Rapamycin [60] and 4-hydroxytamoxifen (4-HT). Moreover, researchers have also engineered light responsive Cas9 systems, with one approach using a fusion of the light inducible heterodimerizing proteins CRY2 and CIB1 [61], and the other using a photoinducible dimerization system called magnets [62].

Additionally, a number of organizations have also sought to increase Cas9 accuracy. Studies on SpCas9 have shown that while it tolerates mismatches in the residues after the first 8–12 residues upstream of the PAM, it has excellent selectivity in those residues. By adjusting the concentration of SpCas9 and sgRNA as well as by using shortened guide RNAs, off target mutagenesis has been reduced [63]. Furthermore, altering either of the Cas9 endonuclease domains' catalytic residues (RuvC: D10A; HNH: H840A) results in the production of nickases (Cas9n), which cause single-stranded DNA (ssDNA) breaks [14]. Off target effects can be minimized by combining Cas9n with two gRNAs to produce two single stranded breaks on opposing strands [64]. Similar to this, high specificity is also obtained by fusing the FokI nuclease to dCas9 to create a programmable nuclease that is similar to ZFN/TALEN but is only catalytically active when two dCas9-FokI complexes are within 15–25 nucleotides of one another [65]. Recently, Cas9 has undergone rational engineering, resulting in a version that stabilizes ssDNA and shows less tolerance for gRNA-DNA mismatches [66]. A modular, chimeric nuclease with a wider sequence-targeting range and better accuracy has been created as a result of the addition of a programmable DNA binding domain (pDBD) to a mutant Cas9 with diminished DNA binding affinity. This nuclease may be utilized to modify practically any genomic region [67].

Base editing has been emerged as another feather in the cap of CRISPR/Cas system. It has been used efficiently for editing of the DNA sequence at targeted sites with precision of a single base modification [55, 56]. The applications of the base editing system are various ranging from correction, point mutation, herbicide resistance, etc. Another CRISPR variant, prime editing, is also getting much fame in "copy and paste" fashion to modify the target genomes. The history of discovery and application of the technology as new breeding technique has been given in Figures 6.2 and 6.3.

6.5 DEVELOPMENT OF CLIMATE SMART CROPS BY USING NEW PLANT BREEDING STRATEGIES

Conventionally, number of crop varieties have been developed by the formers but for the propagation of different crop varieties breeders used different basic techniques like plant selection on the basis desirable characteristics. Advance plant breeding techniques that are based on the genome modification and marker-based selection techniques (MAS) are used for genetics improvement of the crops. These techniques have been reported frequently [68]. Development of a GM crop variety is briefly followed by – (i) identification of the hereditary material or DNA that linked with the desirable trait; (ii) cloning of the GOI into the carrier (vector) or plasmid;

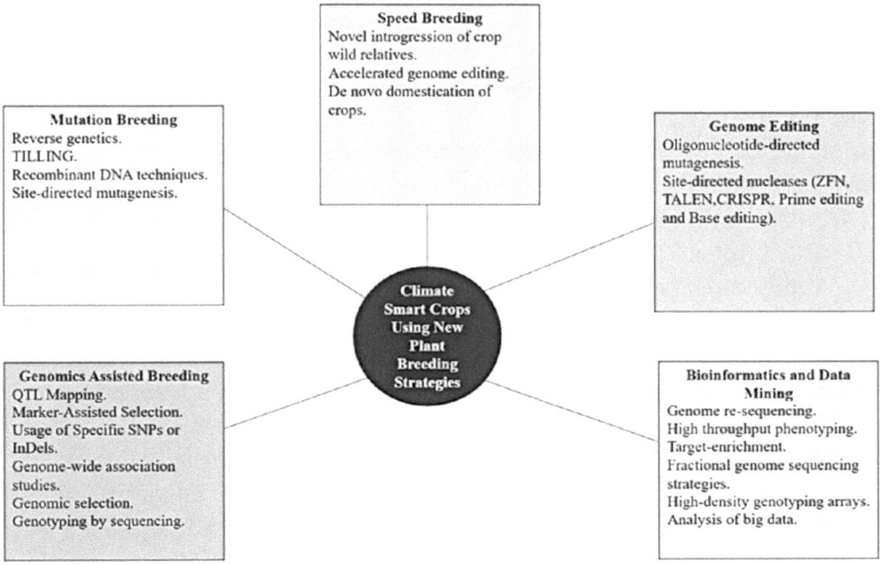

FIGURE 6.2 Development of climate smart crops using new plant breeding strategies.

(iii) transformation of the GOI to the target plant by using different type of transformation techniques; and (iv) multiplication of desired plants contains trait of interest. GM crops have been criticized because of agricultural policies and less public understanding [69]. The one of the major concerns related to GM crops is the economic and social issues such as ownership, regulation of GM product and its marketing. One of the major concerns is the antibiotic resistance gene flow into the non-GM varieties. Due to these concerns the regulation of GM crops is stringent, make end product expensive and availability of new varieties to the farmer community is slow. The plant breeders face problems in producing resistant varieties in the climate change scenario.

The use of GM crops remains restricted in the society, but the latest plant breeding strategies (NPBTs) arose as an advance approach that is used to produce new varieties with improved characteristics. NPBTs including GE platforms such as ZFN, TALEN [70]. Mega-nucleases and Clustered Regularly Interspaced Short Palindromic Repeat (CRISPR) or CRISPR-associated protein (Cas). The advance NPBTs include the speed breeding which shorten the breeding cycles of crops [71] next generation sequencing [72] and phenotyping platforms [73]. Those crops which have difficult flowering pattern are improved by using robust and efficient GE approaches [74]. GE approaches such as TALENs and CRISPR/Cas9 different from conventional methods in many ways, i.e., the precise genetic manipulation of DNA without inserting antibiotic resistant gene which reduce the chance of foreign DNA escape [75]. But in conventional method the GM crop is produced by adding the foreign genetic material or DNA (T-DNA) into the Agrobacterium and after the successful transformation modified crop obtained. Now, same in case with the genome-editing technique such as in CRISPR the insertion of T-DNA restrains by using virus-based DNA insertion or ribonucleoprotein (RNP) complex [76].

Additionally, plant breeders successfully developed the transgene free CRISPR crops by using different type of methodologies such as through agroinfiltration transient expression of programmable nucleases, protoplast-based delivery of preassembled CRIS PR-Cas RNPs and the site-specific addition of a CRISPR into other chromosomal locations following subsequent segregation or removal [77]. The NPBTs produce valuable products including browning-resistant mushrooms, false flax with enhanced omega-3 oils and elevated amylopectin waxy *Zea mays* [78], which rapidly reach to the market with reasonable prices. These valuable products are developed via CRISPR/Cas9 approach and certified by US Department of Agriculture (USDA).

6.6 CRISPR-ASSOCIATED GENE EDITING: EVOLUTION AND APPLICATION OF SITE-SPECIFIC NUCLEASES (SSN)

Different types of site-specific nucleases (SSNs) are used in GE because these nucleases can bind to specific site of the DNA/RNA and cut it. These SSNs are designed in such a way, after cutting near to the target site they can creates the double-stranded breaks (DSBs) in sequence [79]. ZFNs, Mega-nucleases, TALENs, and CRISPR/Cas9 are the four major classes of SSNs [79, 80]. The above-mentioned SSNs have significant plant breeding potential because they are widely used for the modulating host machinery by altering their genome structure such as gene knock-in (KI) and knock-out, stacking, translation modulation and targeted mutagenesis. In comparison with the conventional plant breeding technique (takes 10 years for developing new variety) the SSNs are time saving and also provide significant economic benefits [81].

Remarkably, in 2013, the CRISPR/Cas technology has been emerged as the important groundbreaking SSNs that was widely used for plant GE and in comparison, with the other NPBTs its application in crop development is increasing rapidly [82], Some important agriculture traits like environmental stresses (heat or cold stress tolerance), plant morphology (increased grain size, weight, length, and yield) and bacterial, viral and fungal resistance in plants can be achieved by using these GE techniques [81]. Significantly, many groups have recently achieved these genome modifications by the use of transgene-free systems.

CRISPR-Cas9 is the advance approach that thoroughly used in genome modification of many crops [83]. Cas9 is firstly discovered in *Streptococcus pyogenes,* kind of Type II adaptive immune system that are used to develop the immunity by developing defensive mechanism against many invading viruses [14]. In the DNA of invading phages, the CRISPR/Cas9 is used to introduce the site-specific DSBs resulting in activation of cellular DNA repair mechanism, non-homologous end-joining (NHEJ), that gives the inexact repair, or if the joining is through homology-directed repair (HDR) it may give the specific repair. These repairs are responsible to cause the insertion deletion (indels) mutation in invading viruses and this mutation can alter the virus genome or make it dysfunctional and, in this way, bacteria can develop the natural defense against the viruses.

The short CRISPR RNA (crRNA) are transcribed with the help of CRISPR locus that further binds to the complementary sequence near the protospacer-adjacent motif site (PAM) on the targeted genome. In *S. pyogenes,* the PAM is used to recognize the target sequence and bind with it. PAM is a trinucleotide sequence that used to bind with targeted viral DNA [84]. After binding with viral DNA, which activate RNA (tracrRNA) and this tracrRNA is now able to bind with crRNA. The crRNA gives guide RNA (gRNA) which binds with Cas9 endonuclease and RNase III resulting in the formation of Cas9 complex. After binding of the gRNA at specific site, it activates the Cas9 nucleases that creates a DSB about three nucleotides upstream from the PAM region of targeted sequences [85]. To get the accurate and size specific GE, CRISPR Cas9 is used because the Cas9 mediated cleavage at specific site is obtain with the help of synthetic single guide (sg) RNAs that gives the precise results. With the advancements in CRISPR-Cas, breeders perform a lot of work to improve crops through the host gene regulation and metabolic engineering for example, is one of several new CRISPR systems Cas12 [86], Cas13a [87], Cas13b [88], and fCas9 [65], are presently used in research purpose. The application of CRISPR is not limited due to the development of few new GE technologies like RNA processing Cas9 (RCas9), deactivated Cas9 (dCas9) and its fusion with other proteins such as (cytidine deaminase fusion). All the approaches are providing a wide range of applications of CRISPR technology for the scientists.

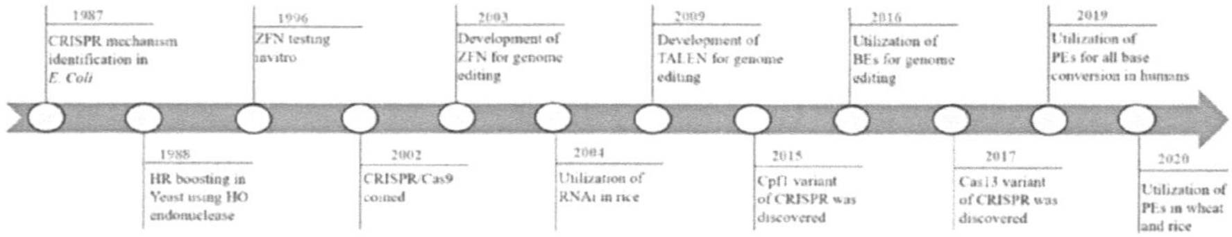

FIGURE 6.3 A brief timeline of different site-specific nuclease with historical events.

6.7 APPLICATIONS OF CRISPR FOR ENGINEERING THE DISEASE-RESISTANT PLANTS

As mentioned above, the wide range of CRISPR-based application have introduced. In this part, we categories the data at how CRISPR has been used successfully to develop the disease-resistant plants [89].

Indel mutation is a coding sequence of genome is responsible to the trigger gene disruption and the CRISPR-Cas9 is most commonly used for this purpose. NHEJ DNA-repair process is responsible to introduce the indel mutation that give rise to a frame shift mutation and disrupt the gene development at target site. So, this methodology has been used to incorporate desirable traits into a variety of crops, including basic cereals including rice and wheat. By disrupting a plant susceptibility (S) gene breeder are able to alter plant–pathogen interaction and resulting in decreased pathogen suitability on the host plant [90] or developed the disease free plants, e.g., CRISPR-Cas9 is used to introduce indels mutation that affecting the eukaryotic translation initiation factor 4E proteins (eIF4Es), that is responsible develop the resistance against numerous RNA viruses in Arabidopsis and cucumber (*Cucumis sativus*) [91].

6.7.1 INDELS IN PROMOTER REGIONS CAUSE GENE DISRUPTION

Indels can be introduced in the promoter region of a plant gene rather than the coding region. Two types of the CRISPR-based promoter editing can be used: (i) in order to destroy or completely inhibit the gene expression disrupt the promotor sequence; and (ii) prevent the pathogen effectors proteins from binding with the promotor sequence to interrupt the effector binding site.

Some other methods were used to break the connection between a bacterial blight pathogen's effector and the promoter region of the rice sugar transporter gene (OsSWEET14), resulting in blight resistance [92]. CRISPR/Cas9 have been used to change gene regulation by editing the cis-regulatory elements and targeting upstream open reading frame (ORF) regions in addition to promoter editing [93].

6.7.2 GENE DELETION CAUSED BY MULTIPLEX SGRNAS

For the insertion of DSBs at specific location in a genome multiple sgRNA can be used in CRISPR system. Start and stop codon are presents near the GOI, two sgRNAs are binds before the start and after the stop codons, e.g., after binding with start and stop codons these sgRNAs are responsible to trigger the DSBs at specific position. Before the cellular repair NHEJ mechanism can repair the DSBs the DNA fragment that having the GOI can be removed. Whereas, for removing the large chromosomal fragment sgRNA can be used. The sgRNA having the PAM (trinucleotide sequence) that can be used to remove large as well as the single fragment of GOI at any genomic region [94]. Additionally, the formation of SpCas9-NG (rationally engineered Cas9), that can recognize NG PAMs, which have a higher specificity than standard NGG PAMs [95]. Gene clusters make this method especially useful with respect to the engineered pathogen resistance. For developing the long-term resistance against the target phages, several S genes are found in close proximity on the chromosome [96], that can be used for removing the chromosomal fragment.

6.7.3 HOMOLOGY-DIRECTED REPAIR MECHANISM AND GENE INSERTION

According to the above-mentioned data it is clear that the CRISPR/Cas can be used to modify the *S* gene to produce disease resistance(s). All plant proteins, like those proteins that are generated by *S* genes, are essential and multifunctional, so disrupting them can affect the of plant health and productivity. Previously, it is clearly mentioned that a lot of alternative approaches are presents that can be used to alter the gene expression disrupting the gene, e.g., promoter editing and cis-regulatory factor. While, in case of the plant pathogen interaction where *S* genes have not been thoroughly investigated it is important to use *R* gene against different pathogens. In that condition, for inserting the *R* genes the CRISPR toolkit can be used when the Cas9 have been developed the sgRNAs are responsible to develop the DSBs and these sgRNA-dictated DSBs are used an alternative route for CRISPR-mediated gene insertion. In this pathway, HDR mechanism is used for insertion instead of the NHEJ machinery. Cas9 and the sgRNAs are added to a fragment that contains R gene surrounded by a homologous sequence to the ends of the DSB. Now, the R gene is inserted in between the two DSB sites by using HDR machinery and this methodology have been successfully used for inserting the one or more than one gene at specific location in a genome [97]. Still, in plants, the efficiency and application HDR is very poor and less reported [98], that can make the gene insertion very difficult in plants. While different type of new strategies is being developed that can be used improve the mechanism of gene insertion in plants [99]. CRISPR-associated gene insertion has been used to investigate basic role of S gene in addition to producing disease resistance. Some researchers are using HDR to incorporate and fuse the green fluorescent protein (GFP) with glutathione S-transferase loci in rice [100]. This feature may be used to investigate S gene functions including regulation (spatiotemporal) of S gene expression and protein localization. During the infection, the host proteins are responsible to cause the pathogenicity so it can be labeled with the GFP. This labeling is used to investigating the S gene expression and localization of host proteins. The approach is extremely efficient, but the main drawback of this approach could be the possible change in the expression and localization due to modification in the protein.

6.7.4 BIOMIMETICS VIA PROMOTER/GENE REPLACEMENT

The use of CRISPR for gene insertion is likely to face extensive regulatory processes and possibly public rejection due to concerns about inserting the foreign DNA into crops. CRISPR provides an alternative solution that operates on the basis of biomimicking to solve this issue. For decades, scientists have determined that inherent species and genetic diversity of cultured crops contain a diverse genetic pool, particularly when it comes to resistance to biotic and abiotic stresses [101]. Several reports have described multiple R genes in wild relatives of cultivated species and established effective resistance transmission through the identification of a R gene to the cultivated crop species [102]. Using multiplexed HDR technique, CRISPR might be used to replace the defective R gene in a cultivated crop variety. The application of CRISPR-mediated mutations in such a way that the sequence of the targeted gene is translated to the sequence of a disease resistant variety is referred to as biomimicking. As a result, rather than deleting the entire gene, the scientist only integrates the unique mutations sites related with the disease resistant traits, indicating that the variations of nucleotide between the concern gene in cultivated and wild varieties are not really important to plant productivity. That was largely accomplished by the use of a CRISPR technology that allows for variations in targeted sequence. By fusing nuclease dead Cas9 domain with the cytidine deaminase can be used for on target mutations with more accuracy which is used for gene modifications and replacement in number of organisms [103]. The N176K replacement encoded by the *eIF4E* virus-resistance allele (*eIF4E1*) of *Pisum sativum* was introduced into the *Arabidopsis eIF4E1* gene using the same method to produce *Arabidopsis* plants resistant to the Clover yellow vein virus (*ClYVV*) [104]. Recent research has shown that synthetic oriented evolution can be used to develop gene modifications that are resistant herbicides [105]. Under biotic selective stress, the same strategy could be used to develop gene variants that are resistant to the specific pathogens.

6.8 CRISPR-ASSOCIATED DISEASE TOLERANCE

CRISPR technology and its derivatives are being used in plant biology for a variety of purposes, from studying gene and protein annotation and localization to introducing desirable traits including tolerance to drought and increased grain quantity and quality. Recent review papers have discussed these intriguing applications [79]. The emphasis of this chapter is on engineering of CRISPR-associated disease resistance in plants. The infection of three major plant pathogens such as bacteria, viruses, and fungi are the important cause of plant diseases and crop failure.

CRISPR-mediated plant virus resistance: focusing on virus genomes improved agricultural practices and the use of plant varieties that are disease-resistant. The majority of studies focusing on virus-resistance using CRISPR/Cas9 technology. There are two ways to achieve virus resistance in plants including targeting the viral genome directly or by targeting the virus *S* genes that confer the viral diseases in crop plants.

The first attempts at engineered virus resistance relied on directly attacking the virus genome inside plant cells [106] focused on several ssDNA geminiviruses in two separate studies: Targeting the sgRNA efficiently developed viral resistance against the beet curly top virus (BCTV), merremia mosaic virus (MeMV), tomato yellow leaf curl virus (TYLCV), cotton leaf curl Kokhran virus (CLCuKoV), and cotton leaf curl Multan betasatellite (CLCuMuB) [107]. There are three sites in the viral genome that are targeted two which encodes replication protein and capsid protein, and one is an intergenic region (IR). By targeting these sites in viral genome resulted in insertion and deletion is being made in the virus genome, that confer in a lower virus accumulation and disease symptoms. Baltes et al. and Ji et al. shows similar results by targeting the geminiviruses such as Bean yellow dwarf virus (BeYDV) and beet severe curly top virus (BSCTV) [108]. These studies showed the minimal achievement in attaining viral resistant in transgenic crops, besides the CRISPR/Cas9 virus targeting [109]. CRISPR-based direct targeting of virus genomes may also be used to engineer resistance to RNA viruses. For targeting of RNA viruses Cas9 RNA-binding variants used instead of DNA binding Cas9 that are used for DNA viruses. Zhang et al. created *Arabidopsis* and *Nicotiana benthamiana* plants that encodes a FnCas9 gene that are used to targets the tobacco mosaic virus (TMV) and cucumber mosaic virus (CMV). That showed minimal virus accumulation and symptom production [110]. They used similar approach to develop resistance to southern rice black streak dwarf virus which is an RNA virus that attack rice. Aman et al. used an RNA targeting LwaCas13a (previously known as C2c2) to interact with the RNA genome of turnip mosaic virus (*TuMV*) in two separate experiments and demonstrated efficient virus RNA targeting in both the *N. benthamiana* and *Arabidopsis* [111]. The LwaCas13a contain modest targeting efficiency against the *TuMV* virus in the studies of Aman et al. these studies showed the huge potential of Cas13 in engineering transgenic crops against the RNA based viruses. This prompted Mahas et al. to explore Cas13-mediated virus immunity by screening various Cas13 variants and found Cas13d system most efficient against the RNA viruses such as *TMV*, *TuMV*, and potato virus X (*PVX*) [112]. Cas13 remarkably targeting the RNA viruses in crops that provides protection against the viriods. But its leftovers to be understood if Cas13 can successfully target and degrade the circular, highly compact, and secondary structure-rich RNA viroid genomes [113]. To confirm these results, further research on economically important plants and field trials will be done. While alternative approaches such as multiplex editing search to resolve the virus accumulation in plants. But certain limitations are associated with this method such as virus escape and the evolution of CRISPR-mediated resistant virus production [114].

6.8.1 PLANT VIRUS RESISTANCE: TARGETING THE PLANT S GENE(S) USING CRISPR/CAS9

The direct targeting of viral genome is avoided due to high copy number and recombination rate, so by using CRISPR-mediated targeting of S gene(s) is preferred because it target the plant genome rather than the viral genome. CRISPR-mediated targeting of *S* genes considered to be an easiest way to engineer viral immunity when *S* genes are well identified defined in crop plants. Additionally, the transgene free CRISPR strategy may be used to target the *S* gene. In various RNA potyviruses eIF4E genes considered as *S* gene that encodes cap-binding proteins that are necessary for the cellular infection cycle of various RNA viruses. CRISPR/Cas9 targeting the eIF4E genes to create virus resistance against potyviruses. In many plant organisms, the 5′-terminal protein VPg in potyvirus interacts with the *eIF4Es* so, by blocking this interaction using CRISPR/Cas9 enhance plant immunity against the potyviruses. In two different studies, CRISPR-associated eIF4E targeting found a successful resistance to multiple viruses in *Arabidopsis* and cucumber, including TuMV, cucumber vein yellowing virus (CVYV), zucchini yellow mosaic virus (ZYMV), and Papaya ring spot mosaic virus-W (PRSV-W). In many later experiments other economically important crops' eIF4E isoforms targeting by CRISPR resulted in broad spectrum resistance in number of hosts. Gomez et al. found that by inhibiting the eIF4E isoforms novel cap-binding protein-1 (nCBP-1) and nCBP-2 in cassava substantially reduced the symptoms of the cassava brown streak virus (CBSV) disease [115]. Besides that, a biomimicking approach resulted in CRISPR-mediated mutations in Arabidopsis eIF4E1 that transformed its sequence to fit the *P. sativum* eIF4E virus-resistance allele, resulting in the successful introduction of ClYVV resistance. Similarly, Macovei et al. used biomimicking of eIF4G alleles to show resistance in rice to the rice Tungro spherical virus [116]. In conclusion, *S* genes have been used by various CRISPR systems to achieve virus resistance in a variety of crop organisms.

6.8.2 BACTERIAL RESISTANCE USING CRISPR/CAS9

CRISPR-mediated bacterial resistance is created by targeting the OsSWEET gene/s that activates immunity against bacterial blight caused by Xanthomonas oryzae pv. In rice, the genes in SWEET clade III family such as OsSWEET11, OsSWEET13, and OsSWEET14 involves in sugar transporters such as glucose and sucrose export. So, induction of SWEET gene by transcription activator-like effectors (TALEs) activates sugar release into the apoplast, that provides the energy for the pathogen and considered as S gene.

GE tool such as TALENs used to knock-outs one or more *OsSWEET* in rice to confer bacterial blight resistance [92]. Furthermore, TALE-binding elements (EBEs) in the *OsSWEET* promoters have been identified as key virulence factors in recessive *OsSWEET*-mediated resistance [117]. Recent study showed that by creating mutations in the promoters of all three *OsSWEET* genes by CRISPR-Cas9 confer resistance to bacterial blight in rice.

Bacterial speck disease is caused by *Pseudomonas syringae* pv in tomato (*Solanum lycopersicum*). It results in production of coronatine, involving the co-receptor JAZ2, which stimulates stomatal opening with facilitating bacterial colonization. *SlJAZ2* was edited with CRISPR-Cas9 which encoded a variant of JAZ2 lacking the C-terminal domain, resulting in resistance to bacterial speck disease by stoppage of stomatal reopening [118]. *DMR6* (downy mildew resistant 6) is another *Arabidopsis* gene that is closely linked to salicylic acid control and create pathogen infections in crop plants. CRISPR has been used to target SlDmr6-1, a tomato ortholog that confers wider resistance to pathogens such as *P. syringae, Phytophthora capsici*, and *Xanthomonas* spp. [119].

Citrus canker is one of the another most important bacterial disease caused by *Xanthomonas citri subsp. Citri.* And damages citrus yield loss in all over the world. A citrus canker susceptibility gene PthA4, the bacterium's main TALE, binds to the effector-binding element in the promoter of LATERAL ORGAN BOUNDARIES 1 (*CsLOB1*) that activates the expression to endorse citrus canker formation [120]. In two different experiments, the symptom of citrus canker is reduced by targeting *CsLOB1* using CRISPR in two citrus varieties, grapefruit (*Citrus paradisi*) [121] and Wanjincheng orange (*Citrus sinensis* Osbeck) [120].

6.8.3 FUNGAL RESISTANCE USING CRISPR/CAS9

Approximately 30% of evolving plant diseases are due to the fungal pathogen [122] and infect broad range of economically important crops. CRISPR systems targets the plant *S* genes that are susceptible for fungal pathogen and confer resistance in plants.

Several studies demonstrated that Mutation in Mlo enhance plant immunity against fungal disease such as powdery mildew [123]. Fungal pathogen penetration into the host epidermal cells is facilitated by the membrane associated protein that encodes by Mlo [124]. Powdery mildew resistance was also created in wheat and tomato by CRISPR-mediated nutation in Mlo. In wheat powdery mildew resistance is also created by targeting Edr1 (enhanced disease resistance 1) that encodes Raf-like mitogen-activated protein [101]. Recently, in tomato plants powdery mildew disease symptoms reduced by CRISPR-mediated targeting of Powdery mildew resistance 4 (Pmr4) S gene [125].

OsERF922 that are negative regulator of blast resistance in rice and encodes a special ethylene responsive factor that involves in biotic stress regulation are targeted by CRISPR/Cas9.

In the grape (Vitis vinifera) cultivar Thompson, Wang et al. used CRISPR to target and disrupt VvWRKY52, a transcription factor gene that plays a key role in plant response to biotic stress and seedless, with improved resistance to Botrytis cinerea, a fungal pathogen [52]. Zhang et al. showed that disrupting the gene encoding tomato mitogen-activated protein kinase 3 (SlMAPK3), which controls the accumulation of ROS, with CRISPR, produced tomato plants immune to *B. cinerea* [126].

6.8.4 OOMYCETE RESISTANCE USING CRISPR/CAS9

Other biotic stress such as oomycete infection is also control by using CRISPR/Cas9 system in papaya *Phytophthora palmivora* is a damaging oomycete pathogen. A papaya plant that are mutant for functional cysteine protease inhibitor (PpalEPIC8) was created using CRISPR-Cas9 system that confer enhance resistance to *P. palmivora* [127]. Similarly, in tropical tree Theobroma cacao that grows cocoa beans, resistance to another oomycete pathogen, *Phytophthora tropicalis*, has also been engineered [128] (Table 6.1).

6.9 REGULATION OF TRANSGENIC CROPS

Since 1995, when the GMO was firstly introduced the strict regulation and nationwide bans have hindered the commercial growth and widespread acceptance of GM crops, especially in Europe [142]. Non-transgenic genome edited crop research and development may be equally delayed if regulatory

TABLE 6.1 Examples of Genome Editing for Crop Improvement

Crops	Editing System	Type of Edit	Target Gene(s)	Target Traits	Results	References
Brassica oleracea	CRISPR/Cas9	Gene disruption	*Bolc.GA4.a*	Dwarfing and fruit dehiscence.	Significant dwarfing was observed in all transformed lines relative to wild-type, as well as decreased fruit dehiscence.	[129]
Citrus (orange)	CRISPR/Cas9	Promotor disruption.	CsLOB1 (promoter)	Citrus canker resistance.	When compared to the wild-type, the frequency of citrus canker disease was reduced by 83.2–98.3%.	[121]
Cucumber	CRISPR/Cas9	Gene disruption	eIF4E	Broad virus resistance.	Resistance to cucumber vein yellowing virus infection and potyviruses zucchini yellow mosaic virus and papaya ring spot mosaic virus-W.	[130]
Maize	CRISPR/Cas9	Promoter swap.	*ARGOS8*	Drought tolerance.	Under flowering drought stress, the yield was 3.8% higher than the control.	[131]
Potato	TALEN	Gene disruption.	*VInv*	Low concentration of reducing sugars that lead to low palatability and possible health risks.	Reducing sugar concentrations that are undetectable.	[132]
Rice	CRISPR/Cas9	Promoter disruption.	*OsSWEET11*, *OsSWEET14*	Bacterial blight resistance.	The blight susceptibility gene's promoter was disturbed.	[133]
Rice	CRISPR/Cas9	Gene disruption.	Gn1a, GS3, and DEP1.	Grain number, grain size, panicle architecture.	Panicles with more seeds, bigger grains, and denser, upright panicles.	[134]
Soybean	TALEN	Gene disruption.	FAD2-1A, FAD2-1B	Low polyunsaturated fats.	The monounsaturated oleic acid content raised from 20% to 80%, while the polyunsaturated linoleic acid content reduced from 50% to under 4%.	[135]

TABLE 6.1 *(Continued)*

Crops	Editing System	Type of Edit	Target Gene(s)	Target Traits	Results	References
Tobacco	CRISPR/Cas9	Viral gene disruption	43 regions in the viral genome.	Resistance to the geminivirus beet severe curly top virus.	In systemic leaves, no serious leaf curling symptoms were found, and virus prevalence in local leaves was decreased by 90–97%.	[136]
Tobacco	CRISPR/Cas9	Viral gene disruption.	Six regions in the viral genome.	Resistance to the geminivirus bean yellow dwarf virus.	Transgenic plants that were challenged with geminivirus and expressed CRISPR-Cas reagents had a 5–87% lower virus load and less infections.	[108]
Tomato	CRISPR/Cas9	Viral gene disruption.	Three regions in the viral genome.	Resistance to the geminivirus tomato yellow leaf curl virus (TYLCV).	The accumulation of viral DNA was reduced, preventing, or greatly reducing infection symptoms.	[107]
Tomato	CRISPR/Cas9	Gene disruption/ knockdown.	SP5G	Time to harvest.	Plants could be harvested two weeks before controls, resulting in a lower overall yield but a higher harvest index.	[137]
Tomato	CRISPR/Cas9	Gene disruption	RIN	Fruit ripening (shelf life).	Ripening was delayed and indefinitely inhibited, likely without the negative side effects seen with traditional RIN mutants.	[138]
Tomato	CRISPR/Cas9	Gene disruption	*SlIAA9*	Parthenocarpy (leading to seedless fruit).	Before fertilization, fruit production was triggered, resulting in seedless tomato fruits.	[139]
Tomato	CRISPR/Cas9	Gene disruption	*SlMlo1*	Resistance to powdery mildew.	Rapid development of a powdery mildew-resistant tomato without the use of transgenes.	[140]
Wheat	TALEN and CRISPR/Cas9	Gene disruption.	*TaMLO-A1*, *TaMLO-B1*, and *TaMLO-D1*	Resistance to powdery mildew.	On the leaves of edited plants, the number of mildews microcolonies produced on the leaves was substantially decreased, and fungal growth was observed on the leaves.	[141]

authorities regard them as GM crops. In comparison to transgenic GM crops, GM crops may be indistinguishable from conventional crops. To distinguish between traditional mutagenesis breeding and GE, regulations may seek to govern the method of crop enhancement rather than the end product. Fortunately, advancements in GE machinery distribution, such as the use of RNP complexes, which do not need even the temporary insertion of foreign DNA into the genome, could quickly undermine crop improvement legislation [143]. In 2018, the European Court of Justice will determine whether genome-edited crops can be classified as GM crops. Many other countries' legislation lags beyond scientific innovation, but the status of genome-edited crops is still undefined. Regulation of genome-edited crops must account for rapid technological advancements and differentiate between transgenic GM crops and non-transgenic genome-edited crops. Canada exemplifies a realistic approach to biotechnology enforcement by analyzing plants on an individual basis, focusing on novel characteristics rather than the breeding process [144].

6.10 CONCLUSION AND FUTURE PROSPECTS

Global climate change is anticipated to impose a severe risk to agrarian profitability around the globe, and consequently challenge food as well as nutritional security. Transgene based and molecular breeding technologies have though greatly assisted in the development of elite crop varieties with enhanced resistance to abiotic stresses. But due to the complex inheritance as well as high G x E associated with these traits, these conventional approaches do not bear the potential for improving such traits that too in the face of increasing global population and climate change. However, in spite of the fact that impact of climate change on the crop's resistance is hard to foresee and is likely to be variable depending on the crop and climate, GE could contribute significantly to reduce the impact of climate change on future cropping scenarios.

In the last three decades, molecular breeding has encountered two big bottlenecks. The use of marker genes during genetic transformation was the source of the first concern. Since transgenes have a low genetic transformation efficiency, researchers used marker genes to separate the transformed cells from the grid of untransformed cells. The majority of the marker genes were antibiotic or herbicide resistance genes, which caused controversy among social workers and other regulatory agencies. Several strategies for overcoming this have been released, but transgenic material's acceptance in most parts of the world remains questionable. The second problem was the transformation methodology, which resulted in the transgene being randomly inserted at every locus along every chromosome. This can disrupt a resident gene or result in an undesirable phenotype. Another issue was the transgene's multicopy and partial integration, which could result in gene expression silencing. In addition to addressing all of the above challenges, genome engineering has four main applications in plant molecular breeding. The gene disruption/knockout using NHEJ is the first and perhaps simplest use of CRISPR technology. As the recent explosion of NGS data and advances in computational power continue to classify and predict novel genes and gene networks, biologists face a significant challenge in confirming these gene functions *in vivo*. CRISPR-mediated gene knockout will speed up the process of validating these findings and identifying genes that have a direct impact on plant abiotic stress tolerance. The ability to restore (allelic improvement) or substitute (allelic substitution) the mutated or inferior copies of agronomically significant genes by HDR is one of the most desirable expectations from CRISPR technology. With point mutations or modified sequence fragments, it can be used to construct effective gene KI or knockout lines. It may also be used to add a restriction site to a region of interest or even a small tag to a gene or protein. By permanently replacing the target gene's promoter or inserting an enhancer to a dCas9 enzyme designed to bind near the target gene's promoter, the CRISPR/Cas system may also be used to reprogram the target gene's expression. The third application involves inserting a transgene at a specific locus or hotspot in the host genome. Although, the CRISPR/Cas T-DNA can be segregated out (meiotic purification) to produce improved varieties without bacterial genome cuts, introducing a transgene into a specific locus would result in stable expression without affecting any other genes. The fourth application is to use base editors to establish artificial genetic diversity in rice cultivars. Furthermore, Epigenetic modulation and genome imaging are two applications of CRISPR technology. Although the risks of using CRISPR are real, the risks associated with climate change to food security are even greater. As a result, genome engineering using the CRISPR-Cas method for selective mutagenesis holds enormous promise for producing elite cultivars of crop plants with improved and long-lasting climate tolerance. It would be appropriate to use CRISPR technology to develop climate-resilient crop varieties that can withstand extreme weather in order to feed the world's ever-increasing population.

KEYWORDS

- **climate change**
- **climate-smart crops**
- **CRISPR/Cas**
- **fungal resistance**
- **gene editing**
- **genetically modified organism**
- **new breeding techniques**
- **plant breeding**

REFERENCES

1. Ackerly, D., Loarie, S., Cornwell, W., Weiss, S., Hamilton, H., Branciforte, R., & Kraft, N., (2010). The geography of climate change: Implications for conservation biogeography. *Diversity and Distributions, 16*(3), 476–487.
2. Gornall, J., Betts, R., Burke, E., Clark, R., Camp, J., Willett, K., & Wiltshire, A., (2010). Implications of climate change for agricultural productivity in the early 21st century. *Philosophical Transactions of the Royal Society B: Biological Sciences, 365*(1554), 2973–2989.
3. Rosenzweig, C., Iglesius, A., Yang, X. B., Epstein, P. R., & Chivian, E., (2001). *Climate Change and Extreme Weather Events-Implications for Food Production, Plant Diseases, and Pests, 2*(2), 24.
4. (a) Nuruddin, M. M., Madramootoo, C. A., & Dodds, G. T., (2003). Effects of water stress at different growth stages on greenhouse tomato yield and quality. *HortScience, 38*(7), 1389–1393; (b) Greven, M., Raw, V., & West, B., (2009). Effects of timing of water stress on yield and berry size. *Water Science and Technology, 60*(5), 1249–1255.
5. Porter, J. R., & Gawith, M., (1999). Temperatures and the growth and development of wheat: A review. *European Journal of Agronomy, 10*(1), 23–36.
6. Zhao, C., Piao, S., Wang, X., Huang, Y., Ciais, P., Elliott, J., Huang, M., et al., (2016). Plausible rice yield losses under future climate warming. *Nature Plants, 3*(1), 1–5.
7. Bassu, S., Brisson, N., Durand, J. L., Boote, K., Lizaso, J., Jones, J. W., Rosenzweig, C., et al., (2014). How do various maize crop models vary in their responses to climate change factors? *Global Change Biology, 20*(7), 2301–2320.
8. Deryng, D., Conway, D., Ramankutty, N., Price, J., & Warren, R., (2014). Global crop yield response to extreme heat stress under multiple climate change futures. *Environmental Research Letters, 9*(3), 034011.
9. Zhao, C., Liu, B., Piao, S., Wang, X., Lobell, D. B., Huang, Y., Huang, M., et al., (2017). Temperature increase reduces global yields of major crops in four independent estimates. *Proceedings of the National Academy of Sciences, 114*(35), 9326–9331.

10. Irigoyen, J. J., Goicoechea, N., Antolín, M., Pascual, I., Sánchez-Díaz, M., Aguirreolea, J., & Morales, F., (2014). Growth, photosynthetic acclimation, and yield quality in legumes under climate change simulations: An updated survey. *Plant Science, 226*, 22–29.
11. Roussi, A., (2020). Why gigantic locust swarms are challenging governments and researchers. *Nature, 579*(7799), 330–330.
12. Varshney, R. K., Bansal, K. C., Aggarwal, P. K., Datta, S. K., & Craufurd, P. Q., (2011). Agricultural biotechnology for crop improvement in a variable climate: Hope or hype? *Trends in Plant Science, 16*(7), 363–371.
13. (a) Cardi, T., (2016). Cisgenesis and genome editing: Combining concepts and efforts for a smarter use of genetic resources in crop breeding. *Plant Breeding, 135*(2), 139–147; (b) Rinaldo, A. R., & Ayliffe, M., (2015). Gene targeting and editing in crop plants: A new era of precision opportunities. *Molecular Breeding, 35*(1), 1–15.
14. Jinek, M., Chylinski, K., Fonfara, I., Hauer, M., Doudna, J. A., & Charpentier, E., (2012). A programmable dual-RNA–guided DNA endonuclease in adaptive bacterial immunity. *Science, 337*(6096), 816–821.
15. Curtin, S. J., Zhang, F., Sander, J. D., Haun, W. J., Starker, C., Baltes, N. J., Reyon, D., et al., (2011). Targeted mutagenesis of duplicated genes in soybean with zinc-finger nucleases. *Plant Physiology, 156*(2), 466–473.
16. Feng, Z., Mao, Y., Xu, N., Zhang, B., Wei, P., Yang, D. L., Wang, Z., et al., (2014). Multigeneration analysis reveals the inheritance, specificity, and patterns of CRISPR/Cas-induced gene modifications in *Arabidopsis*. *Proceedings of the National Academy of Sciences, 111*(12), 4632–4637.
17. (a) Mickelbart, M. V., Hasegawa, P. M., & Bailey-Serres, J., (2015). Genetic mechanisms of abiotic stress tolerance that translate to crop yield stability. *Nature Reviews Genetics, 16*(4), 237–251; (b) Ahmad, S., Sheng, Z., Jalal, R. S., Tabassum, J., Ahmed, F. K., Hu, S., Shao, G., et al., (2021). CRISPR–Cas technology towards improvement of abiotic stress tolerance in plants. In: *CRISPR and RNAi Systems* (pp. 755–772). Elsevier; (c) Ahmad, S., Tang, L., Shahzad, R., Mawia, A. M., Rao, G. S., Jamil, S., Wei, C., et al., (2021). CRISPR-based crop improvements: A way forward to achieve zero hunger. *J. Agric. Food. Chem.*
18. Pereira, A., (2016). Plant abiotic stress challenges from the changing environment. *Frontiers in Plant Science, 7*, 1123.
19. Bhat, J. A., Ali, S., Salgotra, R. K., Mir, Z. A., Dutta, S., Jadon, V., Tyagi, A., et al., (2016). Genomic selection in the era of next generation sequencing for complex traits in plant breeding. *Frontiers in Genetics, 7*, 221.
20. Sander, J. D., & Joung, J. K., (2014). CRISPR-Cas systems for editing, regulating, and targeting genomes. *Nature Biotechnology, 32*(4), 347–355.
21. (a) Sasano, Y., Nagasawa, K., Kaboli, S., Sugiyama, M., & Harashima, S., (2016). CRISPR-PCS: A powerful new approach to inducing multiple chromosome splitting in *Saccharomyces cerevisiae*. *Scientific Reports, 6*(1), 1–11; (b) Monsur, M. B., Shao, G., Lv, Y., Ahmad, S., Wei, X., Hu, P., & Tang, S., (2020). Base editing: The ever expanding clustered regularly interspaced short palindromic repeats (CRISPR) tool kit for precise genome editing in plants. *Genes, 11*(4), 466.
22. (a) Wang, L., Chen, L., Li, R., Zhao, R., Yang, M., Sheng, J., & Shen, L., (2017). Reduced drought tolerance by CRISPR/Cas9-mediated SlMAPK3 mutagenesis in tomato plants. *Journal of Agricultural and Food Chemistry, 65*(39), 8674–8682; (b) Nieves-Cordones,

M., Mohamed, S., Tanoi, K., Kobayashi, N. I., Takagi, K., Vernet, A., Guiderdoni, E., Périn, C., et al., (2017). Production of low-Cs+ rice plants by inactivation of the K+ transporter Os HAK 1 with the CRISPR-Cas system. *The Plant Journal, 92*(1), 43–56; (c) Tabassum, J., Ahmad, S., Hussain, B., Mawia, A. M., Zeb, A., & Ju, L., (2021). Applications and potential of genome-editing systems in rice improvement: Current and future perspectives. *Agronomy, 11*(7), 1359.

23. Miao, C., Xiao, L., Hua, K., Zou, C., Zhao, Y., Bressan, R. A., & Zhu, J. K., (2018). Mutations in a subfamily of abscisic acid receptor genes promote rice growth and productivity. *Proceedings of the National Academy of Sciences, 115*(23), 6058–6063.
24. Takagi, H., Tamiru, M., Abe, A., Yoshida, K., Uemura, A., Yaegashi, H., Obara, T., et al., (2015). MutMap accelerates breeding of a salt-tolerant rice cultivar. *Nature Biotechnology, 33*(5), 445–449.
25. Zhang, A., Liu, Y., Wang, F., Li, T., Chen, Z., Kong, D., Bi, J., et al., (2019). Enhanced rice salinity tolerance via CRISPR/Cas9-targeted mutagenesis of the OsRR22 gene. *Molecular Breeding, 39*(3), 1–10.
26. Li, X. M., Chao, D. Y., Wu, Y., Huang, X., Chen, K., Cui, L. G., Su, L., et al., (2015). Natural alleles of a proteasome α2 subunit gene contribute to thermotolerance and adaptation of African rice. *Nature Genetics, 47*(7), 827–833.
27. Sauer, N. J., Narváez-Vásquez, J., Mozoruk, J., Miller, R. B., Warburg, Z. J., Woodward, M. J., Mihiret, Y. A., et al., (2016). Oligonucleotide-mediated genome editing provides precision and function to engineered nucleases and antibiotics in plants. *Plant Physiology, 170*(4), 1917–1928.
28. Svitashev, S., Schwartz, C., Lenderts, B., Young, J. K., & Cigan, A. M., (2016). Genome editing in maize directed by CRISPR–Cas9 ribonucleoprotein complexes. *Nature Communications, 7*(1), 1–7.
29. Butler, N. M., Baltes, N. J., Voytas, D. F., & Douches, D. S., (2016). Geminivirus-mediated genome editing in potato (*Solanum tuberosum* L.) using sequence-specific nucleases. *Frontiers in Plant Science, 7*, 1045.
30. Shimatani, Z., Kashojiya, S., Takayama, M., Terada, R., Arazoe, T., Ishii, H., Teramura, H., et al., (2017). Targeted base editing in rice and tomato using a CRISPR-Cas9 cytidine deaminase fusion. *Nature Biotechnology, 35*(5), 441–443.
31. Tang, N., Ma, S., Zong, W., Yang, N., Lv, Y., Yan, C., Guo, Z., et al., (2016). MODD mediates deactivation and degradation of OsbZIP46 to negatively regulate ABA signaling and drought resistance in rice. *The Plant Cell, 28*(9), 2161–2177.
32. Chen, Y., Ma, J., Zhang, X., Yang, Y., Zhou, D., Yu, Q., Que, Y., et al., (2017). A novel non-specific lipid transfer protein gene from sugarcane (NsLTPs), obviously responded to abiotic stresses and signaling molecules of SA and MeJA. *Sugar Tech, 19*(1), 17–25.
33. Tang, L., Mao, B., Li, Y., Lv, Q., Zhang, L., Chen, C., He, H., et al., (2017). Knockout of OsNramp5 using the CRISPR/Cas9 system produces low Cd-accumulating indica rice without compromising yield. *Sci. Rep., 7*(1), 1–12.
34. Ribeiro, C. W., Korbes, A. P., Garighan, J. A., Jardim-Messeder, D., Carvalho, F. E., Sousa, R. H., Caverzan, A., et al., (2017). Rice peroxisomal ascorbate peroxidase knockdown affects ROS signaling and triggers early leaf senescence. *Plant Science, 263*, 55–65.
35. Wu, J., Yang, R., Yang, Z., Yao, S., Zhao, S., Wang, Y., Li, P., et al., (2017). ROS accumulation and antiviral defense control by microRNA528 in rice. *Nature Plants, 3*(1), 1–7.

36. Sinharoy, S., Liu, C., Breakspear, A., Guan, D., Shailes, S., Nakashima, J., Zhang, S., et al., (2016). A *Medicago truncatula* cystathionine-β-synthase-like domain-containing protein is required for rhizobial infection and symbiotic nitrogen fixation. *Plant Physiology, 170*(4), 2204–2217.
37. Zafar, S. A., Patil, S., Uzair, M., Fang, J., Zhao, J., Yuan, S., Uzair, M., et al., (2019). Degenerated panicle and partial sterility 1 (DPS1) encodes a CBS domain containing protein required for anther cuticle and panicle development in rice. *New Phytologist*.
38. Zhao, J., Zhao, L., Zhang, M., Zafar, S. A., Fang, J., Li, M., Zhang, W., & Li, X., (2017). *Arabidopsis* E3 ubiquitin ligases PUB22 and PUB23 negatively regulate drought tolerance by targeting ABA receptor PYL9 for degradation. *International Journal of Molecular Sciences, 18*(9), 1841.
39. Fang, H., Meng, Q., Xu, J., Tang, H., Tang, S., Zhang, H., & Huang, J., (2015). Knock-down of stress inducible OsSRFP1 encoding an E3 ubiquitin ligase with transcriptional activation activity confers abiotic stress tolerance through enhancing antioxidant protection in rice. *Plant Molecular Biology, 87*(4, 5), 441–458.
40. Ning, Y., Jantasuriyarat, C., Zhao, Q., Zhang, H., Chen, S., Liu, J., Liu, L., et al., (2011). The SINA E3 ligase OsDIS1 negatively regulates drought response in rice. *Plant Physiology, 157*(1), 242–255.
41. (a) Zhang, M., Zhao, J., Li, L., Gao, Y., Zhao, L., Patil, S. B., Fang, J., et al., (2017). The *Arabidopsis* U-box E3 ubiquitin ligase PUB30 negatively regulates salt tolerance by facilitating BRI1 kinase inhibitor 1 (BKI1) degradation. *Plant, Cell & Environment, 40*(11), 2831–2843; (b) De Toledo, T. D. P., Brail, Q., Dahlbeck, D., & Staskawicz, B., (2016). CRISPR-Cas9 mediated mutagenesis of a DMR6 ortholog in tomato confers broad-spectrum disease resistance. *BioRxiv*, 064824.
42. He, L., Shi, X., Wang, Y., Guo, Y., Yang, K., & Wang, Y., (2017). *Arabidopsis* ANAC069 binds to C[A/G]CG[T/G] sequences to negatively regulate salt and osmotic stress tolerance. *Plant Molecular Biology, 93*(4, 5), 369–387.
43. Wang, C. T., Ru, J. N., Liu, Y. W., Li, M., Zhao, D., Yang, J. F., Fu, J. D., & Xu, Z. S., (2018). Maize WRKY transcription factor ZmWRKY106 confers drought and heat tolerance in transgenic plants. *International Journal of Molecular Sciences, 19*(10), 3046.
44. Gao, P., Bai, X., Yang, L., Lv, D., Pan, X., Li, Y., Cai, H., et al., (2011). osa-MIR393: A salinity-and alkaline stress-related microRNA gene. *Molecular Biology Reports, 38*(1), 237–242.
45. Chiou, T. J., Aung, K., Lin, S. I., Wu, C. C., Chiang, S. F., & Su, C. L., (2006). Regulation of phosphate homeostasis by microRNA in *Arabidopsis*. *The Plant Cell, 18*(2), 412–421.
46. Chandrasegaran, S., & Carroll, D., (2016). Origins of programmable nucleases for genome engineering. *Journal of Molecular Biology, 428*(5), 963–989.
47. Bortesi, L., & Fischer, R., (2015). The CRISPR/Cas9 system for plant genome editing and beyond. *Biotechnol. Adv., 33*(1), 41–52.
48. Zheng, Y., Shen, W., Zhang, J., Yang, B., Liu, Y. N., Qi, H., Yu, X., et al., (2018). CRISPR interference-based specific and efficient gene inactivation in the brain. *Nature Neuroscience, 21*(3), 447–454.
49. Piatek, A., Ali, Z., Baazim, H., Li, L., Abulfaraj, A., Al-Shareef, S., Aouida, M., & Mahfouz, M. M., (2015). RNA-guided transcriptional regulation in planta via synthetic dC as9-based transcription factors. *Plant Biotechnology Journal, 13*(4), 578–589.

50. Klap, C., Yeshayahou, E., Bolger, A. M., Arazi, T., Gupta, S. K., Shabtai, S., Usadel, B., et al., (2017). Tomato facultative parthenocarpy results from Sl AGAMOUS-LIKE 6 loss of function. *Plant Biotechnology Journal, 15*(5), 634–647.
51. Miao, J., Guo, D., Zhang, J., Huang, Q., Qin, G., Zhang, X., Wan, J., et al., (2013). Targeted mutagenesis in rice using CRISPR-Cas system. *Cell Research, 23*(10), 1233–1236.
52. Wang, X., Tu, M., Wang, D., Liu, J., Li, Y., Li, Z., Wang, Y., & Wang, X., (2018). CRISPR/Cas9-mediated efficient targeted mutagenesis in grape in the first generation. *Plant Biotechnology Journal, 16*(4), 844–855.
53. Char, S. N., Neelakandan, A. K., Nahampun, H., Frame, B., Main, M., Spalding, M. H., Becraft, P. W., et al., (2017). An Agrobacterium-delivered CRISPR/Cas9 system for high-frequency targeted mutagenesis in maize. *Plant Biotechnology Journal, 15*(2), 257–268.
54. Gao, W., Long, L., Tian, X., Xu, F., Liu, J., Singh, P. K., Botella, J. R., & Song, C., (2017). Genome editing in cotton with the CRISPR/Cas9 system. *Frontiers in Plant Science, 8*, 1364.
55. Rodríguez-Leal, D., Lemmon, Z. H., Man, J., Bartlett, M. E., & Lippman, Z. B., (2017). Engineering quantitative trait variation for crop improvement by genome editing. *Cell, 171*(2), 470–480. e8.
56. Eid, A., Alshareef, S., & Mahfouz, M. M., (2018). CRISPR base editors: Genome editing without double-stranded breaks. *Biochemical Journal, 475*(11), 1955–1964.
57. (a) Mali, P., Yang, L., Esvelt, K. M., Aach, J., Guell, M., DiCarlo, J. E., Norville, J. E., & Church, G. M., (2013). RNA-guided human genome engineering via Cas9. *Science, 339*(6121), 823–826; (b) Cong, L., Ran, F. A., Cox, D., Lin, S., Barretto, R., Habib, N., Hsu, P. D., et al., (2013). Multiplex genome engineering using CRISPR/Cas systems. *Science, 339*(6121), 819–823.
58. González, F., Zhu, Z., Shi, Z. D., Lelli, K., Verma, N., Li, Q. V., & Huangfu, D., (2014). An iCRISPR platform for rapid, multiplexable, and inducible genome editing in human pluripotent stem cells. *Cell Stem Cell, 15*(2), 215–226.
59. Dow, L. E., Fisher, J., O'rourke, K. P., Muley, A., Kastenhuber, E. R., Livshits, G., Tschaharganeh, D. F., et al., (2015). Inducible *in vivo* genome editing with CRISPR-Cas9. *Nature Biotechnology, 33*(4), 390–394.
60. Zetsche, B., Volz, S. E., & Zhang, F., (2015). A split-Cas9 architecture for inducible genome editing and transcription modulation. *Nature Biotechnology, 33*(2), 139–142.
61. Polstein, L. R., & Gersbach, C. A., (2015). A light-inducible CRISPR-Cas9 system for control of endogenous gene activation. *Nature Chemical Biology, 11*(3), 198–200.
62. Nihongaki, Y., Kawano, F., Nakajima, T., & Sato, M., (2015). Photoactivatable CRISPR-Cas9 for optogenetic genome editing. *Nature Biotechnology, 33*(7), 755–760.
63. Fu, Y., Sander, J. D., Reyon, D., Cascio, V. M., & Joung, J. K., (2014). Improving CRISPR-Cas nuclease specificity using truncated guide RNAs. *Nature Biotechnology, 32*(3), 279–284.
64. (a) Ran, F. A., Hsu, P. D., Lin, C. Y., Gootenberg, J. S., Konermann, S., Trevino, A. E., Scott, D. A., et al., (2013). Double nicking by RNA-guided CRISPR Cas9 for enhanced genome editing specificity. *Cell, 154*(6), 1380–1389; (b) Mali, P., Aach, J., Stranges, P. B., Esvelt, K. M., Moosburner, M., Kosuri, S., Yang, L., & Church, G. M., (2013). Cas9 transcriptional activators for target specificity screening and paired nickases for cooperative genome engineering. *Nature Biotechnology, 31*(9), 833–838.

65. Guilinger, J. P., Thompson, D. B., & Liu, D. R., (2014). Fusion of catalytically inactive Cas9 to FokI nuclease improves the specificity of genome modification. *Nature Biotechnology, 32*(6), 577.
66. (a) Slaymaker, I. M., Gao, L., Zetsche, B., Scott, D. A., Yan, W. X., & Zhang, F., (2016). Rationally engineered Cas9 nucleases with improved specificity. *Science, 351*(6268), 84–88; (b) Kleinstiver, B. P., Prew, M. S., Tsai, S. Q., Topkar, V. V., Nguyen, N. T., Zheng, Z., Gonzales, A. P., et al., (2015). Engineered CRISPR-Cas9 nucleases with altered PAM specificities. *Nature, 523*(7561), 481–485.
67. Bolukbasi, M. F., Gupta, A., Oikemus, S., Derr, A. G., Garber, M., Brodsky, M. H., Zhu, L. J., & Wolfe, S. A., (2015). DNA-binding-domain fusions enhance the targeting range and precision of Cas9. *Nature Methods, 12*(12), 1150–1156.
68. (a) Tester, M., & Langridge, P., (2010). Breeding technologies to increase crop production in a changing world. *Science, 327*(5967), 818–822; (b) Bailey-Serres, J., Parker, J. E., Ainsworth, E. A., Oldroyd, G. E., & Schroeder, J. I., (2019). Genetic strategies for improving crop yields. *Nature, 575*(7781), 109–118.
69. Ezezika, O. C., Daar, A. S., Barber, K., Mabeya, J., Thomas, F., Deadman, J., Wang, D., & Singer, P. A., (2012). Factors influencing agbiotech adoption and development in sub-Saharan Africa. *Nature Biotechnology, 30*(1), 38.
70. Zaidi, S. S. E. A., Vanderschuren, H., Qaim, M., Mahfouz, M. M., Kohli, A., Mansoor, S., & Tester, M., (2019). New plant breeding technologies for food security. *Science, 363*(6434), 1390, 1391.
71. (a) Ghosh, S., Watson, A., Gonzalez-Navarro, O. E., Ramirez-Gonzalez, R. H., Yanes, L., Mendoza-Suárez, M., Simmonds, J., et al., (2018). Speed breeding in growth chambers and glasshouses for crop breeding and model plant research. *Nature Protocols, 13*(12), 2944–2963; (b) Watson, A., Ghosh, S., Williams, M. J., Cuddy, W. S., Simmonds, J., Rey, M. D., Hatta, M. A. M., et al., (2018). Speed breeding is a powerful tool to accelerate crop research and breeding. *Nature Plants, 4*(1), 23–29.
72. Le Nguyen, K., Grondin, A., Courtois, B., & Gantet, P., (2019). Next-generation sequencing accelerates crop gene discovery. *Trends in Plant Science, 24*(3), 263–274.
73. Barabaschi, D., Tondelli, A., Desiderio, F., Volante, A., Vaccino, P., Valè, G., & Cattivelli, L., (2016). Next generation breeding. *Plant Science, 242*, 3–13.
74. Zorrilla-Fontanesi, Y., Pauwels, L., Panis, B., Signorelli, S., Vanderschuren, H., & Swennen, R., (2020). Strategies to revise agrosystems and breeding to control *Fusarium* wilt of banana. *Nature Food, 1*(10), 599–604.
75. He, Y., & Zhao, Y., (2020). Technological breakthroughs in generating transgene-free and genetically stable CRISPR-edited plants. *aBIOTECH, 1*(1), 88–96.
76. Liang, Z., Chen, K., Zhang, Y., Liu, J., Yin, K., Qiu, J. L., & Gao, C., (2018). Genome editing of bread wheat using biolistic delivery of CRISPR/Cas9 *in vitro* transcripts or ribonucleoproteins. *Nature Protocols, 13*(3), 413.
77. (a) Demirer, G. S., Zhang, H., Matos, J. L., Goh, N. S., Cunningham, F. J., Sung, Y., Chang, R., et al., (2019). High aspect ratio nanomaterials enable delivery of functional genetic material without DNA integration in mature plants. *Nature Nanotechnology, 14*(5), 456–464; (b) Kwak, S. Y., Lew, T. T. S., Sweeney, C. J., Koman, V. B., Wong, M. H., Bohmert-Tatarev, K., Snell, K. D., et al., (2019). Chloroplast-selective gene delivery and expression in planta using chitosan-complexed single-walled carbon nanotube carriers. *Nature Nanotechnology, 14*(5), 447–455.

78. Waltz, E., (2016). CRISPR-edited crops free to enter market, skip regulation. *Nature*. Publishing Group.
79. Pickar-Oliver, A., & Gersbach, C. A., (2019). The next generation of CRISPR–Cas technologies and applications. *Nature Reviews Molecular Cell Biology, 20*(8), 490–507.
80. Stella, S., & Montoya, G., (2016). The genome editing revolution: A CRISPR-Cas TALE off-target story. *Inside the Cell, 1*(1), 7–16.
81. Chen, K., Wang, Y., Zhang, R., Zhang, H., & Gao, C., (2019). CRISPR/Cas genome editing and precision plant breeding in agriculture. *Annual Review of Plant Biology, 70*, 667–697.
82. Li, J. F., Norville, J. E., Aach, J., McCormack, M., Zhang, D., Bush, J., Church, G. M., & Sheen, J., (2013). Multiplex and homologous recombination-mediated genome editing in *Arabidopsis* and *Nicotiana benthamiana* using guide RNA and Cas9. *Nat. Biotechnol., 31*(8), 688–691.
83. Hanna, R. E., & Doench, J. G., (2020). Design and analysis of CRISPR–Cas experiments. *Nature Biotechnology, 38*(7), 813–823.
84. Wright, A. V., Nuñez, J. K., & Doudna, J. A., (2016). Biology and applications of CRISPR systems: Harnessing nature's toolbox for genome engineering. *Cell, 164*(1, 2), 29–44.
85. O'Connell, M. R., Oakes, B. L., Sternberg, S. H., East-Seletsky, A., Kaplan, M., & Doudna, J. A., (2014). Programmable RNA recognition and cleavage by CRISPR/Cas9. *Nature, 516*(7530), 263–266.
86. Zaidi, S. S. E. A., Mahfouz, M. M., & Mansoor, S., (2017). CRISPR-Cpf1: A new tool for plant genome editing. *Trends in Plant Science, 22*(7), 550–553.
87. Aman, R., Ali, Z., Butt, H., Mahas, A., Aljedaani, F., Khan, M. Z., Ding, S., & Mahfouz, M., (2018). RNA virus interference via CRISPR/Cas13a system in plants. *Genome Biol., 19*(1), 1–9.
88. Smargon, A. A., Cox, D. B., Pyzocha, N. K., Zheng, K., Slaymaker, I. M., Gootenberg, J. S., Abudayyeh, O. A., et al., (2017). Cas13b is a type VI-B CRISPR-associated RNA-guided RNase differentially regulated by accessory proteins Csx27 and Csx28. *Molecular Cell, 65*(4), 618–630. e7.
89. Ahmad, S., Wei, X., Sheng, Z., Hu, P., & Tang, S., (2020). CRISPR/Cas9 for development of disease resistance in plants: Recent progress, limitations, and future prospects. *Briefings in Functional Genomics, 19*(1), 26–39.
90. Zaidi, S. S. E. A., Mukhtar, M. S., & Mansoor, S., (2018). Genome editing: Targeting susceptibility genes for plant disease resistance. *Trends in Biotechnology, 36*(9), 898–906.
91. Pyott, D. E., Sheehan, E., & Molnar, A., (2016). Engineering of CRISPR/Cas9-mediated potyvirus resistance in transgene-free *Arabidopsis* plants. *Mol. Plant Pathol., 17*(8), 1276–1288.
92. Blanvillain-Baufumé, S., Reschke, M., Solé, M., Auguy, F., Doucoure, H., Szurek, B., Meynard, D., et al., (2017). Targeted promoter editing for rice resistance to *Xanthomonas oryzae* pv. oryzae reveals differential activities for SWEET 14-inducing TAL effectors. *Plant Biotechnology Journal, 15*(3), 306–317.
93. Si, X., Zhang, H., Wang, Y., Chen, K., & Gao, C., (2020). Manipulating gene translation in plants by CRISPR–Cas9-mediated genome editing of upstream open reading frames. *Nature Protocols, 15*(2), 338–363.

94. Cai, Y., Chen, L., Sun, S., Wu, C., Yao, W., Jiang, B., Han, T., & Hou, W., (2018). CRISPR/Cas9-mediated deletion of large genomic fragments in soybean. *International Journal of Molecular Sciences, 19*(12), 3835.
95. Nishimasu, H., Shi, X., Ishiguro, S., Gao, L., Hirano, S., Okazaki, S., Noda, T., et al., (2018). Engineered CRISPR-Cas9 nuclease with expanded targeting space. *Science, 361*(6408), 1259–1262.
96. Ashfield, T., Egan, A. N., Pfeil, B. E., Chen, N. W., Podicheti, R., Ratnaparkhe, M. B., Ameline-Torregrosa, C., et al., (2012). Evolution of a complex disease resistance gene cluster in diploid *Phaseolus* and tetraploid glycine. *Plant Physiology, 159*(1), 336–354.
97. Van, V. T., Sung, Y. W., Kim, J., Doan, D. T. H., Tran, M. T., & Kim, J. Y., (2019). Challenges and perspectives in homology-directed gene targeting in monocot plants. *Rice, 12*(1), 1–29.
98. Liu, M., Rehman, S., Tang, X., Gu, K., Fan, Q., Chen, D., & Ma, W., (2019). Methodologies for improving HDR efficiency. *Frontiers in Genetics, 9*, 691.
99. Butt, H., Eid, A., Ali, Z., Atia, M. A., Mokhtar, M. M., Hassan, N., Lee, C. M., et al., (2017). Efficient CRISPR/Cas9-mediated genome editing using a chimeric single-guide RNA molecule. *Frontiers in Plant Science, 8*, 1441.
100. Wang, M., Lu, Y., Botella, J. R., Mao, Y., Hua, K., & Zhu, J. K., (2017). Gene targeting by homology-directed repair in rice using a geminivirus-based CRISPR/Cas9 system. *Molecular Plant, 10*(7), 1007–1010.
101. Zhang, H., Mittal, N., Leamy, L. J., Barazani, O., & Song, B. H., (2017). Back into the wild—Apply untapped genetic diversity of wild relatives for crop improvement. *Evolutionary Applications, 10*(1), 5–24.
102. Arora, S., Steuernagel, B., Gaurav, K., Chandramohan, S., Long, Y., Matny, O., Johnson, R., et al., (2019). Resistance gene cloning from a wild crop relative by sequence capture and association genetics. *Nature Biotechnology, 37*(2), 139–143.
103. Komor, A. C., Kim, Y. B., Packer, M. S., Zuris, J. A., & Liu, D. R., (2016). Programmable editing of a target base in genomic DNA without double-stranded DNA cleavage. *Nature, 533*(7603), 420–424.
104. Bastet, A., Zafirov, D., Giovinazzo, N., Guyon-Debast, A., Nogué, F., Robaglia, C., & Gallois, J. L., (2019). Mimicking natural polymorphism in eIF 4E by CRISPR-Cas9 base editing is associated with resistance to potyviruses. *Plant Biotechnology Journal, 17*(9), 1736–1750.
105. Butt, H., Zaidi, S. S. E. A., Hassan, N., & Mahfouz, M., (2020). CRISPR-based directed evolution for crop improvement. *Trends in Biotechnology, 38*(3), 236–240.
106. Zaidi, S. S. E. A., Mansoor, S., Ali, Z., Tashkandi, M., & Mahfouz, M. M., (2016). Engineering plants for geminivirus resistance with CRISPR/Cas9 system. *Trends in Plant Science, 21*(4), 279–281.
107. Ali, Z., Abulfaraj, A., Idris, A., Ali, S., Tashkandi, M., & Mahfouz, M. M., (2015). CRISPR/Cas9-mediated viral interference in plants. *Genome Biology, 16*(1), 1–11.
108. Baltes, N. J., Hummel, A. W., Konecna, E., Cegan, R., Bruns, A. N., Bisaro, D. M., & Voytas, D. F., (2015). Conferring resistance to geminiviruses with the CRISPR–Cas prokaryotic immune system. *Nature Plants, 1*(10), 1–4.
109. Tripathi, J. N., Ntui, V. O., Ron, M., Muiruri, S. K., Britt, A., & Tripathi, L., (2019). CRISPR/Cas9 editing of endogenous banana streak virus in the B genome of *Musa* spp. overcomes a major challenge in banana breeding. *Communications Biology, 2*(1), 1–11.

110. Zhang, T., Zheng, Q., Yi, X., An, H., Zhao, Y., Ma, S., & Zhou, G., (2018). Establishing RNA virus resistance in plants by harnessing CRISPR immune system. *Plant Biotechnology Journal, 16*(8), 1415–1423.
111. Aman, R., Mahas, A., Butt, H., Ali, Z., Aljedaani, F., & Mahfouz, M., (2018). Engineering RNA virus interference via the CRISPR/Cas13 machinery in *Arabidopsis*. *Viruses, 10*(12), 732.
112. Mahas, A., Aman, R., & Mahfouz, M., (2019). CRISPR-Cas13d mediates robust RNA virus interference in plants. *Genome Biology, 20*(1), 1–16.
113. Hadidi, A., (2019). Next-generation sequencing and CRISPR/Cas13 editing in viroid research and molecular diagnostics. *Viruses, 11*(2), 120.
114. Rybicki, E. P., (2019). CRISPR–Cas9 strikes out in cassava. *Nature Biotechnology, 37*(7), 727–728.
115. Gomez, M. A., Lin, Z. D., Moll, T., Luebbert, C., Chauhan, R. D., Vijayaraghavan, A., Kelley, R., et al., (2017). Simultaneous CRISPR/Cas9-mediated editing of cassava eIF4E isoforms nCBP-1 and nCBP-2 confers elevated resistance to cassava brown streak disease. *bioRxiv,* 209874.
116. Macovei, A., Sevilla, N. R., Cantos, C., Jonson, G. B., Slamet-Loedin, I., Čermák, T., Voytas, D. F., et al., (2018). Novel alleles of rice eIF4G generated by CRISPR/Cas9-targeted mutagenesis confer resistance to rice tungro spherical virus. *Plant Biotechnol. J., 16*(11), 1918–1927.
117. Zhou, J., Peng, Z., Long, J., Sosso, D., Liu, B., Eom, J. S., Huang, S., et al., (2015). Gene targeting by the TAL effector PthXo2 reveals cryptic resistance gene for bacterial blight of rice. *The Plant Journal, 82*(4), 632–643.
118. Ortigosa, A., Gimenez-Ibanez, S., Leonhardt, N., & Solano, R., (2019). Design of a bacterial speck resistant tomato by CRISPR/Cas9-mediated editing of Sl JAZ 2. *Plant Biotechnology Journal, 17*(3), 665–673.
119. Thomazella, D., Brail, Q., Dahlbeck, D., & Staskawic, Z., (2016). *CRISPR-Cas9-Mediated Mutagenesis of a DMR6 Ortholog in Tomato Confers Broad-Spectrum Disease Resistance*. bioRxiv Preprint.
120. Jia, H., Zhang, Y., Orbović, V., Xu, J., White, F. F., Jones, J. B., & Wang, N., (2017). Genome editing of the disease susceptibility gene Cs LOB 1 in citrus confers resistance to citrus canker. *Plant Biotechnology Journal, 15*(7), 817–823.
121. Peng, A., Chen, S., Lei, T., Xu, L., He, Y., Wu, L., Yao, L., & Zou, X., (2017). Engineering canker-resistant plants through CRISPR/Cas9-targeted editing of the susceptibility gene Cs LOB 1 promoter in citrus. *Plant Biotechnology Journal, 15*(12), 1509–1519.
122. Giraud, T., Gladieux, P., & Gavrilets, S., (2010). Linking the emergence of fungal plant diseases with ecological speciation. *Trends in Ecology & Evolution, 25*(7), 387–395.
123. Acevedo-Garcia, J., Kusch, S., & Panstruga, R., (2014). Magical mystery tour: MLO proteins in plant immunity and beyond. *New Phytologist, 204*(2), 273–281.
124. Freisleben, R., & Lein, A., (1942). Über die auffindung einer mehltauresistenten mutante nach röntgenbestrahlung einer anfälligen reinen Linie von sommergerste. *Naturwissenschaften, 30*(40), 608–608.
125. Martínez, M. I. S., Bracuto, V., Koseoglou, E., Appiano, M., Jacobsen, E., Visser, R. G., Wolters, A. M. A., & Bai, Y., (2020). CRISPR/Cas9-targeted mutagenesis of the tomato susceptibility gene PMR4 for resistance against powdery mildew. *BMC Plant Biology, 20*(1), 1–13.

126. Zhang, S., Wang, L., Zhao, R., Yu, W., Li, R., Li, Y., Sheng, J., & Shen, L., (2018). Knockout of SlMAPK3 reduced disease resistance to *Botrytis cinerea* in tomato plants. *Journal of Agricultural and Food Chemistry, 66*(34), 8949–8956.
127. Gumtow, R., Wu, D., Uchida, J., & Tian, M., (2018). A *Phytophthora palmivora* extracellular cystatin-like protease inhibitor targets papain to contribute to virulence on papaya. *Molecular Plant-Microbe Interactions, 31*(3), 363–373.
128. Fister, A. S., Landherr, L., Maximova, S. N., & Guiltinan, M. J., (2018). Transient expression of CRISPR/Cas9 machinery targeting TcNPR3 enhances defense response in *Theobroma cacao*. *Frontiers in Plant Science, 9*, 268.
129. Lawrenson, T., Shorinola, O., Stacey, N., Li, C., Østergaard, L., Patron, N., Uauy, C., & Harwood, W., (2015). Induction of targeted, heritable mutations in barley and *Brassica oleracea* using RNA-guided Cas9 nuclease. *Genome Biology, 16*(1), 1–13.
130. Chandrasekaran, J., Brumin, M., Wolf, D., Leibman, D., Klap, C., Pearlsman, M., Sherman, A., et al., (2016). Development of broad virus resistance in non-transgenic cucumber using CRISPR/Cas9 technology. *Mol. Plant Pathol., 17*(7), 1140–1153.
131. Shi, J., Gao, H., Wang, H., Lafitte, H. R., Archibald, R. L., Yang, M., Hakimi, S. M., et al., (2017). ARGOS 8 variants generated by CRISPR-Cas9 improve maize grain yield under field drought stress conditions. *Plant Biotechnology Journal, 15*(2), 207–216.
132. Clasen, B. M., Stoddard, T. J., Luo, S., Demorest, Z. L., Li, J., Cedrone, F., Tibebu, R., et al., (2016). Improving cold storage and processing traits in potato through targeted gene knockout. *Plant Biotechnology Journal, 14*(1), 169–176.
133. Jiang, W., Zhou, H., Bi, H., Fromm, M., Yang, B., & Weeks, D. P., (2013). Demonstration of CRISPR/Cas9/sgRNA-mediated targeted gene modification in *Arabidopsis*, tobacco, sorghum, and rice. *Nucleic Acids Research, 41*(20), e188–e188.
134. Li, M., Li, X., Zhou, Z., Wu, P., Fang, M., Pan, X., Lin, Q., et al., (2016). Reassessment of the four yield-related genes Gn1a, DEP1, GS3, and IPA1 in rice using a CRISPR/Cas9 system. *Frontiers in Plant Science, 7*, 377.
135. Demorest, Z. L., Coffman, A., Baltes, N. J., Stoddard, T. J., Clasen, B. M., Luo, S., Retterath, A., et al., (2016). Direct stacking of sequence-specific nuclease-induced mutations to produce high oleic and low linolenic soybean oil. *BMC Plant Biology, 16*(1), 1–8.
136. Ji, X., Zhang, H., Zhang, Y., Wang, Y., & Gao, C., (2015). Establishing a CRISPR–Cas-like immune system conferring DNA virus resistance in plants. *Nature Plants, 1*(10), 1–4.
137. Soyk, S., Müller, N. A., Park, S. J., Schmalenbach, I., Jiang, K., Hayama, R., Zhang, L., et al., (2017). Variation in the flowering gene SELF PRUNING 5G promotes day-neutrality and early yield in tomato. *Nature Genetics, 49*(1), 162–168.
138. Ito, Y., Nishizawa-Yokoi, A., Endo, M., Mikami, M., & Toki, S., (2015). CRISPR/Cas9-mediated mutagenesis of the RIN locus that regulates tomato fruit ripening. *Biochemical and Biophysical Research Communications, 467*(1), 76–82.
139. Ueta, R., Abe, C., Watanabe, T., Sugano, S. S., Ishihara, R., Ezura, H., Osakabe, Y., & Osakabe, K., (2017). Rapid breeding of parthenocarpic tomato plants using CRISPR/Cas9. *Scientific Reports, 7*(1), 1–8.
140. Nekrasov, V., Wang, C., Win, J., Lanz, C., Weigel, D., & Kamoun, S., (2017). Rapid generation of a transgene-free powdery mildew resistant tomato by genome deletion. *Sci. Rep., 7*(1), 1–6.

141. Wang, Y., Cheng, X., Shan, Q., Zhang, Y., Liu, J., Gao, C., & Qiu, J. L., (2014). Simultaneous editing of three homoeoalleles in hexaploid bread wheat confers heritable resistance to powdery mildew. *Nature Biotechnology, 32*(9), 947–951.
142. Ahmad, S., Shahzad, R., Jamil, S., Tabassum, J., Chaudhary, M. A. M., Atif, R. M., Iqbal, M. M., et al., (2021). Regulatory aspects, risk assessment, and toxicity associated with RNAi and CRISPR methods. In: *CRISPR and RNAi Systems* (pp. 687–721). Elsevier.
143. Wolter, F., & Puchta, H., (2017). Knocking out consumer concerns and regulator's rules: Efficient use of CRISPR/Cas ribonucleoprotein complexes for genome editing in cereals. *Genome Biology, 18*(1), 1–3.
144. Livingston, N. M., (2016). *The Livingston Tomato Report 2016: The Philosophical Environmentalists Guide to Justice in the Global Food System.* Masters' Dissertation; York University, North York, Ontario, Canada.
145. Chen, L., Ren, F., Zhong, H., Jiang, W., & Li, X. (2010). Identification and expression analysis of genes in response to high-salinity and drought stresses in Brassica napus. *Acta Biochim Biophys Sin, 42*(2), 154–164.
146. Gong, P., Zhang, J., Li, H., Yang, C., Zhang, C., Zhang, X., ... & Ye, Z. (2010). Transcriptional profiles of drought-responsive genes in modulating transcription signal transduction, and biochemical pathways in tomato. *Journal of experimental botany, 61*(13), 3563–3575.
147. Liu, C., Wang, B., Li, Z., Peng, Z., & Zhang, J. (2018). TsNAC1 is a key transcription factor in abiotic stress resistance and growth. *Plant physiology, 176*(1), 742–756.
148. Zuo, B., Zheng, X., He, P., Wang, L., Lei, Q., Feng, C., ... & Kong, J. (2014). Overexpression of MzASMT improves melatonin production and enhances drought tolerance in transgenic Arabidopsis thaliana plants. *Journal of Pineal Research, 57*(4), 408–417.
149. Byeon, Y., & Back, K. (2016). Low melatonin production by suppression of either serotonin N-acetyltransferase or N-acetylserotonin methyltransferase in rice causes seedling growth retardation with yield penalty, abiotic stress susceptibility, and enhanced coleoptile growth under anoxic conditions. *Journal of Pineal Research, 60*(3), 348–359.
150. Antoniou, C., Chatzimichail, G., Xenofontos, R., Pavlou, J. J., Panagiotou, E., Christou, A., & Fotopoulos, V. (2017). Melatonin systemically ameliorates drought stress-induced damage in M edicago sativa plants by modulating nitro-oxidative homeostasis and proline metabolism. *Journal of Pineal Research, 62*(4), e12401.
151. Jung, C., Seo, J. S., Han, S. W., Koo, Y. J., Kim, C. H., Song, S. I., ... & Cheong, J. J. (2008). Overexpression of AtMYB44 enhances stomatal closure to confer abiotic stress tolerance in transgenic Arabidopsis. *Plant physiology, 146*(2), 623.

CHAPTER 7

APPLICATIONS AND EXPLORATIONS OF CRISPR-BASED GENOME EDITING TOOL-BOX IN ABIOTIC STRESS TOLERANCE IN PLANTS

DEEPU PANDITA

Government Department of School Education, Jammu, Jammu and Kashmir, India

ABSTRACT

By the year 2050, global population is expected to increase equal to 9.7 billion. To feed so many mouths, production of food needs a colossal escalation. The abiotic stresses, for instance, water scarcity, salinity, high temperature, low temperature, water submergence and heavy metal contamination have significantly restricted the universal agricultural and crop produce and threaten future food security. The threat to the food security is further proceeding in scenarios of global warming and climate fluctuations. The revolutionary tools of CRISPR/Cas genome editing (GE) platforms are game-changing, indispensable, faster, low-priced, more precise, potential, and most promising technologies to alleviate complex machinery of abiotic stress tolerance in plants and in designing novel climate smart abiotic stress-tolerant crops. CRISPR/Cas mediated GE facilitates abiotic stress tolerance in a number of crop plants.

CRISPR-Based Gene Editing: A Solution for Global Food Security. Shakeel Ahmad, Asif Ali Khan, and Muhammad Arshad Javed (Eds.)

7.1 INTRODUCTION

Firstly, the global population has been estimated to increase up to 9.7 billion. As such the production of agriculture needs to be increased by 70–85% till the year 2050 [1]. Secondly, global warming results in climate change which decline crop growth, crop yields and agriculture by greater than 50% every year worldwide. The climatic factors such as change in water distribution, daily temperatures, salinity, and soil nutrition influence crop production [3, 4]. Human food demand will almost double by 2050s [5] and increase in crop production remains a challenge. Modern breeding technologies and genetic engineering approaches will help in enhancing abiotic stress tolerances and develop climate-resilient crops. Improvement of crops by conventional breeding does not produce climate smart resilience because of complex inheritance and high genotype plus environmental interaction [6]. Undesirable genes also get transmitted together with gene of Interest (GOI) in plant breeding (Figure 7.1(a)). Recent advancements of genetic engineering and omics technologies have revolutionized crop research and act as suitable strategies for the integration of foreign DNA and new nucleotides, deletion of DNA segments and epigenetic alterations. These directly allow the study of genotype interaction with phenotype and modify genetic material to produce superior cultivars [7]. Genetically modified (GM) crops have improved traits through recombinant DNA technology but with limitations and regulatory and ethical issues because of safety and efficacy alarms in commercialization [8, 9]. GMOs have externally introduced gene constructs having selectable marker genes. Recombinant DNA machinery is costly and slow (8 to 10 years are requisite for releasing improved varieties) and non-targeted transgene integration in plant genome may interrupt endogenous gene expression. In GMOs, only genes of Interest (GOI) are transferred to recipient plant (Figure 7.1(b)). Clustered Regularly Interspaced Short Palindromic Repeats (CRISPR)/Cas system has developed as a promising and revolutionary tool of genome editing (GE) to efficiently introduce hereditary mutations by removing, addition or modification of DNA sequences in the plant genome [10] (Figure 7.1(c)). Future applications of CRISPR/Cas in molecular breeding will be key approach to improve abiotic stress tolerance, elevate crop improvement, and sustain global food security during climate change. CRISPR/Cas is delivered as a protein-RNA complex in target plant and degrades after edition. Thus, non-genetically/non-transgenic modified crops with desirable characters can be produced without regulations on genetic modifications. The way of gene transfer in plant breeding, GMOs, and GE are depicted in Figure 7.1.

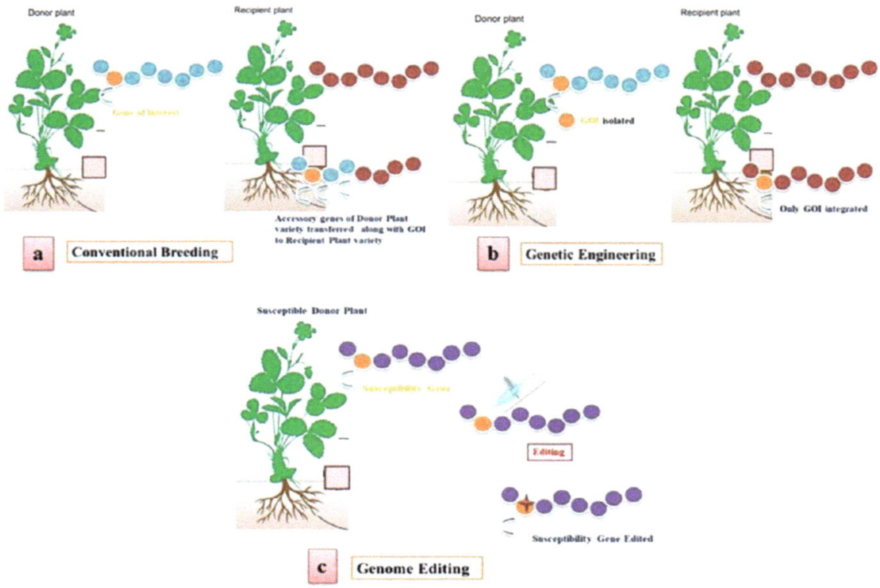

FIGURE 7.1 Gene transfer in (a) conventional breeding; (b) GMOs; and (c) genome editing.

7.2 HISTORY OF CRISPR

CRISPR term was coined by Jansen et al. [11]. Discovery of bacterial CRISPR array (repeat sequences type I-E CRISPR/Cas system) was identified by Ishino et al. [12]. CRISPR families show ubiquitous nature in prokaryotic organisms (2000). CRISPR/Cas9 was identified in 2002 and spacers in CRISPR as foreign bacteriophage DNA in 2005. CRISPR/Cas provides resistance to bacteriophages and their role in bacterial adaptive immunity system was characterized in 2006–2007 [2]. In year 2008, prospacers were discovered and RNA mediated DNA targeting mechanism was deduced. CR ISPR/Cas9 was designed and characterized as GE tool in 2012 [13] and regulation of genes through CRISPR interference (CRISPRi) and CRISPR activation (CRISPRa) in 2013 [14]. CRISPR/Cas12a was discovered in 2015 and CRISPR/Cas13a in 2016. CRISPR/Cas mediated GE was reported in plants [15, 16] and animals in the year 2013 [17].

7.3 CRISPR/CAS SYSTEM

CRISPR are specialized DNA of 28–37 base pairs in locus of bacterial systems which provides them protection against invading and foreign viruses.

Between the repeat sequences in CRISPR array are located the spacer DNA sequences assimilated by bacteria from the DNA of bacteriophage that has invaded bacteria earlier which serve as the genetic memory of previous bacteriophage infection (Figure 7.2). On invasion by the same bacteriophage, the host bacterial cell defends itself by cleavage and degradation of attacking sequences of virus complementary to the spacer sequences. Bacterial cells protect itself from invading viral genome by succeeding steps (Figure 7.2).

7.3.1 ADAPTIVE IMMUNITY SYSTEM BY CRISPR/CAS

CRISPR/Cas mediated acquired immunity in in bacteria is broadly classified into 3 major evolutionary stages adaptation, expression, and interference (Figure 7.2).

7.3.1.1 CRISPR ACQUISITION

Adaptation (spacer acquisition) involves cleavage, processing or spacer acquisition and integration of foreign DNA of the invading bacteriophage into locus of CRISPR as new spacers labeled as protospacers. Cas1 enzymes possess nuclease activity and assist in the process of acquisition [18].

7.3.1.2 CRRNA EXPRESSION

Transcription of CRISPR array and insertion of spacers forms pre-CRISPR RNA. Pre-CRISPR RNA on maturation forms mature crRNA. Processing of pre-crRNA and integration with Cas effector nucleases lead to formation of crRNA effector complex [18, 19].

7.3.1.3 CRISPR INTERFERENCE (CRISPRI)

CRISPR/Cas effector complex helps in recognition of 2–6 bp long protospacer adjacent motif (PAM) of target viruses and cleaves target DNA [20] (Figure 7.2).

7.4 CLASSIFICATION

A CRISPR/Cas system has been categorized into Class 1 and Class 2. The Class 1 and Class 2 are again sectioned into 6 types and 34 subtypes [21]. Class 1 has types of I, III, and IV whereas Class 2 has II, V, and VI types [22]. GE is predominantly done by class 2 of CRISPR/Cas system. The types

II and V are used in genome engineering and type VI for RNA engineering [23, 24]. The diagrammatic representation of the classification of CRISPR/Cas system is given in Figure 7.3.

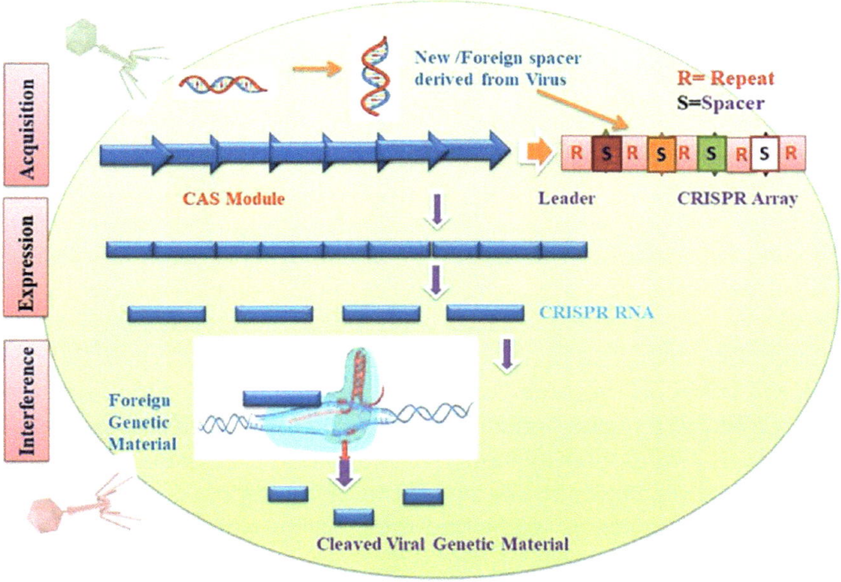

FIGURE 7.2 Bacterial adaptive immunity by CRISPR/Cas.

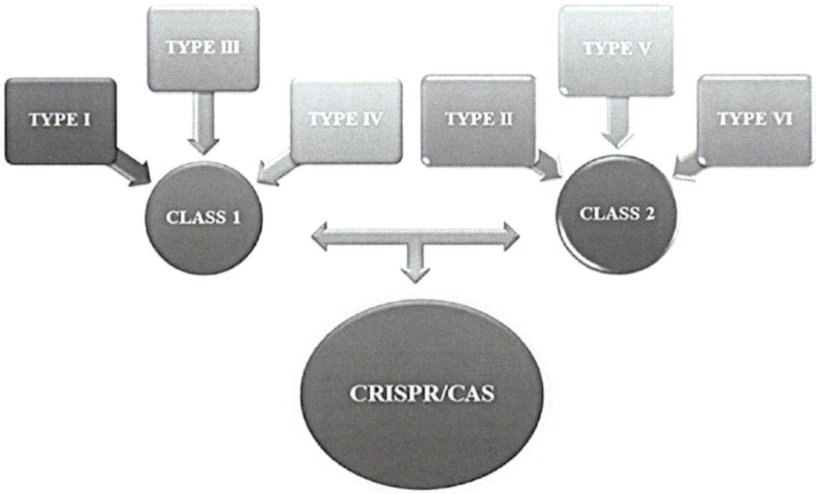

FIGURE 7.3 CRISPR/Cas system classification.

7.5 CRISPR/CAS-MEDIATED GENOME EDITING (GE) FOR PLANT ABIOTIC STRESS TOLERANCE

The yield of crop plants is influenced by various abiotic stresses such as drought, flooding, salinity, cold, and high temperature with higher risks under the scenario of climate changes [25, 26]. Drought, salinity, low/high temperature and heavy metal abiotic stresses pose key challenge to agriculture. Drought and salinity occur at the global levels and cause yield loss in majority of crops [27, 28]. CRISPR has enabled heat/high temperature, cold/low temperature, herbicide tolerance and resistance against viruses, bacteria, and fungi in *Oryza sativa* L., *Nicotiana tabacum* L., *Triticum aestivum* L., *Solanum tuberosum* L., *Zea mays* L., *Solanum lycopersicum* L., *Gossypium* spp., *Glycine max* (L.) Merr. and brassicas [18]. Some sensitivity (S) genes which may act as potential targets for improving abiotic stress tolerance through GE are *RGLG2* targets *AtERF53* [29], *ARR1*, *ARR10*, *ARR12* [30], *AtPUB19* [31] in Arabidopsis, *OsDIS1* in Rice [32] for drought tolerance, *OsERF922* in rice for salt tolerance [33] and *PIF3* and *AtVDAC1* in *Arabidopsis thaliana* for tolerance to cold [34, 35]. Brassinazole resistant 1 (BZR1) gene regulates the brassinosteroid (BR) response and BR-mediated developmental processes and thermo-tolerance through regulation of Feronia (*FER*) genes by CRISPR-bzr1- and BZR1-overexpression lines [36].

7.5.1 CRISPR/CAS-MEDIATED GENOME EDITING (GE) IN ORYZA SATIVA L.

CRISPR/Cas9 technology helps in development of elite cultivars of drought and salt tolerance (DST) and grain yield in indica mega rice cultivar MTU1010 by mutating (OsDST) gene. Cas9-free dstΔ184–305 demonstrated adequate tolerance to osmotic and high salt stress in the seedling stage [37]. CRISPR/Cas9 mediated knocked-out generation of triple mutant of cold tolerance endogenous gene MYB30 and two yield related genes improved cold tolerance and yield in rice [38]. CRISPR/Cas9 generated rice mutants to understand mechanism of ABA-stress-activated protein kinase 2 [39]. SNF1-related protein kinase 2 (SnRK2) regulates osmotic stress response and ABA signaling. CRISPR/Cas9 edited SAPK2 mutants were ABA insensitive due to non-functional SAPK2 and more sensitive to oxidative and drought stress in rice [39]. CRISPR/Cas9 targeted OsRR22 mutants showed improvement of salinity tolerance in rice seedlings [40]. CRISPR/Cas9-targeted mutagenesis of enhanced response to ABA1 (OsERA1) which regulates ABA signaling and dehydration response confers increased response to abscisic acid (ABA) and tolerance to drought in osera 1 mutant lines of rice. OsERA1 negatively regulates primary root growth in normal conditions [41]. Three CRISPR/Cas9 knockout mutants (ospqt3) of *Oryza sativa* PARAQUAT TOLERANCE 3 (OsPQT3) which encodes E3 ubiquitin ligase enhance resistance to oxidative and salinity abiotic stresses and increases grain yield in field. The ospqt3 mutants' show enhanced OsGPX1, OsAPX1, and OsSOD1 expression [42]. CRISPR/Cas9 edited Semi-Dwarfl (SD1) in elite landraces Kasalath and TeTePu (TTP), show shorter plant height and improved resistance against lodging [43]. CRISPR/Cas9 designed OsNAC45 knock-out lines of rice were extra sensitive to salt stress [44]. Cas12a mediated OsEPFL9 gene knock-out significantly reduces stomatal count in leaves and increases water use efficiency under situations of stress in rice [45]. In *Oryza sativa* L., the modification of abiotic tolerance genes has been done using CRISPR/Cas9, e.g., *OsMYB1* (MYB family transcription factor), *OsPDS* (albino), *OsMYB5* (MYB family transcription factor), *OsEPSPS* (lethal), *OsPMS3* (photoperiod sensitive male sterile), *OsDERF1* (drought tolerance), *OsYSA* (albino young seedling), *OsMSH1* (pleiotropic phenotype), *OsROC5* (abaxial leaf rolling), and *OsSPP* (early seedling leaf chlorosis) genes [46, 47]. Based on CRISPR/Cas9 editing in Nipponbare rice, a variety of *TIFY1b* (a transcription factor) mutant lines with some desirable changes in the cold-tolerant gene were generated [48]. CRISPR-based RNA editing tools target miRNAs, e.g., OsMIR528 gene encoding miRNA edited at DNA level helps in salt stress tolerance in rice [49].

7.5.2 CRISPR/CAS-MEDIATED GENOME EDITING (GE) IN ARABIDOPSIS THALIANA L. HEYNH.

CRISPR/Cas9 mediated GE was for the first time done in jasmonate-zim-domain protein 1 (*JAZ1*), brassinosteroid insensitive 1 (*BRI1*), and gibberellic acid insensitive (*GAI*) genes of *Arabidopsis* [50]. CRISPR/Cas9 in *Arabidopsis* generated mutants (cbfs double and triple) to determine role of C-repeat binding factors under cold stress [51]. Knockout of entire family of *AITR* (ABA-induced transcription repressors) (*AITR3* and *AITR4*) conserved transcription factors (TFs) family by CRISPR/Cas9 increased drought and salinity abiotic stress tolerance in Arabidopsis. AITRs act as negative regulators of ABA signaling and abiotic stresses. Thus, AITRs can be targeted for the molecular breeding of crop plants [52, 53].

7.5.3 CRISPR/CAS-MEDIATED GENOME EDITING (GE) IN SOLANUM LYCOPERSICUM L.

Mitogen-activated protein kinase (MAPK) signaling molecules respond to drought in *Solanum lycopersicum* L. CRISPR/Cas9 designed mitogen-activated protein kinases 3 (slmapk3) mutants enhanced defense response to water scarcity in tomato and show severe wilting symptoms [54]. C-repeat binding factor 1 (CBF1) protects cold-sensitive tomato, because CRISPR/Cas9 generated cbf1 mutants show more severe chilling-injury [55]. Long noncoding RNAs (lncRNAs) regulate expression of genes. CRISPR/Cas9 mediated editing of lncRNA1459 generates tomato mutants with delayed fruit ripening [56].

7.5.4 CRISPR/CAS-MEDIATED GENOME EDITING (GE) IN ZEA MAYS L.

Auxin regulated gene involved in organ size (*ARGOS*) gene family act as negative regulators of ethylene response and play significant role in the ethylene signal transduction pathways. CRISPR/Cas9 mediated *ARGOS8* gene overexpression in maize enhances drought tolerance. Mutants showed enhanced yields under drought [57]. CRISPR/Cas9 facilitated GE produced new allelic variants of *ARGOS8* (*ARGOS8-v1* and *ARGOS8-v2*) for improved yield under stress [58].

7.5.5 CRISPR/CAS-MEDIATED GENOME EDITING (GE) IN TRITICUM AESTIVUM L.

In wheat protoplasts, CRISPR/Cas9 enabled GE of two abiotic stress-connected genes viz., wheat dehydration responsive element binding protein 2 (*TaDREB2*) and wheat ethylene responsive factor 3 (*TaERF3*) [59]. The other chief models where CRISPR/Cas facilitated GE has been effectively done for attaining plant abiotic stress tolerance are enlisted in Table 7.1.

7.6 CONCLUSION AND FUTURE DIRECTIONS

The GE tool of CRISPR/Cas has enabled designing of climate resilient and abiotic stress resistant plants. These will be beneficial in scenarios of the upcoming escalation in human population and climate change. CRISPR/Cas

TABLE 7.1 Applications of CRISPR/Cas-Mediated Genome Editing Tools in Abiotic Stress Tolerance

CRISPR/Cas Type	Genes Targeted	Impact Produced	Transgenic Plants	References
CRISPR/Cas9	Auxin-regulated gene involved in organ size 8 (*ARGOSS*).	Drought tolerance	*Zea mays* L.	[58]
	Arabidopsis thaliana vacuolar H^+ pyrophosphatase (AVP1).	Drought stress tolerance	*Arabidopsis thaliana* L. Heynh.	[60]
	Drb2a and *Drb2b*	Salinity and drought tolerance	*Glycine max* L. Merr.	[61]
	NPRI	Drought tolerance	*Solanum lycopersicum* L.	[62]
	Open stomata 2 knockout.	Drought tolerance	*Arabidopsis thaliana* (L.) Heynh.	[63]
	Oryza sativa bilateral blade senescence 1 (*BBS1*).	Sensitivity to drought	*Oryza sativa* L.	[64]
	OsDERF1, *OsPMS3*, *OsMSH1*, *OsMYB5*	Drought tolerance	*Oryza sativa* L.	[65]
	OsMPK2, *OsPDS*, *OsBADH2*	Multiple-stress tolerance	*Oryza sativa* L.	[66]
	OsNAC041	Salt tolerance	*Oryza sativa* L.	[67]
	OsRR22	Salt tolerance	*Oryza sativa* L.	[68]
	PYL1, *PYL4*, *PYL6*	Regulation of stress responses	*Oryza sativa* L.	[69]
	SAPK1 and *SAPK2*	Salt tolerance	*Glycine max* (L.) Merr.	[70]
	SIAGL6	Heat stress tolerance	*Solanum lycopersicum* L.	[71]
	TaDREB-2 and *3*	Drought tolerance	*Triticum aestivum* L.	[59]
	ZmHKT1	Salt tolerance	*Zea mays* L.	[72]
	Heat shock protein 90.	Heat tolerance	*Tetraselmis suecica* Kylin (Butcher).	[73]

TABLE 7.1 *(Continued)*

CRISPR/Cas Type	Genes Targeted	Impact Produced	Transgenic Plants	References
CRISPR/Cas9 knock out/CRISPRi	*OsSIT1*	Salt tolerance	*Oryza sativa* L.	[74]
	OsMODD	Drought tolerance	*Oryza sativa* L.	[75]
	OsSRL1, *OsSRL2*	Drought tolerance	*Oryza sativa* L.	[76]
	OsSAPK2	Drought tolerance	*Oryza sativa* L.	[70]
	TIFY1a, TIFY1b	Cold tolerance	*Oryza sativa* L.	[48]
	SIMAPK3	Drought tolerance	*Solanum lycopersicum* L.	[54]
	SINPR1	Drought tolerance	*Solanum lycopersicum* L.	[77]
CRISPR/Cas9-HDR or C>T and A>G editors	*OsCOLD1*	Cold tolerance	*Oryza sativa* L.	[78]
	OsTT1	Heat tolerance	African rice	[79]
CRISPR/dCas9HAT	*AREB1*	Drought tolerance	*Arabidopsis thaliana* (L.) Heynh.	[80]
CRISPR/Cpf1/ LbCas12a	*SlHKT 1; 2 HDR*	Multiple-stress tolerance/salt tolerance.	*Solanum lycopersicum* L.	[81]
CRISPR/Cas12a	*OsEPFL9*	Regulation of stomata density.	*Oryza sativa* L.	[45]

mediated GE system has fundamentally influenced crop improvement. Plant GE has no ethical issues. CRISPR/Cas-mediated GE has great potential in the area of stress biology due to greater efficiency and specificity. CRISPR/Cas facilitates editing abiotic stress tolerance in various plants. Scientific research using CRISPR/Cas systems has been reported to improve effectively the abiotic stress tolerance in a number of plants as discussed above and has effective roles in future applications of molecular breeding. Different versions of CRISPR/Cas system have emerged as proficient tools of GE for plant abiotic stress tolerance. Plants originating from CRISPR GE technology described above are non-GM crops and do not face the GMO regulatory regimes. So can be subjected to commercialization and marketing. This will enable designing of crop plants with superior genotypes and phenotypes.

KEYWORDS

- **abiotic stress tolerance**
- **adaptive immunity system**
- **Arabidopsis thaliana**
- **CRISPR/Cas**
- **food security**
- **genome editing**
- **global warming**
- **plant abiotic stress tolerance**

REFERENCES

1. Ray, D. K., Mueller, N. D., West, P. C., & Foley, J. A., (2013). Yield trends are insufficient to double global crop production by 2050. *PLoS One, 8*, e66428.
2. Barrangou, R., Fremaux, C., Deveau, H., Richards, M., Boyaval, P., Moineau, S., Romero, D. A., & Horvath, P., (2007). CRISPR provides acquired resistance against viruses in prokaryotes. *Science, 315*, 1709–1712.
3. Gusta, L., (2012). Abiotic stresses and agricultural sustainability. *Journal of Crop Improvement, 26*, 415–427.
4. Verma, A. K., & Deepti, S., (2016). Abiotic stress and crop improvement: Current scenario. *Advances in Plants & Agriculture Research, 4*, 345, 346.

5. Tilman, D., Balzer, C., Hill, J., & Befort, B. L., (2011). Global food demand and the sustainable intensification of agriculture. *Proceedings of the National Academy of Sciences of the United States of America, 108*, 20260–20264.
6. Bhat, J. A., Ali, S., Salgotra, R. K., Mir, Z. A., Dutta, S., Jadon, V., Tyagi, A., et al., (2016). Genomic selection in the era of next generation sequencing for complex traits in plant breeding. *Frontiers in Genetics, 7*, 221.
7. Tester, M., & Langridge, P., (2010). Breeding technologies to increase crop production in a changing world. *Science, 327*, 818–822.
8. Prado, J. R., Segers, G., Voelker, T., Carson, D., Dobert, R., Phillips, J., Cook, K., et al., (2014). Genetically engineered crops: From idea to product. *Annual Review of Plant Biology, 65*, 769–790.
9. Raman, R., (2017). The impact of genetically modified (GM) crops in modern agriculture: A review. *GM Crops & Food, 8*, 195–208.
10. Bortesi, L., & Fischer, R., (2015). The CRISPR/Cas9 system for plant genome editing and beyond. *Biotechnology Advances, 33*, 41–52.
11. Jansen, R., Embden, J. D. V., Gaastra, W., & Schouls, L. M., (2002). Identification of genes that are associated with DNA repeats in prokaryotes. *Mol. Microboil., 43*, 1565–1575.
12. Ishino, Y., Shinagawa, H., Makino, K., Amemura, M., & Nakata, A., (1987). Nucleotide sequence of the iap gene, responsible for alkaline phosphatase isozyme conversion in *Escherichia coli*, and identification of the gene product. *J. Bacteriol., 169*, 5429–5433.
13. Jinek, M., Chylinski, K., Fonfara, I., Hauer, M., Doudna, J. A., & Charpentier, E., (2012). A programmable dual-RNA-guided DNA endonuclease in adaptive bacterial immunity. *Science, 337*, 816–821.
14. Bikard, D., Jiang, W., Samai, P., Hochschild, A., Zhang, F., & Marraffini, L. A., (2013). Programmable repression and activation of bacterial gene expression using an engineered CRISPR-Cas system. *Nucleic Acids Res., 41*(15), 7429–7437. doi: 10.1093/nar/gkt520.
15. Li, J. F., Norville., J. E., Aach., J., McCormack, M., Zhang, D., Bush, J., Church, G. M., & Sheen, J., (2013). Multiplex and homologous recombination-mediated genome editing in *Arabidopsis* and *Nicotiana benthamiana* using guide RNA and Cas9. *Nature Biotechnology, 31*, 688–691. doi: 10.1038/nbt.2654.
16. Shan, Q., Wang, Y., Li, J., Zhang, Y., Chen, K., Liang, Z., Zhang, K., et al., (2013). Targeted genome modification of crop plants using a CRISPR-Cas system. *Nat. Biotechnol., 31*, 686.
17. Mali, P., Yang, L., Esvelt, K. M., Aach, J., Guell, M., DiCarlo, J. E., Norville, J. E., & Church, G. M., (2013). RNA-guided human genome engineering via Cas9. *Science, 339*, 823–826.
18. Chen, K., Wang, Y., Zhang, R., Zhang, H., & Gao, C., (2019). CRISPR/Cas genome editing and precision plant breeding in agriculture. *Annu. Rev. Plant Biol., 70*, 667–697.
19. Terns, M. P., (2018). CRISPR-based technologies: Impact of RNA-targeting systems. *Mol. Cell, 72*(3), 404–412. doi: 10.1016/j.molcel.2018.09.018.
20. Knott, G. J., & Jennifer, A. D., (2018). CRISPR-Cas guides the future of genetic engineering. *Science, 361*(6405), 866–869. https://doi.org/10.1126/science.aat5011.
21. Makarova, K. S., Wolf, Y. I., & Koonin, E. V., (2018). Classification and nomenclature of CRISPR-Cas systems: Where from here? *CRISPR J, 1*, 325–336.

22. Manghwar, H., Lindsey, K., Zhang, X., & Jin, S., (2019). CRISPR/Cas system: Recent *Advances and Future Prospects for Genome Editing. Trends in Plant Science, 24*(12), 1102–1125. https://doi.org/10.1016/j.tplants.2019. 09.006.
23. Moon, S. B., Kim, D. Y., Jeong-Heon, K., Kim, & Yong-Sam, K., (2019). Recent advances in the CRISPR genome editing tool set. *Experimental & Molecular Medicine, 51*, 130 https://doi.org/10.1038/s12276-019-0339-7.
24. Burmistrz, M., Krakowski, K., & Krawczyk-Balska, A., (2020). RNA-targeting CRISPR–Cas systems and their applications. *Int. J. Mol. Sci., 21*, 1122. doi: 10.3390/ijms21031122.
25. Wang, W., Vinocur, B., & Altman, A., (2003). Plant responses to drought, salinity, and extreme temperatures: Towards genetic engineering for stress tolerance. *Planta., 218*, 1–14.
26. Mushtaq, M., Bhat, J. A., Mir, Z. A., Sakina, A., Ali, S., Singh, A. K., Tyagi, A., et al., (2018). CRISPR/Cas approach: A new way of looking at plant-abiotic interactions. *J. Plant Physiol., 224, 225*, 156–162. doi: 10.1016/j.jplph.2018.04.001.
27. Boyer, J. S., (1982). Plant productivity and environment. *Science, 218*, 443–448.
28. Ghosh, D., & Xu, J., (2014). Abiotic stress responses in plant roots: A proteomics perspective. *Front. Plant Sci., 5*, 6.
29. Cheng, M. C., Hsieh, E. J., Chen, J. H., Chen, H. Y., & Lin, T. P., (2012). *Arabidopsis* RGLG2, functioning as a RING E3 ligase, interacts with AtERF53 and negatively regulates the plant drought stress response. *Plant Physiology, 158*, 363–375.
30. Nguyen, K. H., Ha, C. V., Nishiyama, R., et al., (2016). *Arabidopsis* type B cytokinin response regulators ARR1, ARR10, and ARR12 negatively regulate plant responses to drought. *Proceedings of the National Academy of Sciences, USA, 113*, 3090–3095.
31. Liu, Y. C., Wu, Y. R., Huang, X. H., Sun, J., & Xie, Q., (2011). AtPUB19, a U-box E3 ubiquitin ligase, negatively regulates abscisic acid and drought responses in *Arabidopsis* thaliana. *Molecular Plant, 4*, 938–946.
32. Ning, Y., Jantasuriyarat, C., Zhao, Q., et al., (2011). The SINA E3 ligase OsDIS1 negatively regulates drought response in rice. *Plant Physiology, 157*, 242–255.
33. Liu, D., Chen, X., Liu, J., Ye, J., & Guo, Z., (2012). The rice ERF transcription factor OsERF922 negatively regulates resistance to *Magnaporthe oryzae* and salt tolerance. *Journal of Experimental Botany, 63*, 3899–3911.
34. Jiang, B., Shi, Y., Zhang, X., Xin, X., Qi, L., Guo, H., Li, J., & Yang, S., (2017). PIF3 is a negative regulator of the CBF pathway and freezing tolerance in *Arabidopsis. Proceedings of the National Academy of Sciences, USA, 114,* E6695–E6702.
35. Li, Z. Y., Xu, Z. S., He, G. Y., Yang, G. X., Chen, M., Li, L. C., & Ma, Y., (2013). The voltage dependent anion channel 1 (AtVDAC1) negatively regulates plant cold responses during germination and seedling development in *Arabidopsis* and interacts with calcium sensor CBL1. *International Journal of Molecular Sciences, 14*, 701–713.
36. Yin, Y., Qin, K., Song, X., Zhang, Q., Zhou, Y., Xia, X., & Yu, J., (2018). BZR1 transcription factor regulates heat stress tolerance through FERONIA receptor-like kinase-mediated reactive oxygen species signaling in tomato. *Plant Cell Physiol., 59*(11), 2239–2254. doi: 10.1093/pcp/pcy146. PMID: 30107607.
37. Kumar, S. V. V., Verma, R. K., Yadav, S. K., et al., (2020). CRISPR-Cas9 mediated genome editing of drought and salt tolerance (OsDST) gene in indica mega rice cultivar MTU1010. *Physiol. Mol. Biol. Plants., 26*, 1099–1110. https://doi.org/10.1007/s12298-020-00819-w.

38. Zeng, Y., Wen, J., Zhao, W., Wang, Q., & Huang, W., (2020). Rational improvement of rice yield and cold tolerance by editing the three genes OsPIN5b, GS3, and OsMYB30 with the CRISPR-Cas9 system. *Front. Plant Sci., 10*, 1663. https://doi.org/10.3389/fpls.2019.01663.
39. Lou, D., Wang, H., Liang, G., & Yu, D., (2017). OsSAPK2 confers abscisic acid sensitivity and tolerance to drought stress in rice. *Front. Plant Sci., 8*, 993. https://doi.org/10.3389/fpls.2017.00993.
40. Zhang, A., Liu, Y., Wang, F., Li, T., Chen, Z., Kong, D., et al., (2019). Enhanced rice salinity tolerance via CRISPR/Cas9-targeted mutagenesis of the OsRR22 gene. *Mol. Breed., 39*, 47. https://doi.org/10.1007/s11032-019-0954-y.
41. Ogata, T., Ishizaki, T., Fujita, M., & Fujita, Y., (2020). CRISPR/Cas9-targeted mutagenesis of OsERA1 confers enhanced responses to abscisic acid and drought stress and increased primary root growth under nonstressed conditions in rice. *PLoS One, 15*(12), e0243376. https://doi.org/10.1371/journal.pone.0243376.
42. Alfatih, A., Wu, J., Jan, S. U., Zhang, Z. S., Xia, J. Q., & Xiang, C. B., (2020). Loss of rice PARAQUAT TOLERANCE 3 confers enhanced resistance to abiotic stresses and increases grain yield in field. *Plant Cell Environ., 43*(11), 2743–2754. https://doi.org/10.1111/pce.13856.
43. Hu, X., Cui, Y., Dong, G., et al., (2019). Using CRISPR-Cas9 to generate semi-dwarf rice lines in elite landraces. *Sci. Rep., 9*, 19096. https://doi.org/10.1038/s41598-019-55757-9.
44. Zhang, X., Long, Y., Huang, J., et al., (2020). OsNAC45 is involved in ABA response and salt tolerance in rice. *Rice, 13*, 79. https://doi.org/10.1186/s12284–020-00440-1.
45. Yin, X., Anand, A., Quick, P., & Bandyopadhyay, A., (2019). Editing a stomatal developmental gene in rice with CRISPR/Cpf1. In: Qi, Y., (ed.), *Plant Genome Editing with CRISPR Systems; Methods in Molecular Biology* (Vol. 1917). Humana Press, New York, NY. https://doi.org/10.1007/978-1-4939–8991-1_19.
46. Zhang, H., Zhang, J., Wei, P., et al., (2014). The CRISPR/Cas9 system produces specific and homozygous targeted gene editing in rice in one generation. *Plant Biotechnology Journal, 12*, 797–807.
47. Xu, G., Yuan, M., Ai, C., Liu, L., Zhuang, E., Karapetyan, S., Wang, S., & Dong, X., (2017). uORF-mediated translation allows engineered plant disease resistance without fitness costs. *Nature, 545*, 491–494.
48. Huang, X. Z., Zeng, X. F., Li, J. R., & Zhao, D. G., (2017). Construction and analysis of tify1a and tify1b mutants in rice (*Oryza sativa*) based on CRISPR/Cas9 technology. *J. Agric. Biotechnol., 25*, 1003–1012. 10.3969/j.issn.1674-7968.2017.06.015.
49. Zhou, J., Deng, K., Cheng, Y., Zhong, Z., Tian, L., Tang, X., Tang, A., Zheng, X., Zhang, T., Qi, Y., et al., (2017). CRISPR-Cas9 based genome editing reveals new insights into microRNA function and regulation in rice. *Front. Plant Sci., 8*, 1–12.
50. Feng, Z., Zhang, B., Ding, W., Liu, X., Yang, D. L., Wei, P., et al., (2013). Efficient genome editing in plants using a CRISPR/Cas system. *Cell Res., 23*, 1229–1232. https://doi.org/10.1038/cr.2013.114.
51. Jia, Y., Ding, Y., Shi, Y., Zhang, X., Gong, Z., & Yang, S., (2016). The cbfs triple mutants reveal the essential functions of CBF s in cold acclimation and allow the definition of CBF regulons in *Arabidopsis*. *New Phytol., 212*, 345.
52. Tian, H., Chen, S., Yang, W., Wang, T., Zheng, K., Wang, Y., et al., (2017). A novel family of transcription factors conserved in angiosperms is required for ABA signaling. *Plant Cell Environ., 40*, 2958–2971.

53. Chen, S., Zhang, N., Zhou, G., Hussain, S., Ahmed, S., Tian, H., & Wang, S., (2021). Knockout of the entire family of AITR genes in *Arabidopsis* leads to enhanced drought and salinity tolerance without fitness costs. *BMC Plant Biology, 21*, 137. https://doi.org/10.1186/s12870-021-02907-9.
54. Wang, L., Chen, L., Li, R., Zhao, R., Yang, M., Sheng, J., & Shen, L., (2017). Reduced drought tolerance by CRISPR/Cas9-mediated SlMAPK3 mutagenesis in tomato plants. *J. Agric. Food Chem., 65*, 8674–8682. https://doi.org/10.1021/acs.jafc.7b027 45.
55. Li, R., Zhang, L., Wang, L., Chen, L., Zhao, R., Sheng, J., & Shen, L., (2018). Reduction of tomato-plant chilling tolerance by CRISPR-Cas9-mediated SlCBF1 mutagenesis. *J. Agric. Food Chem., 66*(34), 9042–9051. doi: 10.1021/acs.jafc.8b02177.
56. Li, R., Fu, D., Zhu, B., Luo, Y., & Zhu, H., (2018). CRISPR/Cas9-mediated mutagenesis of lncRNA1459 alters tomato fruit ripening. *Plant J., 94*(3), 513–524. doi: 10.1111/tpj.13872.
57. Hirai, M. Y., Sugiyama, K., Sawada, Y., Tohge, T., Obayashi, T., Suzuki, A., & Goda, H., (2007). Omics based identification of *Arabidopsis* Myb transcription factors regulating aliphatic glucosinolate biosynthesis. *Proc. Natl. Acad. Sci. USA, 104*, 6478.
58. Shi, J., Gao, H., Wang, H., Lafitte, H. R., Archibald, R. L., Yang, M., et al., (2017). ARGOS8 variants generated by CRISPR-Cas9 improve maize grain yield under field drought stress conditions. *Plant Biotechnol. J., 15*, 207–216. https ://doi.org/10.1111/pbi.12603.
59. Kim, D., Alptekin, B., & Budak, H., (2018). CRISPR/Cas9 genome editing in wheat. *Funct. Integr. Genomics, 18*, 31–41. 10.1007/s10142-017-0572-x.
60. Park, J. J., Dempewolf, E., Zhang, W., & Wang, Z. Y., (2017). RNA-guided transcriptional activation via CRISPR/dCas9 mimics overexpression phenotypes in *Arabidopsis*. *PLoS One, 12*, e0179410.
61. Curtin, S. J., Xiong, Y., Michno, J. M., Campbell, B. W., Stec, A. O., ˇCermák, T., Starker, C., et al., (2018). CRISPR/Cas9 and TALENs generate heritable mutations for genes involved in small RNA processing of glycine max and *Medicago truncatula*. *Plant Biotechnol. J., 16*, 1125–1137.
62. Erpen-Dalla, C. L., Mahmoud, L. M., Moraes, T. S., Mou, Z., Grosser, J. W., & Dutt, M., (2019). Development of improved fruit, vegetable, and ornamental crops using the CRISPR/Cas9 genome editing technique. *Plants, 8*, 601.
63. Osakabe, Y., Watanabe, T., Sugano, S. S., Ueta, R., Ishihara, R., Shinozaki, K., & Osakabe, K., (2016). Optimization of CRISPR/Cas9 genome editing to modify abiotic stress responses in plants. *Sci. Rep., 6*, 26685. https ://doi.org/10.1038/srep2 6685.
64. Zeng, D. D., Yang, C. C., Qin, R., Alamin, M., Yue, E. K., Jin, X. L., & Shi, C. H., (2018). A guanine insert in OsBBS1 leads to early leaf senescence and salt stress sensitivity in rice (*Oryza sativa* L.). *Plant Cell Reports, 37*(6), 933–946. doi:10.1007/s00299-018–2280-y.
65. Hsu, P. D., Lander, E. S., & Zhang, F., (2014). Development and applications of CRISPR-Cas9 for genome engineering. *Cell, 157*(6), 1262–1278.
66. Korotkova, A. M., Gerasimova, S. V., & Khlestkina, E. K., (2019). Current achievements in modifying crop genes using CRISPR/Cas system. *Vavilovskii Zhurnal Genet. Selektsii, 23*, 29–37.
67. Bo, W., Zhaohui, Z., Huanhuan, Z., Xia, W., Binglin, L., Lijia, Y., Xiangyan, H., et al., (2019). Targeted mutagenesis of NAC transcription factor gene, OsNAC041, leading to salt sensitivity in rice. *Rice Sci., 26*, 98–108.

68. Zhang, A., Liu, Y., Wang, F., Li, T., Chen, Z., Kong, D., Bi, J., Zhang, F., Luo, X., & Wang, J., (2019). Enhanced rice salinity tolerance via CRISPR/Cas9-targeted mutagenesis of the OsRR22 gene. *Mol. Breed., 39*, 47.
69. Miao, C., Xiao, L., Hua, K., Zou, C., Zhao, Y., Bressan, R. A., et al., (2018). Mutations in a subfamily of abscisic acid receptor genes promote rice growth and productivity. *PNAS, 115*, 6058–6063.
70. Lou, D., Wang, H., Liang, G., & Yu, D., (2017). OsSAPK2 confers abscisic acid sensitivity and tolerance to drought stress in rice. *Front Plant Sci., 8*, 993. https ://doi.org/10.3389/fpls.2017.00993.
71. Klap, C., Yeshayahou, E., Bolger, A. M., Arazi, T., Gupta, S. K., Shabtai, S., Usadel, B., et al., (2017). Tomato facultative parthenocarpy results from SlAGAMOUS-LIKE 6 loss of function. *Plant Biotechnol. J., 15*, 634–647.
72. Zhang, M., Cao, Y., Wang, Z., Wang, Z. Q., Shi, J., Liang, X., Song, W., et al., (2018). A retrotransposon in an HKT1 family sodium transporter causes variation of leaf Na+ exclusion and salt tolerance in maize. *New Phytol., 217*, 1161–1176.
73. Xu, J., Soni, V., Chopra, M., & Chan, O., (2020). Genetic modification of the HSP90 GENE using CRISPR-Cas9 to enhance thermotolerance in *T. Suecica. URNCST Journal., 4*(4). https://doi.org/10.26685/urncst.178.
74. Li, C. H., Wang, G., Zhao, J. L., Zhang, L. Q., Ai, L. F., Han, Y. F., et al., (2014). The receptor-like kinase SIT1 mediates salt sensitivity by activating MAPK3/6 and regulating ethylene homeostasis in rice. *The Plant Cell, 26*(6), 2538–2553. doi:10.1105/tpc.114.125187.
75. Tang, N., Ma, S., Zong, W., Yang, N., Lv, Y., Yan, C., Guo, Z., Li, J., Li, X., Xiang, Y., et al., (2016). MODD mediates deactivation and degradation of OsbZIP46 to negatively regulate ABA signaling and drought resistance in rice. *The Plant Cell., 28*(9), 2161–2177. doi:10.1105/tpc.16.00171.
76. Liao, S., Qin, X., Luo, L., Han, Y., Wang, X., Usman, B., Nawaz, G., et al., (2019). CRISPR/Cas9-induced mutagenesis of semi-rolled leaf 1,2 confers curled leaf phenotype and drought tolerance by influencing protein expression patterns and ROS scavenging in rice (*Oryza sativa* L.). *Agronomy, 9*, 728. https ://doi.org/10.3390/agronomy91 10728.
77. Li, R., Liu, C., Zhao, R., Wang, L., Chen, L., Yu, W., Zhang, S., Shen, J., & Shen, L., (2019). CRISPR/Cas9-mediated SlNPR1 mutagenesis reduces tomato plant drought tolerance. *BMC Plant Biol., 19*, 38. https ://doi.org/10.1186/s1287 0-018-1627-4.
78. Ma, Y., Dai, X., Xu, Y., Luo, W., Zheng, X., Zeng, D., Pan, Y., Lin, X., Liu, H., Zhang, D., et al., (2015). COLD1 confers chilling tolerance in rice. *Cell, 160*(6), 1209–1221. 10.1016/j.cell.2015.01.046.
79. Li, X. M., Chao, D. Y., Wu, Y., Huang, X., Chen, K., Cui, L. G., et al., (2015). Natural alleles of a proteasome α2 subunit gene contribute to thermotolerance and adaptation of African rice. *Nature Genetics, 47*(7), 827–833. doi: 10.1038/ng.3305.
80. Roca-Paixão, J. F., Gillet, F. X., Ribeiro, T. P., Bournaud, C., Lourenco-Tessutti, T., Noriega, D. D., Melo, B., et al., (2019). Improved drought stress tolerance in *Arabidopsis* by CRISPR/dCas9 fusion with a histone acetyl transferase. *Sci. Rep., 9*, 8080. https://doi.org/10.1038/s4159 8-019-44571-y.
81. Vu, T. V., Sivankalyani, V., Kim, E. J., Doan, D. T. H., Tran, M. T., Kim, J., Sung, Y. W., et al., (2020). Highly efficient homology-directed repair using CRISPR/Cpf1-geminiviral replicon in tomato. *Plant Biotechnol. J.,* 1–11. doi:10.1111/pbi.13373.

CHAPTER 8

CRISPR-CAS-MEDIATED MUTAGENESIS FOR DEVELOPING RESISTANCE AGAINST PLANT PATHOGENS

KHADIJA AALIYA,[1] SEHRISH IFTIKHAR,[2] MUHAMMA NAEEM SATTAR,[3] FARAH SAEED,[4] FARAH K. AHMED,[5] SAJID FAIZ,[6] ADEL A. REZK,[7] and KAMEL A. ABD-ELSALAM[8]

[1]*Center of Excellence in Molecular Biology, University of the Punjab, Lahore, Pakistan*

[2]*Microfluidics Innovation Center, Elvesys, Paris, France*

[3]*Central Laboratories, King Faisal University, Al-Ahsa, Kingdom of Saudi Arabia*

[4]*Institute of Agricultural Sciences (IAGS), University of the Punjab, Quaid-e-Azam Campus, Lahore, Pakistan*

[5]*Biotechnology English Program, Faculty of Agriculture, Cairo University, Giza, Egypt*

[6]*Department of Plant Breeding and Genetics, The University of Haripur, Khyber Pakhtunkhwa, Pakistan*

[7]*Department of Biotechnology, College of Agricultural and Food Sciences, King Faisal University, Al-Ahsa, Kingdom of Saudi Arabia*

[8]*Plant Pathology Research Institute, Agricultural Research Center (ARC), Giza, Egypt*

CRISPR-Based Gene Editing: A Solution for Global Food Security. Shakeel Ahmad, Asif Ali Khan, and Muhammad Arshad Javed (Eds.)

ABSTRACT

CRISPR technology for plant genome editing (GE) unlocks new possibilities for engineering genes for growth and development, biofortification, disease resistance, better tolerance to various biotic and abiotic stresses. In this context, we review the latest advances in CRISPR technology for fungal, bacterial, viral, nematode, and insect resistance engineering. Finally, we emphasize emerging technological challenges and technology loopholes and unresolved issues for future study. Finally, the future perspective of plant GE and breeding for resistance are discussed. The development of disease-resistant crop types with targeted CRISPR mutagenesis may be a good way to address obstacles to disease resistance in crop breeding.

8.1 INTRODUCTION

The world's population is projected to reach 10 billion by 2050, while the available crop and water supplies decline, with a rise of about 20–70% from current production level in the demand for food [1–3]. Therefore, it is a great challenge to ensure food safety for this vast population, particularly under climatic resilient context. The increase of human population and climate change is one of the critical issues worldwide to be addressed in the development of high-yielding crops with their evolved adaptation to the environment [4]. CRISPR (clustered regularly interspaced short palindromic repeats), which was discovered in 1993, is an easy and efficient method for genome editing (GE) [5, 6]. CRISPR/Cas9 technology could become basic protocols for the acquisition of plants with unique appealing characteristics to help achieve the international zero-hunger objectives [7]. The main causative agents that cause biotic stress and reduce plant yield, are plant pathogens (viruses, bacteria, fungi, and nematodes) and plant insects either causing directing crop losses or indirectly by acting as vectors for certain pathogens. The continuing growth of many new strains of deadly pests also makes it very complicated to combat these pathogens. Worldwide crop losses have been recorded due to ~137 pathogens associated with wheat, rice, maize, potato, and soybean [8]. Biotic stresses have negative impacts on various plant hosts caused by other living organisms therefore, it is important to consider the plant pathogens to defend farming against the destructive effects of biotic stressors [9]. Currently, CRISPR techniques have been utilized to strengthen the world food chain and to show how the CRISPR toolbox can be used to further enhance food security by conferring resistance to plant pathogens and food-borne pathogens or by enhancing the functionality of beneficial microorganisms in cultivation and probiotics [10, 11]. CRISPR/Cas9 has led to crop protection strategies through the development of mutated variants of disease susceptibility genes that suppress/eliminate pathogenic effector interaction [12]. The use of CRISPR/Cas offers a time-saving approach to functional genomics to produce resistance cultivars against new fungal strains that may also be used to promote plant pathogens as potential competitors and/or in plant defense [3]. The use of CRISPR/Cas9 as a GE tool has been extended to support the classical breeding of horticultural and field crops. The main aim of the chapter is to discuss CRISPR/Cas9 system applications to safeguard crop plants against fungal, bacterial, and viral pathogens.

8.2 PLANT-PATHOGEN INTERACTIONS AND PLANT DEFENSE MECHANISM AGAINST PATHOGENS

The complex interactions between plants and their associated microbes positively or negatively affect the plant ecology [9]. Plants are susceptible to a variety of economically important pathogens including bacteria, fungi, nematodes, phytoplasma, and viruses [13]. More than 40% of losses in yield in plants are caused by biotic stresses accounting for 15% of global food production drop [14, 15]. Meanwhile, the recurrent yield losses due to plant pathogens create a continuous challenge to the global food production and to feed the global population, which is projected to be doubled by 2050 [14, 15].

The intensive and non-judicial use of chemical-based insecticides and pesticides to control the plant pathogens, sustain plant health and overcome yield losses could not bring a fair control over plant pathogens [16, 17]. Rather, applications of chemicals have contaminated cultivated soils, water reservoirs, vegetation, food chains and impose risk to end-users [16, 17].

In plants, the defense mechanism is mediated by pathogen as well as plant-derived molecules. The synergy between plant and the pathogen involves a complex plethora of molecular interactions [18]. The defense-related interacting molecules generally rely upon the interacting pathogen for example fungi, viruses, or bacteria. For instance, bacteria use different secretary systems (type II, III, and IV) to release virulence-related biomolecules, which interact with the host plant [19]. Nevertheless, fungi secrete numerous enzymes to degrade the plant cell wall, which is the first defense line of the host plant, followed by the development of haustorium or discharge of infectious biomolecules to initiate the infection. While viruses have adopted a different mechanism of entry either by directly entering mechanically through wounding or biologically using vectors to initiate infection. Contrarily, plants have also developed defense strategies where they have some receptors called wall-associated kinases (WAKs) or trans-membrane protein recognition receptors (PRRs) [20] to initiate the defense responses. Host receptors sense the microbes-associated molecular patterns or pathogen-associated molecular patterns (MAMPs or PAMPs) resulting in defense response to fight with the infectious pathogen known as PAMP-triggered immunity (PTI) [21]. While pathogens have developed effectors-triggered susceptibility (ETS) in response to the host's defense system, where various pathogens excrete some effectors to overcome PTI by accelerating susceptibility proteins and to fight against host plant immunity. On the contrary, the plants activate effectors-triggered immunity (ETI), as a secondary response to the pathogens, where the effectors or avirulence (avr) proteins recognize resistance (R) genes. Quantitatively, ETI response is much quicker and stronger as compared to PTI and cause localized cell death or hypersensitive response (HR) to simultaneously eradicate the pathogen and the invaded plant cells. Generally, PTI is known as a nonspecific type of immunity, which is conserved in some pathogens and activated by the PAMPs detection. On the other hand, ETI is considered as a highly specific form of immunity, which is activated through sensing the pathogen effectors and simultaneous HR response leading to programmed cell death (PCD) ultimately hindering the pathogen invasion at the infection site [22]. As a consequence of PTI and ETI activation, the expression of pathogenesis-related genes is induced upon pathogen attack and mitogen-activated protein kinase (MAPK) is activated, which modify transcription factors (TFs) or plant hormones. As a result, the downstream events are also affected such as cell-wall adjustments, closed stomata, HR, generation of anti-microbial compounds (chitinases, phytoalexins, defensins) or generation of reactive oxygen species (ROS) [21]. The plants detect and eradicate the invading pathogens using the aforementioned strategies as their active defense mechanism.

8.3 COMPLEXITIES OF DISEASE RESISTANCE IN PLANTS

Plants and their pathogens are in constant combat where plants are always trying to defend themselves and pathogens are in a race to overcome the defense system of the plants. As a result, plants must equip themselves with the necessary defense mechanisms to cope with the attacking pathogens. Prichard and birch described in their zig-zag model that plants have adopted various defense strategies resulting from the selection pressure during evolution while pathogens have adopted numerous evasion capabilities [23]. This phenomenon was further illustrated by Anderson, describing plants with developed physical barriers to restrict the entry of invading pathogens as well as they also use some receptors either in plasma membrane or inter/intracellular space, which can initiate the defense mechanism once pathogen invades or by detecting pathogen-derived compounds. Plants can secrete some antimicrobial compounds to recognize the pathogen or their associated compounds and subsequently detoxify and inhibit their spreading [15]. In the same way, pathogens have also adopted some strategies to overcome plant defense system and then to initiate and spread the disease. For example, various fungal and bacterial pathogens have evolved to secrete the cell wall-degrading enzymes, e.g., cellulase, xylanase, etc., with the help of which the pathogen can contravene the cell membrane of plants and can invade within the host [24]. Some pathogens have adopted the other methods of invasion where they secrete some effector molecules and can silence the host defense mechanism, resulting in disease progression [25]. Deep understandings of these molecular mechanisms between plants and pathogens have enabled the researchers to develop disease-resistant or less susceptible plant varieties for sustainable agriculture. The rapid advancement in molecular biology techniques allowed changing the host genome either by inserting some genes, to restrict the activity of cell wall-degrading enzyme, disintegrate the toxins, and secrete some antimicrobial compounds, or deleting susceptible genes to pathogen invasion [26]. Advances in high-throughput molecular techniques allowed the in-depth understanding of infection mechanism of pathogens and immunity responses in plants providing opportunities for GE to develop resistance against plant pathogens. In this background, CRISPR based GE technology is applied to modify the plant traits, improve resistance, and consequently improve crop production.

8.4 CRISPR-CAS9 SYSTEM AS BACTERIAL IMMUNE SYSTEM AND MAGNIFICENT TOOL FOR PLANT EDITING

In bacteria and archaea, immunity against bacteriophages and transposons is attained with the help of RNA mediated CRISPR/Cas systems [8]. CRISPR/Cas9 system is comprised of two components: Cas9 nucleases and CRISPR array. CRISPR array is constituted with clustered short palindromic repeats, which are regularly interspaced with spacer sequences of the same length and Cas genes [27]. Spacer sequences are typically the full or partial genetic materials of the invading pathogens, which have previously caused infections.

In response to pathogen attack, foreign genetic material is integrated into CRISPR array. This event persuades to development a type of memory in a bacterial genome. As a result, transcription of CRISPR array leads to develop mature crRNA which is called crRNA biogenesis [28]. crRNA biogenesis is the transcription of pre-crRNA from CRISPR array and then its final endonucleolytic cleavage to form crRNA. This crRNA is responsible for recognition and degradation of pathogenic genetic material (Figure 8.1).

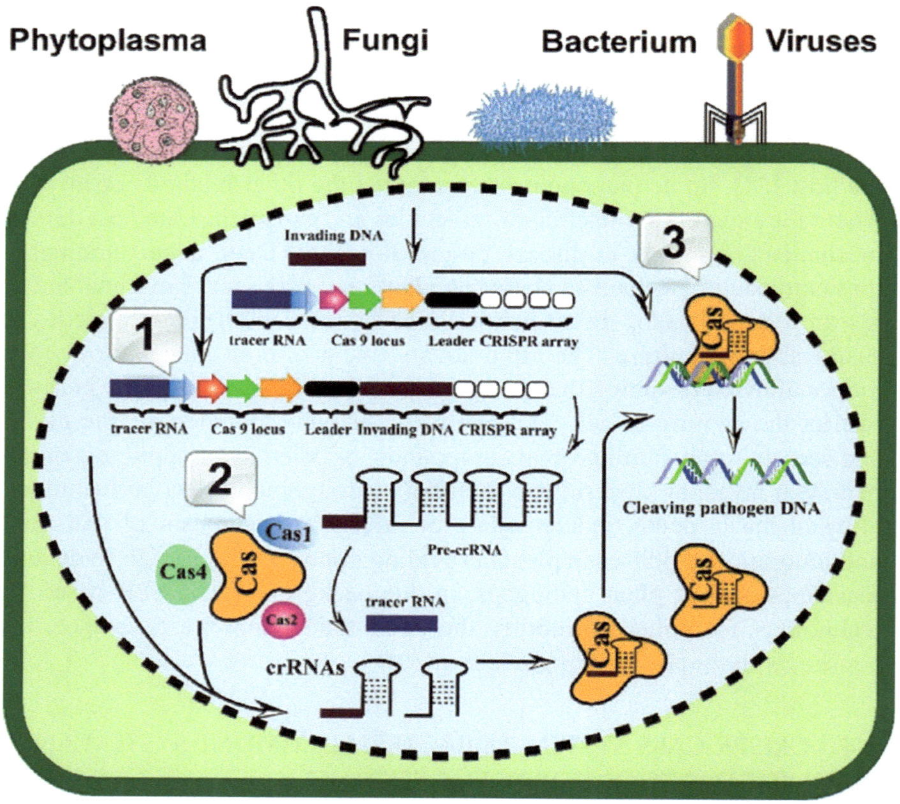

FIGURE 8.1 A three-step CRISPR/Cas9-based resistance model showing: (1) Identification and recruitment of foreign nucleic acid (fungal, bacterial, viral or phytoplasmic). The dark brown bar depicting the invaded nucleic acid, which is integrated into CRISPR-array shown in white rectangles and call new spacers. (2) The CRISPR machinery and components are accelerated inside the infected cell. And (3) the mature crRNA biosynthesis started with hybridization of crRNA to the invaded genome and this combination is recognized by Cas proteins. Coordination between Cas helicase and CRISPR cleave the foreign nucleic acid, respectively.

Source: Modified from: Sattar et al. [29].

The CRISPR/Cas systems can be grouped into two distinct systems based on structure, function, and sequences of Cas proteins. The class 1 CRISPR/Cas systems include types I, III, and IV, which uses a multi-protein effector complex. On the contrary, class 2 of CRISPR/Cas systems contains types II, V, and VI, which uses a single effector protein [30–32]. As a result, the class 2 CRISPR/Cas systems have been extensively applied for the manipulation and detection of nucleic acid [33, 34]. The Cas proteins of types II and V are usually applied to target DNA while the Cas proteins of type VI are used to target RNA. Cas9 enzyme has the ability to snip DNA 3-pb upstream of the PAM. It has 2 distinct domains: HNH-like nuclease domain, which is responsible for the cleavage of DNA strand complementary to target strand and the RuvC-like nuclease domain, responsible for the cleavage of non-complementary strand [35], respectively. Cas9 nucleases introduce double-stranded DNA (dsDNA) breaks (DSBs) right after the protospacer adjacent motifs (PAM). These PAMs are the short strings of nucleotides (2–5 bases) attached to the 3′-end of the spacer sequence provide the feasibility for Cas nucleases to identify the target sequence [36]. The CRISPR mediated DSB is repaired in two ways: a non-homologous end joining (NHEJ) mechanism or the homology-directed repair (HDR) mechanism [37]. Among these, the fallible NHEJ mechanism repairs the DSBs through inserting sequences or deleting sequences (collectively called InDels), which results in loss of function and genetically modified (GM) crops. Until now, CRISPR-Cas9 GE has been successfully exploited to edit the genome of both monocot and dicot plants species [38]. The genome of model plants such as *Nicotiana benthamiana* and *Arabidopsis thaliana* along with cash crops including, wheat, rice, barley, maize, cotton, tomato, soybean, potato, grape, citrus, etc., have been modified irrespective of their complex polyploid [36, 39].

8.5 CRISPR/CAS9-BASED HOST-MEDIATED RESISTANCE AGAINST FUNGAL PATHOGENS

In some circumstances, biotic factors such as fungal infections have a negative effect not only on the growth and production of the plant but also deteriorate the quality of the product. While transgenic technology has been used to insert certain genes to produce transgenic plants with high pathogen tolerances, various transgenic technologies techniques has been limited progress as none of gene played an important role in the plants' response to diseases [28]. Since CRISPR/Cas9 has been used as a GE tool to produce new plant resistance genotypes to various fungal pathogens [7,

40–42]. Numerous methods were introduced to enhance tolerance of plants to phytopathogenic agents, based on molecular biology and genetic modification. *Several resistance genes* have been inserted to combat harmful fungal pathogens and incorporated into plants. Today, the CRISPR/Cas as a modern molecular biology technique is used to help these genetic manipulations [43]. CRISPR/Cas9 applied to the gene *VvWRKY52* (a transcription factor) transcription factor of *Agrobacterium* with four guided RNAs (gRNAs) resulted in the generation of the transgenic plant and the destruction of *VvWRKY52* magnified resistance against *Botrytis cinerea* in grapevine [44]. Today a wide variety of crops are being improved through CRISPR-mediated disease resistance, but editing the epigenome by targeting epigenetic markers to specific genes may be more effective than altering the genome [9, 45]. CRISPR/Cas system was applied effectively in different species of Phytophthora, including *Phytophthora sojae* [46], *Phytophthora capsici* [47], and *Ustilago trichophora* [48].

Another approach to increase crop resilience is to alter or signal pathways to plant hormone biosynthesis [26]. In order to increase resistance to *Magnaporthe oryzae* induced blast disease, for example, a particular ethylene sensible r*ice* ERF transcription factor *OsERF 922* had been discovered and mutated using CRISPR-Cas9 technology [49]. DsDNA disrupts in the phytoene desaturase (PDS) gene of rice and mildew resistance locus O (MLO) in wheat were engineered by designing a sgRNA-driven CRISPR/Cas9 cassette [50]. The production of wheat is seriously impaired by pathogens such as *Blumeria graminis f.* sp. *tritici*. Targeted mutagenesis was initially applied via the three wheat Mlo (*TaMLO*) knockout [39]. The knockout resulted in an improvement in the frequency of mutation, enhanced resistance against *B. graminis* f. sp. *tritici* in mutant wheat plants. The silencing of *TaLpx-1*, which encodes a 9-LOX with homology to *LOX1* and *LOX5*, resulted in an increased tolerance to *F. graminearum* in wheat, implying that 9-LOXs play a conserved function as sensitivity factors in disease caused by this causal organism in Arabidopsis and wheat host [51]. Studies in 4 genes, *ZmPDS*, *ZmIPK1A*, *ZmIPK*, and *ZmMRP4* from *Zea mays*, have shown that the CRISPR/Cas also has high efficacy in the maize GE against *Ustilaginoidea virens* [52]. Zheng and his research group investigated the participation of the *Mlo* genes in endophytic fungus *L. taurica* susceptibility. A loss of function in the SlMlo1 gene causes resistance to powdery mildew diseases caused by Oidium neolycopersici in tomato (*Solanum lycopersicum*) [53]. Non-transgenic plants such as tomatoes have shown strong resistance to *O. neolycopersici* infection obtained by the use of CRISPR/Cas9, which targets *SlMlo-1*. Further research has shown that changes in the *SlDMR6-1* tomato gene increase resistance to *Phytophthora capsici* infection [54]. Enhanced gene resistance to diseases 1 (*EDR1*) negatively regulates plant fungal pathogen resistance. Mediating mutation through CRISPR/Cas9 of the *EDR1* wheat gene led to resistant plants against fungal infection [55]. Powdery mildew in tomato is caused by the fungal isolate *O. neolycopersici*. A systematic characterization of the MLO (mildew locus O) family in tomato was investigated to monitor Powdery Mildew Fungus caused by Odium neolycopersici by studying their genomic composition, expression profile, and predicted protein motifs. There are 17 *MLO* genes in tomato, which can be classified into six clades. *MLO* gene expression can vary by tissue, and maybe some *MLO* genes have alternate splicing variants in different plant tissues [56]. CRISPR/Cas9 technology has been used by Nekrasov and his team for the generation of loss of function mutations in the *SLMLO1* locus [57]. The two chosen targets within *SlMlo1* were placed on opposite DNA strands and separated by 42 bp within the *SlMLO1* locus, resulting in a unique form of mutation rather than indels. CRISPR/Cas mediated GE mechanism applied in plant breeding for resistance. Examples of developing disease-resistant cultivars via the CRISPR/Cas9 mechanism by targeting susceptibility genes (S). S-gene disorders and disease-resistant variety production through CRISPR/Cas9 system are more stable, time-saving, and functional. Flowchart of CRISPR gene editing for disease resistance has been shown in Figure 8.2.

Resistance to *Verticillium dahliae* infestation was recently obtained by a change in the *Gh14-3-3d* gene [58]. The transgenic plants that resulted were found to be extremely fungal resistant [59]. This technology is now being used to attack particular domains of certain virulent organisms, as well as in the study of various factors involved in Phytophthora infection on various plants such as papaya [60, 61]. *CRISPR, with a focus on Phytophthora, is also used to investigate the function of particular genes, fungicide resistance mechanisms, and plant/host-pathogen interactions in various organisms [44]. CRISPR was also used to efficiently and precisely target and disrupt the virulent gene AVR4/6 in Phytophthora sojae [46]. TcNPR-3 gene targeting in Theobroma cacao enhances disease-resistance to P. tropicalis* [62]. Disruption of the *PpalEPIC8* gene in Papaya and the *BnWRKY70* gene in *Brassica napus* increases tolerance to *P. palmivora* and *Sclerotinia sclerotiorum* infections, respectively [61, 63].

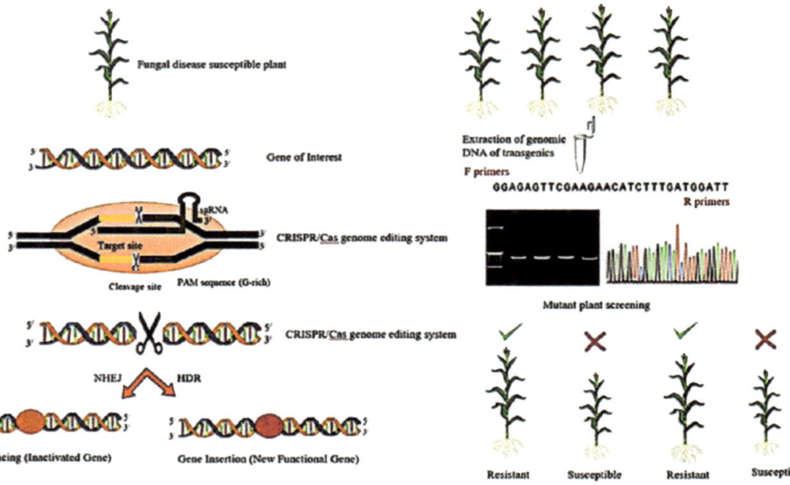

FIGURE 8.2 Schematic description of CRISPR/Cas system application for generating disease resistant germplasm. (1) Selection of elite cultivar susceptible to disease; (2) selection of negative regulator gene with predicted/known function for disease resistance; (3) employing CRISPR/Cas system for targeted genome modification for gene of interest; (4) genome editing through CRISPR/Cas system and cleavage of target site to generate mutation(s); (5) mutation through non-homologs end joining (NHEJ) for gene silencing or homology direct repair (HDR) for gene insertion; (6) generation of mutant plants through agrobacterium transformation and selection among T_0 for mutation detection; (7) designing of primers to detect mutation and Sanger sequencing to identify homozygous mutant plants; and (8) selection of homozygous mutant plants and selection of resistant plants after disease inoculation.

In another study on *Phytophthora sojae*, creating multiple mutations in PsORP1 using CRISPR-Cas9 demonstrated different resistance levels to oxathiapiprolin (OXPT) fungicide.

Another research on *P. sojae* found that causing several mutations in *PsORP1* using CRISPR-Cas9 resulted in various degrees of resistance to the OXPT fungicide [47]. CRISPR-Cas9 has been used effectively in *Phytophthora* species such as *P. infestans* and *P. sojae* to target and disrupt virulent genes that serve a role in the development of *effector* proteins that cause various infections in plants [64]. Furthermore, CRISPR is being successfully used in plants to modify various susceptible genes that appear to react during the infection period of some Phytophthora species, thus increasing host immunity and resistance to such disease (Table 8.1). Recently, Hahn & Scalliet identified the critical features of the CRISPR scheme, including how it opens new avenues for unraveling the virulence mechanisms of Botrytis and other plant pathogenic fungi [65], as well as how it can speed up research for the discovery of new antifungal compounds.

TABLE 8.1 CRISPR-Cas-Mediated Technology for Enhanced Fungal Diseases Resistance and Genetic Engineering in Plant Pathogenic Fungi

Plant Host	Targeted Gene	Fungal Pathogens	Plant Diseases resistance	References
Wheat	*TaMLO*	*Blumeria graminis* f. sp. *Tritici*	Powdery mildew	[39]
Wheat	*TaMLO-A1*, *TaMLO-B1*, and *TaMLO-D1*.	*Blumeria graminis* f. sp. *tritici*	Powdery mildew	[66]
Tomato	*SlMlo*	*Oidium neolycopersici*	Powdery mildew	[56]
Tomato	*SlMlo*	*Oidium neolycopersici*	Powdery mildew	[57]
Tomato	*SlMlo*	*Oidium neolycopersici*	Powdery mildew	[67]
Avocado	*PaNPR2* and *PaNPR4*	*Phytophthora cinnamomi*	Root rot/dieback	[68]
Cacao	*TcNPR3*	*Phytophthora tropicalis*	Black pod	[62]
Lychee	*PAE4* and *PAE5*	*Peronophthora litchii*	Downy blight	[69]
Papaya	*alEPIC8*	*Phytophthora palmivora*	Reduce pathogenicity	[61]
Potato	*Avr1*, *PiTubA2*, and *PiAP5*	*Phytophthora infestans*	Late blight	[70]
Soybean	*Avr4/6*	*Phytophthora sojae*	Damping off	[46]
Tomato	*DMR6*	*Phytophthora capsici*	Resistance to infection.	[54]
Banana	*MaATG8s*	*Fusarium oxysporum* f. sp. *cubense*	Panama disease	[71]
Fruits	*WRKY52*	*Botrytis cinerea*	Grey mold	[72]
Potato	*S-genes*	*Phytophthora infestans*	Potato blight disease	[73]
Rice	*OsERF922*	*Magnaporthe oryzae*	Rice blast	[74]
Rice	*OsSEC3A*	*Magnaporthe oryzae*	Rice blast	[41]

TABLE 8.1 *(Continued)*

Plant Host	Targeted Gene	Fungal Pathogens	Plant Diseases resistance	References
Rice	*OsMPK5*	*Magnaporthe oryzae*	Rice blast	[75]
Sugarcane	*ScGluA1*	*Sporisorium scitamineum*	Smut	[76]
Wheat	*TaEDR1*	*Blumeria graminis* f. sp. *tritici*	Powdery mildew	[55]
Papaya	*alEPIC8*	*Phytophthora palmivora*	Bud-rot of palms	[44]
Watermelon	*Clpsk1*	*Fusarium oxysporum* f. sp. *niveum*	Fusarium wilt	[77]
Canola	*Avr*	*Leptosphaeria maculans*	Blackleg	[78]
Maize	*bE1* and *bW2*	*Ustilago maydis*	Corn smut	[79]
Maize	*Pep1* and *Pit2*	*Ustilago maydis*	Maize smut	[80]
Rice	*USTA* and *UvSLT2*	*Ustilaginoidea virens*	Rice false smut	[52]
Rice	*ALB1*	*Magnaporthe oryzae*	Rice blast	[81]
Rice	*OsERF922*	*Magnaporthe oryzae*	Rice blast	[82]
Tomato	*Mlo*	*Oidium neolycopersici*	Powdery mildew	[82, 83]
Wheat	*TaLpx1*	*Fusarium graminearum*	Fusarium head blight	[51]
Wheat and corn	*tri5* and *tri6*	*Fusarium graminearum*	Fusarium head blight	[3]
Wheat	*TaNFXL1*	*Fusarium graminearum*	Fusarium head blight	[40]
Cotton	*Gh14-3-3d*	*Verticillium dahliae*	*Verticillium wilt*	[84]
Fruits	*Bos1*	*Botrytis cinerea*	Grey mold	[41]

8.6 CRISPR/CAS9-BASED HOST MEDIATED RESISTANCE AGAINST BACTERIAL PATHOGENS

Phytopathogenic bacterial infections may result in severe yield losses in agriculture. Examples of bacterial diseases in plants are local lesions, mosaics, and rots [36]. Phytopathogenic bacteria are classified into three groups depending upon their susceptible hosts. "Crop specific" bacteria cause one type of disease in a single crop however, "polyphagous-specific" bacterial infections result in numerous disease symptoms. While the "Kingdom Crossers" can infect and cause severe symptoms across different kingdoms [28]. Different types of bacterial pathogens categorically produce a diverse range of responses in their host plants and therefore, it is challenging to confer broad-spectrum resistance against them. Moreover, continuous evolution and horizontal gene transfer mechanism of bacterial pathogens make them defiant to induced resistance [85].

Bacterial pathogens proceed into the cell either passively through stomatal or stigmatic openings or actively by secreting bioactive compounds [86]. Host susceptible (*S*) genes help the bacterial pathogens to surpass the plant defense mechanism. Plants can recognize the pathogen entry and in response, the immune system of plants tries to eradicate the invading pathogen. However, some bacteria use complex signaling cascades, which are not recognizable by plant defense machinery. Moreover, *S* genes also play a vital role in the plant defense mechanism resulting in disease development [87]. GE ensuing in knock out of these *S* genes can improve the plant immunity; however, there are a few examples of utilizing CRISPR-Cas9 system to generate bacterial disease-resistant plants.

Bacterial Blight of rice is considered among the most devastating diseases in terms of yield loss worldwide. Bacterial Blight is caused by *Xanthomonas oryzae* pv. *oryzae* also known as *Xoo*. Sugar transporter genes of the SWEET family are induced by TALE proteins secreted by *Xoo* into the host [88]. This induced sucrose transportation promotes bacterial infection by providing sugars for pathogen growth. These TALE proteins have their effecter binding sites (EBEs) in *SWEET11*, *SWEET13*, and *SWEET14* genes especially in promoter regions [89–91]. In different studies, EBEs are targeted by CRISPR/Cas9 system to disrupt the interaction of TALEs with sugar transporter genes [92]. Targeted mutagenesis in the promoter region of *Os8N* and coding region of *Xa13* through CRISPR/Cas9 system also has proven improved resistance against the bacterial blight of rice. Similarly knocking out the mitogen-activated protein kinase 5 (*OsMPK5*) gene through target mutation by CRISPR/Cas9 system-induced resistance against bacterial pathogen *Burkholderia glumae* in rice [75].

Erwinia amylovora is the causal bacterial agent for the fire blight of apple and other horticultural crops. Pathogen effector proteins (DspE) are induced by bacteria, which interacts with DIMP (*DspE*-interacting proteins of Malus) genes of apple and results in compromised plant defense mechanism.

CRISPR/Cas9 system was employed for the targeted mutations in *DIMP-1, DIMP-2*, and *DIMP-4* to confer resistance against the pathogen in apple [93]. Citrus canker caused by *Xanthomonas citri* sp. *citri* (*Xcc*), is the main cause for the yield loss of citrus. The *lateral organ boundaries 1* (*CsLOB1*) gene has been found in host susceptibility to *X. citri*. Two different studies provide with evidence that disruption of *CsLOB1* gene either in promoter region [94] or in the gene encoding region [95] resulted in enhanced resistance to bacterial infection (Table 8.2).

Broad-spectrum resistance in tomato against bacterial speck, bacterial spot and *Phytophthora* blight was enhanced by knocking out of the downy mildew resistance gene 6 (*DMR6*) through CRISPR/Cas9 system [54]. Coronatine (COR) secreted by the bacteria, interacts with binding site of *Jasmonate ZIM-domain-2* ortholog (*SlJAZ2*) in tomato resulting in wide-open stomata, which facilitates bacterial entry and infection. Targeted mutation in the *SlJAZ2* gene via CRISPR/Cas9 has enhanced resistance against the bacteria [96]. Identification of more *S* genes in different economically important plants and their target mutations through GE tools like CRISPR/Cas systems can be promising approach towards sustainable agriculture.

8.7 CRISPR/CAS9-BASED RESISTANCE AGAINST PLANT VIRUSES

Plant viruses are known to cause half of all known plant diseases and pose far-reaching threats to agricultural production worldwide. According to an estimation, more than US$ 30 billion losses are linked with viral infection of cultivated crops [100]. Formerly, about 1,500 plant viruses are identified based on genome sequences. All the known are viruses are grouped into 26 families [101]. Plant viruses entirely depend upon the host cellular machinery because they have an obligatory intracellular mode of life to complete their life cycle. In contrast to fungi and bacteria, the availability of efficient antibiotics or other chemicals available to control plant viruses is very rare.

The application of molecular plant breeding and antiviral engineering techniques can be efficiently optimized to confer host-mediated natural resistance against viruses and to develop virus-tolerant plants. The CRISPR/Cas system targeting viral genome or host susceptibility genes can be applied as an effective control measure to combat plant viruses. CRISPR/Cas-mediated gene editing is a very attractive tool that can be easily applied in the field of plant protection.

TABLE 8.2 CRISPR/Cas9 Resistance Against Some Plant Pathogenic Bacteria

Pathogen Name	Disease	Plant Species	Target Gene	Function of Target Genes	Method Used	References
Erwinia amylovora	Fire blight	Apple	*DIPM-1*, *DIPM-2*, and *DIPM-4*	Interact with pathogen effectors.	Knock out	[93]
Xanthomonas citri subsp. Citri	Citrus canker	Citrus	*CsLOB1*/cds region	Help in pathogen infection.	Knock out	[95]
Xanthomonas citri subsp. Citri	Citrus canker	Citrus	*CsLOB1*/promoter	Help in pathogen infection.	Knock out	[94]
Burkholderia glumae	Not diffened	Rice	*OsMPK5*/cds region	Help in pathogen infection.	Knock out	[75]
Xanthomonas oryzae pv. oryzae	Bacterial blight	Rice	*SWEET11*, *SWEET13* and *SWEET14*/promoter	Sucrose transporter gene.	Knock out	[97]
Xanthomonas oryzae pv. oryzae	Bacterial blight	Rice	*OsSWEET11* and *OsSWEET14*/ promoter region.	Sucrose transporter gene.	Knock out	[90]
Xanthomonas oryzae pv. oryzae	Bacterial blight	Rice	*OsSWEET11*, *OsSWEET14*/ promoter	Sucrose transporter gene.	Knock out	[89]
Xanthomonas oryzae pv. oryzae	Bacterial blight	Rice	*OsSWEET13*/cds region	Sucrose transporter gene.	Knock out	[91]
Xanthomonas oryzae pv. oryzae	Bacterial blight	Rice	*Os8N3*/promoter region.	Help in pathogen infection.	Knock out	[98]
Xanthomonas oryzae pv. oryzae	Bacterial blight	Rice	*Xa13*/cds region	Help in pathogen infection.	Knock out	[99]
Phytophthora capsici	Phythophthora blight	Tomato	*SlDMR6–1*/cds region	Regulation of PR genes.	Knock out	[54]
Xanthomonas spp.	Bacterial spot	Tomato	*SlDMR6–1*/cds region.	Regulation of PR genes.	Knock out	[54]
Pseudomonas syringae	Bacterial speck	Tomato	*SlDMR6–1*/cds region	Regulation of PR genes.	Knock out	[54]
Pseudomonas syringae pv. tomato (Pto)	Bacterial speck	Tomato	*SlJAZ2*/cds region	Opens stomata	Knock out	[96]

In general, CRISPR/Cas technology can be applied in two different strategies to combat plant viruses. In the first approach, the viral genome is targeted and destroyed so that replication and infection can be inhibited. In the second strategy, the host susceptibility factors are manipulated, which are crucial for the completion of viral infection or life cycle to inhibit the viral infection (Figure 8.3). Lately, CRISPR/Cas systems have been exclusively applied to develop resistance against plant viruses and other pathogens [102, 103].

8.7.1 RESISTANCE AGAINST DNA VIRUSES

The viruses are usually classified into two broad groups based on their genome type, first group includes DNA viruses and second group comprises RNA viruses. DNA viruses are further classified based upon dsDNA genomes and single-stranded (ssDNA) genomes. The group with ssDNA genome comprises ~485 species and the group with dsDNA genome includes ~85 species. The two most economically important plant virus families are *Geminiviridae* and *Caulimoviridae*. The CRISPR/Cas systems have been extensively applied to precisely target and impair the genomic DNA of these two families [104]. The GE technique has been exclusively used to directly targeting either the viral genome or indirectly inhibiting viral infection by targeting the interacting host factors in plants [105]. Different GE techniques including TALENs (transcription activator-like effector nucleases) and ZFNs (zinc finger nucleases) have already been employed to target ssDNA genome of plant viruses [106, 107]. However, the advantages proposed by CRISPR/Cas systems over ZFN- and TALEN-mediated GE allow it to be readily accepted and providing promising antiviral engineering in plants.

Recently, various researchers have applied CRISPR/Cas9 gene-editing techniques by employing sgRNAs targeting the intergenic region (IR) or viral replication-associated region. This resulted in efficient interference of DNA and consequently conferred resistance against bean yellow dwarf virus (BeYDV) and beet severe curly top virus (BSCTV) in transgenic plants [108, 109]. Tashkandi targeted the replication-associated protein (Rep) and coat protein (CP) encoding sequence of tomato yellow leaf curl virus (TYLCV) using Cas9–sgRNA resulting in development of resistance in the tomato and *N. benthamiana* [110]. Nonetheless, Ali targeted the stem-loop sequence using sgRNAs to confer a broad-spectrum resistance against three different viruses including cotton leaf curl Kokhran virus (CLCuKoV) [111] and TYLCV with monopartite ssDNA genome and merremia mosaic

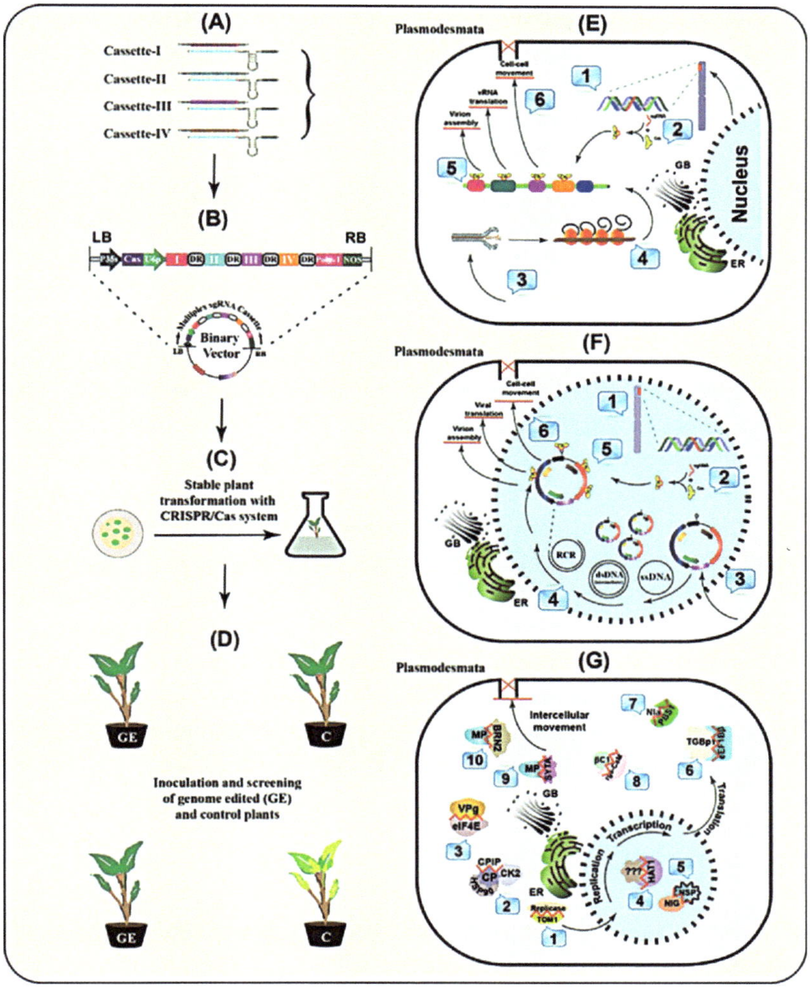

FIGURE 8.3 Detailed description to confer engineered resistance based upon CRISPR/Cas9 model in crop plants. (A) sgRNA designing to target virus or host genes. (B) Expression cassette for multiplexed sgRNAs in a suitable plant transformation binary vector. (C) Stable plant transformation using recombinant binary vector carrying the whole cassette with sgRNAs, promotors, and terminators. (D) Testing of successful plant transformants with genome editing (GE) events against single or multiple plant viruses to confirm the conferred resistance. (E) In planta genetic resistance model against and ssRNA virus with potato virus Y (PVY) as a case study. (F) In planta genetic resistance model against ssDNA virus with cotton leaf curl Kokhran virus (CLCuKoV) as a case study. And (G) in planta genetic resistance model against plant viruses based upon host factors interacting with DNA or RNA genome of plant viruses.

Source: Modified from: Shahid et al. [105].

virus (MeMV) with a bipartite ssDNA genome. Similarly, Iqbal proposed a detailed strategy to collectively target the helper begomoviruses and their associated DNA-satellite components through multiplexed CRISPR/Cas based GE [104]. Yin targeted Rep and IR regions of cotton leaf curl Multan virus (CLCuMuV), which resulted in complete resistance in *N. benthamiana* plants against CLCuMuV [112]. Likewise, CRISPR/Cas9-mediated systems have been applied against plant dsDNA viruses. Liu developed transgenic *Arabidopsis thaliana* plants conferring effective resistance to cauliflower mosaic virus (CaMV) by sgRNAs targeting the CP region of CaMV [113].

In a recent study by Khan, CRISPR-Cas9 system was applied to target βC1 of cotton leaf curl Multan betasatellite (CLCuMuB) and CLCuKoV-encoded Rep and resulted in reduced viral titer in the plants [114]. The rapid advances in CRISPR/Cas9-mediated GE systems have resulted in an increasing number of successful examples with promising resistance against plant DNA viruses. The resistance was developed against banana streak virus (BSV) or wheat dwarf virus (WDV), by targeting the CP region or movement protein (MP) of these viruses, respectively, applying a CRISPR/Cas9 system duplexed with sgRNAs [115, 116]. The same technique can also be applied to engineer the cotton against cotton leaf curl virus (CLCuV), which adversely affects the cotton production globally [117]. These findings justify the application of CRISPR/Cas9-mediated immunity as a potential technology to control economically important ssDNA geminiviruses viruses, like banana bunchy top virus (BBTV) [118].

8.7.2 RESISTANCE AGAINST RNA VIRUSES

RNA plant viruses are comprised of a very diverse group of viruses and are known to cause immense financial losses to crop productivity all over the world. RNA plant viruses are divided into four classes as dsRNA, negative-strand ssRNA (-ssRNA), positive-strand ssRNA (+ssRNA), and retroviruses. Recently developed specific Cas proteins derived as Cas13a (C2c2) from *Leptotrichia shahii* (LshCas13a) and *Francisella novicida* (FnCas9) helped to target the RNA genomes of viruses [32, 119]. Cas13 is an RNAi analog to directly target RNA transcripts of plant viruses [120]. Accordingly, a CRISPR/Cas system with Cas effector proteins can be applied directly to interfere with the genome of RNA viruses [121]. The first-ever study was performed by Zhang where a plant codon-optimized FnCas9 combined with rgRNA was used to develop resistance against ssRNA viruses viz, tobacco mosaic virus (TMV) and cucumber mosaic virus (CMV) in transgenic *A. thaliana* and *N. benthamiana* plants, respectively [59]. Thereafter, in a study by Aman where RNA-targeting CRISPR/Cas system using LshCas13a, was designed to hinder ssRNA genome of turnip mosaic virus (TuMV) in plants [122, 123] (Table 8.3).

The system harboring LshCas13a was used by specifically targeting viral RNA in transgenic rice and tobacco against two-ssRNA viruses, i.e., Southern rice black-streaked dwarf virus (SRBSDV) and rice stripe mosaic virus (RSMV), respectively [124]. Hameed theoretically proposed a broad-spectrum resistance against potato virus Y (PVY) (a positive sense ssRNA virus) [125]. Zhan effectively used the CRISPR/Cas13a system against PVY by targeting the CP-, P3-, or Nib-coding sequences and inhibiting PVY infection [124]. Recently, Zhang, engineered *N. benthamiana* plants by employing CRISPR-Cas13 system to develop resistance against TMV resulting in reduced viral titer [42].

8.7.3 HOST-MEDIATED ENGINEERED RESISTANCE AGAINST VIRUSES

Plant viruses mobilize the cellular translation factors of their host and use them to translate their RNAs as well as to facilitate the infection process. Therefore, the translation-related host factors were originally recognized as pro-viral factors. All these genes have been characterized and used to control plant pathogenic viruses. It was reported that the recessive resistance is caused by the mutation in a recessive gene encoding a host factor, which is usually crucial to a virus for causing infection or completing the life cycle or encoding a negative regulator of plant defense responses [140].

In addition to its application to harness resistance against DNA and RNA viruses, CRISPR/Cas systems can also be applied either to inactivate/mutate the host susceptibility factors (*S*-genes) or to express/activate resistance factors (*R*-genes). The *S*- or *R*-genes can be targeted by CRISPR/Cas-mediated gene engineering and transgene free plants can be yielded by segregating the transgenic sequences from the progeny. The *R*-genes are commonly associated with undesired traits like poor yield, inadequate flavor, or growth-related abnormalities. As result, the *R*-genes have the least potential to be considered for genetic engineering against virus resistance. Likewise, *S*-genes, for example, the *Arabidopsis ssi2* can provide resistance against CMV but are also linked with developmental abnormalities. Even

TABLE 8.3 CRISPR-Cas-Mediated Genome Editing Applied Against Different DNA and RNA Viruses to Different Plant Hosts

Genome	Targeted Viral Pathogens*	Genus	Virus Families	Editing Systems	Plant Host	References
ssDNA	BSCTV	Curtovirus	Geminiviridae	CRISPR-Cas9	*A. thaliana* and *N. benthamiana*	[109]
	CaLCuV	Begomovirus	Geminiviridae		Cabbage	[112]
	BSCTV	Curtovirus	Geminiviridae		*A. thaliana* and *N. benthamiana*	[109]
	BeYDV	Mastrevirus	Geminiviridae		*N. benthamiana*	[108]
	TYLCV	Begomovirus	Geminiviridae		*N. benthamiana*	[111]
	TYLCV, MMV, BCTSV, and BYDV	Begomovirus, Curtovirus, Mastrevirus	Geminiviridae		*A. thaliana* and *N. benthamiana*	[126]
	TYLCV	Begomovirus	Geminiviridae		Tomato	[127]
	ACMV	Begomovirus	Geminiviridae		*N. benthamiana*	[128]
	TYLCV	Begomovirus	Geminiviridae		*N. benthamiana* and tomato	[110]
	WDV	Mastrevirus	Geminiviridae		*Barley*	[115]
	ChiLCV	Begomovirus	Geminiviridae		*N. benthamiana*	[129]
	Geminiviruses	Begomovirus	Geminiviridae	CRISPR-Cas14/a	–	[130]
	Geminiviruses	Begomovirus	Geminiviridae		–	[131]
	Geminiviruses	Begomovirus	Geminiviridae		–	[132]
dsDNA	CaMV	Caulimovirus	Caulimoviridae	CRISPR-Cas9	*A. thaliana*	[113]

TABLE 8.3 *(Continued)*

Genome	Targeted Viral Pathogens*	Genus	Virus Families	Editing Systems	Plant Host	References
ssRNA (+)	CVYV, ZYMV, and PRSV-W	Ipomovirus, potyvirus	Potyviridae	CRISPR-Cas9	*Cucumis sativus*	[133]
	TuMV	Potyvirus	Potyviridae		*A. thaliana*	[134]
	TRV and PEBV	Tobravirus	Virgaviridae		*A. thaliana* and *N. benthamiana*	[122]
	CMV and TMV	Cucumovirus, Tobamovirus	Bromoviridae, Virgaviridae		*A. thaliana* and *N. benthamiana*	[84]
	RTSV	Waikavirus	Sequiviridae		*Oryza sativa*	[135]
	SMV	Potyvirus	Potyviridae		*Soya bean*	[124]
	TRV	Tobravirus	Virgaviridae		*N. benthamiana*	[136]
	PVY	Potyvirus	Potyviridae	CRISPR-Cas13/a	Potato	[137]
	TuMV	Potyvirus	Potyviridae		*A. thaliana*	[123]
	TuMV	Potyvirus	Potyviridae		*N. benthamiana*	[122]
	PVY	Potyvirus	Potyviridae		–	[125]
	RNA viruses	Potyvirus	Potyviridae		*N. benthamiana*	[138]
	RNA viruses	Potyvirus	Potyviridae	CRISPR-CasRx	*N. benthamiana*	[136]
Plant pathogens	–	–	–	CRISPR-Cas13/a	Soybeans	[139]

*Viruses acronyms used are: African cassava mosaic virus (ACMV); bean yellow dwarf virus (BeYDV); beet curly top virus (BCTV); beet severe curly top virus (BSCTV); cabbage leaf curl virus (CaLCuV); chili leaf curl virus (ChiLCV); cauliflower mosaic virus (CaMV); cucumber vein yellowing virus (CVYV); cucumber mosaic virus (CMV); merremia mosaic virus (MMV); papaya ring spot mosaic virus-W (PRSV-W); pea early browning virus (PEBV); potato virus Y (PVY); rice tungro spherical virus (RTSV); soybean mosaic virus (SMV); tobacco mosaic virus (TMV); tobacco rattle virus (TRV); tomato yellow leaf curl virus (TYLCV); turnip mosaic virus (TuMV); wheat dwarf virus (WDV); and zucchini yellow mosaic virus (ZYMV). Virus genomes are denoted as single-stranded (ss) DNA and a positive sense (C) ssRNA.

though, various *S*-genes can be targeted for antiviral engineering and present promising potential. The resistance can be developed by inactivating these susceptibility factors without causing any abnormalities and affecting the general plant health pertaining to the functional redundancy of the isoforms. CRISPR/Cas-mediated GE can be applied to target the recessive gene. Cao divided the *S*-genes into four groups [141]. The candidate genes in the first group include the homeodomain leucine zipper protein 1 [142] and negative regulators of plant defense like rgs-CaM [143]. The second group includes the specific *S*-genes involved in replication/translation/movement of virus, for example, *Sec24a*, *eEF4s*, and *eEF1A* (a COPII coatomer) [144] and tobamovirus multiplication 1 (*TOM1*) and its associated homologs [145]. The third group comprise the S-genes which interact and modify the viral proteins (phosphorylate) including shaggy-related protein kinases (SK4-1, NsAK) [146] and cellular CK2 [21]. The fourth group included the secondary cell wall synthesis factor, which positively interacts with the virus, such as phenylpropanoid metabolism factor (4-coumarate:CoA ligase1, 4CL1) and Bearskin2B, BRN2 [146].

The eukaryotic recessive resistance genes including translation initiation factors (TIFs) and their isoforms are the most commonly exploited genes [147]. These TIFs are a prerequisite for translation processes in plant RNA viruses in various crops. They can develop resistance against a range of viral species [140, 144]. It was confirmed by various studies where CRISPR/Cas9-based cucumber knockout mutants displayed successful resistance against for papaya ringspot mosaic virus-W (PRSV-W), cucumber vein yellowing virus (CVYV), and zucchini yellow mosaic virus (ZYMV) [148]. Macovei [135] applied CRISPR/Cas9-targeted mutagenesis of *eIF4G* (a novel allele of rice) cause rice tungro spherical virus (RTSV) resistance. The CRISPR/Cas9-mediated eIF(iso)4E mutation in cassava and *A. thaliana* showed a fair control over cassava brown streak virus (CBSV) [134, 149]. The CRISPR/SpCas9-mediated GE through mutating the cap-binding protein-1 (nCBP-1 or nCBP-2) confer high tolerance to CBSV in cassava plants [149].

Bastet applied antivirus engineering in Arabidopsis by using a recently developed CRISPR-Cas9n-cytidine base editor [150]. They introduced the N176K mutation, which caused resistance against clover yellow vein virus (ClYVV) by converting *elF4E1* allele as susceptible to resistant allele [150]. Helm modified the cleavage site for the soybean mosaic virus (SMV) on soybean PBS1 decoy protein developing resistance against SMV by triggering cell death in soybean protoplasts [151].

Nevertheless, most of the recently developed strategies based on host-mediated recessive engineered resistance against plant viruses were primarily focused on eIF4s. Moreover, mostly this strategy has been tested successfully against potyviruses and some related plant viruses. Hence, there is a need to identify and characterize more host susceptibility genes to develop resistance against economically important plant viruses (Table 8.4).

The potential recessive targets in light yellow background which are suppressors of plant defenses, the susceptibility factors required for viral translation/replication/movement or other life cycles, the factors in the cellular metabolism, or the modified factors for the benefit of the viral infection or viral protein functions, etc., can be utilized in antiviral engineering by GE technologies. The abbreviations of the virus names in this table are as follows: CVYV, ZYMV, PRSV-W, papaya ringspot virus-W; TuMV, ClYVV, clover yellow vein virus, RTSV, CBSV. The abbreviations of the gene names in this table are listed as follows: ESCRT; endosomal sorting complexes required for transport, GAPDH; glyceraldehyde 3-phosphate dehydrogenase, SYTA1; synaptotagmin 1, CPR5; also referred to as the RYMV2 gene, RIM1; rice dwarf virus multiplication 1, RHD3; root hair defective 3.

8.8 CRISPR/CAS9 RESISTANCE AGAINST INSECTS

Insects are responsible to cause considerable crop losses all over the world. They can cause either direct losses or indirect losses by acting as a vector for transmission of various diseases. It has been reported that an average of 28% losses of annual crop productivity is caused by insect pests. Various pesticides are used to control insects, but appearance of insecticide resistance is a major problem. Also, the non-degradable nature and biomagnification of chemical pesticides necessitate the development of environment friendly and more selective alternative control strategies [153]. Even though, the development of *Bt* crops present compelling environmental and economic prospects and have decreased the application of pesticides, nonetheless, the development of insect resistance has been reported in several studies against *Bt* toxins. The recent advances in recombinant DNA technology have resulted in increased interest in the development of insect-resistant plant varieties. As a result, alternative techniques including RNAi technology were applied to develop resistance in plants against insects. Nonetheless, RNAi technology had many limitations, for example, lack of reproducibility and off-target effects. Recently, Hall, exploited CRISPR/Cas genomics as a promising technology

TABLE 8.4 CRISPR/Cas System-Mediated Resistance by Targeting the Host Factors and Recessive Targets Against Plant Viruses

Targets	Target Gene Function	Viruses	Genome	Genus	Family	Editing Systems	Plant Species	Potential Recessive Resistance Targets (Without Development Defects)	References
Host gene elF4E	Help in viral replication.	CVYV, ZYMV, PRSV-W, TuMV	ssRNA	Ipomovirus, potyvirus, potyvirus, potyvirus	Potyviridae	CRISPR-SpCas9	*Cucumis sativus*	eEFIA, ESCRT, GAPDH, RAB5, SYPK1	[133]
		ClYVV	ssRNA	Potyvirus	Secoviridae	CRISPR/nCaaS-cytidine deaminase	–	SYTA1	[152]
Host gene elF4G	Help in viral replication.	RTSV	ssRNA	Waikavirus	Secoviridae	CRISPR/SpCas9	*Oryza sativa*	CPR5	[135]
Host gene elF(iso)4E	Help in viral replication.	TuMV	ssRNA	Potyvirus	Potyviridae	CRISPR/SpCas9	*A. thaliana*	nCBP-1/ nCBP-2	[149]
Host gene nCBP-l/ nCBP-2	Help in viral replication.	CBSV	ssRNA	Ipomovirus	Potyviridae	CRISPR/SpCas9	*Manihot esculenta*	RHD3	[134]

for the development of resistance against insects in plants [154]. CRISPR/Cas-based system has been applied to identify the essential traits of insect. Huang applied CRISPR in various insects, e.g., silkworms, *Drosophila*, mosquitoes by targeting their cell lines, embryo, or larval stages [155]. A *CRISPR/Cas9* gene drive with a detrimental trait, include distortion in sex ratio, reduction in fertility, and the enhanced sensitivity to chemicals, which can be breed in laboratory followed by their introduction into the open field trials to multiply the engineered GE. Recent advances of CRISPR/Cas9, both in plants and insects, a present potential strategy to develop resistance against insect *via* knockout or insertion of new genes. The CRISPR/Cas9-mediated gene editing can be applied to perform in-depth study of plant response against insects and thus providing alternative strategies to optimize plant defense against insects.

8.9 CRISPR/CAS9 RESISTANCE AGAINST NEMATODES

Plant-parasitic nematodes (PPNs) are one of the important pests, which limit crop production worldwide. Nematodes are obligate sedentary endoparasites, ubiquitous [156]. Nematodes make the second biggest group of animals after insects. In the phylum Nematoda, PPNs are 7% of total species comprising 4,300 species and infect a variety of important crops [157, 158]. The cereal cyst nematodes (CCNs) are considered the most notorious species of PPNs belonging to Heteroderidae. The annual yield losses of up to 3,00,000 tons were reported *H. avenae* in Australia [159]. In northwestern U.S.A., the losses are estimated at 22,000 tons and the average yield reduction was 10% by *H. avenae* and *H. filipjevi* [160].

Various classical breeding approaches have been applied to develop resistance against CCNs. The wheat and barley varieties have been screened to select the suitable parents and subsequently, molecular markers are applied associated with the resistance. CCNs can be effectively controlled by combining host-plant resistance and tolerance against CCNs in order to minimize crop losses [35]. Similarly, biotechnological approaches like gene silencing have also been applied to develop resistance in plants by silencing some parasitism genes of nematodes. Nevertheless, the resistance behavior of different isolates of Heterodera from various regions of the world may differ in response to different cereal crops. In a study by Smiley, the resistant cultivars of wheat, barley, or oat against Heterodera in one region were susceptible to the CCN populations in the other region [160].

The advancement in GE and molecular techniques has resulted in replacement of conventional breeding practices. In recent studies, Dickinson & Goldstein [161] and Paix [162] applied CRISPR/Cas9 system in *Caenorhabditis elegans*. The development paved the path for the characterization of various crucial genes involved in different physiological processes of nematodes. Kang used CRISPR/Cas9-mediated GE to study resistance against soybean cyst nematode in soybean [163]. They knockout two serine hydroxymethyltransferase genes, GmSHMT08 and GmSHMT05, to study the resistance response in soybean hairy roots against *H. glycines*. Despite this, there has not been a lot of research related to resistance responses of the plants against nematodes using the CRISPR/Cas9 system. However, the recently available genome sequences of hosts of CCNs can help to develop non-transgenic cereals by applying CRISPR/Cas9-mediated GE [164].

8.10 LIMITATIONS AND SOLUTIONS

CRISPR/Cas systems have various benefits to induce resistance in agronomic crops due to their easiness, high versatility, and ability to simultaneously targeting multiple traits. These systems ensure substantial potential to minimize the drawbacks of classical breeding practices to generate genetic resistance in plants with high effectiveness and accuracy. Regardless of various benefits and immense applications, CRISPR/Cas9 still has some limitations. However, few significant complications might obstruct the value of CRISPR/Cas9 system to develop resistance against diseases. For example, directly targeting *S* genes may cause certain fitness cost owed to their connection utilizing additional necessary genes mostly the genes monitoring other traits like plant growth and development. Moreover, the interference of any S gene can disrupt a specific product and eventually several other related products might suffer due to mutation in the targeted gene. It may affect the availability of essential micronutrients or negatively correlate to other physiological abnormalities [165]. A workable solution to manage these events is to manipulate and bring in *S* gene variants and knock-in (KI) of preferred traits [152]. *S* genes can be knocked out in pathogen by inducible CRISPR systems that correspond to a specific micronutrient or a sugar promoter and ultimately avoiding fitness consequences. Desired mutations in S genes can be created by GE in most economically important crops without affecting species barriers. In future, discovery of additional S genes may offer new targets for GE. Nowadays, different bioinformatics tools, like CasOFFinder,

CCTop, and various approaches, i.e., SELEX, IDLV capture, Guide-seq, HTGTS, BLESS, Digenome-seq, and DISCOVER, have been developed. The researchers are trying to improvise the existing analytical tools to further support these approaches.

The major challenge in GE applications in many crop plants with polyploidy in their genome is a characterization of the genetic basis of a desirable trait. This is particularly difficult in tree plants in which the application of molecular and genetic tools is always a challenging task and thus exploring their whole genome is critical through next-generation sequencing or other related techniques. It becomes even challenging if an efficient transformation or plant regeneration protocol is not available for that plant species, which is a prerequisite for the efficient application of CRISPR/Cas9 based GE [29]. Another challenging task is to get rid of the genetic footprints of foreign elements in a GE crop plant, such as T-DNA insertions from bacterial vectors or foreign promoter elements. A possible solution is to use backcross breeding methods or high-throughput screening of large plant populations. Moreover, transgene-free plants can also be obtained through expressing ribonucleoprotein (RNP) complexes and their transcripts *in vitro* [56]. Several other factors can also be opted such as size optimization of sgRNAs, GC contents in sgRNAs, co-expression of sgRNAs and Cas protein, highly efficient delivery vectors and designing optimal expression cassettes targeting the right tissue types [105, 166].

Currently, GE is helpful to supplement classical breeding methods by precise genetic modifications and avoiding long cycles of backcrossing and extensive screening. Thus, the plants produced would be indistinguishable from the wild plants and thus can easily tackle the current GMO regulations [167]. The GMO regulatory framework of different countries has different perspectives about current GE plants. In Canada, a pre-market assessment is a pre-requisite for any GE product, showing differences with its wild source [168]. Contrarily, the European Court of Justice has released a strict ruling to declare GE crops under the same GMO regulations [169]. Meanwhile, Australia has allowed any GE crop free of foreign material to be acceptable to pass the current GMO regulations.

8.11 FUTURE PERSPECTIVES

Farmers are now using organic pesticides to keep pests from destroying their crops in the agricultural sector. Government initiatives and non-governmental organizations (NGOs) are actively pressuring the framework to develop safer alternatives. To feed the world's population, which is projected to be 9–10 billion in 2050, a strong breeding strategy, such as speed breeding, that saves time is the need of the hour [170]. To shorten the crop's life cycle, a pace breeding system is being established with the aid of RNAi and CRISPR technologies. CRISPR-Cas and gene-controlled systems dependent on RNAi would be useful to humans in the long run. They will, for example, eradicate disease epidemics, increase agricultural productivity, and monitor pathogen spread by using plant cultivars proven to be resistant to insect pests and herbicide-sensitive pathogens. CRISPR/Cas approaches can become the next generation of diagnostic biosensing tools; they are responsive, precise, and can discriminate at a single DNA base-pair resolution. They can also be multiplexed, allowing a single screening search to recognize different targets. For example, the management of pathogens, including plant cultivars known to resist insect pests and herbicistic pathogens, may help reduce epidemics, enhance farming output, and control the spread of diseases. Several biosafety and social questions about this technology must be addressed by the scientific community to make efficient and durable use of this technology in crop improvement. The CRISPR-powered device could soon be used with pace plant breeding schemes to reshape agri-food supplies and protection. Therefore, in future, we can expect the CRISPR technology to contribute considerably to decomposing plant-pathogen interactions and the development of sustainable and wide-spectrum disease-resistant plants.

8.12 CONCLUSIONS

In terms of breeding resistances, elucidating host-pathogenic interactions, the manufacture of CRISPR-based antimicrobials, and diagnostics for diseases, as well as the reduction of damages from destructive pathogens in agricultural applications, CRISPR/Cas-driven technologies, have emerged as powerful instruments for the transformation into modern agricultural production. Resistance to fungi, bacteria, and viruses has primarily been accomplished by the production of transgenics that target host susceptibility genes. In protecting cultures from infectious, bacterial, and fungal diseases, the CRISPR/Cas method offers a promising solution. CRISPR/Cas is simple to add to phytopathogens because scientists are sufficiently familiar with the molecular processes of the infection of these. The eradication of the virus involves the accuracy and non-damaging of the single targeted gene.

Enhancing disease resistance in plants using the CRISPR/Cas9 system is successful because of three main reasons – (i) availability of exhaustive knowledge on host-pathogen interaction triggered molecular pathways; (ii) single gene control of disease resistance; and (iii) immediate applicability of targeted mutagenesis for disease resistance. However, some critical questions like the following need to be addressed while dealing with CRISPR/Cas9-mediated disease-resistant crops. The use of the CRISPR/Cas9 method to improve disease resistance in plants is effective for three reasons: (a) the existence of extensive information on host-pathogen interaction induced molecular pathways; (b) single gene regulation of disease resistance; and (c) the immediate applicability of selective mutagenesis for disease resistance. However, some important issues, such as the ones mentioned below, must be answered when dealing with CRISPR/Cas9-mediated disease-resistant crops. Hence, it is the right time to make a worldwide consensus about GE crops to be exempted from current GMO regulations so, that we may harvest its full potential in addressing the food security and climate change problems.

KEYWORDS

- **climate change**
- **CRISPR/Cas**
- **disease-resistant crops**
- **food security**
- **fungal pathogens**
- **pathogenic organisms**
- **plant genome editing**
- **resistance genes**

REFERENCES

1. Hunter, M. C., Smith, R. G., Schipanski, M. E., Atwood, L. W., & Mortensen, D. A., (2017). Agriculture in 2050: Recalibrating targets for sustainable intensification. *BioScience, 67*, 386–391.
2. Fiaz, S., Ahmad, S., Noor, M. A., Wang, X., Younas, A., Riaz, A., Riaz, A., & Ali, F., (2019). Applications of the CRISPR/Cas9 system for rice grain quality improvement: Perspectives and opportunities. *Int. J. Mol. Sci., 20*, 888.

3. Muñoz, I. V., Sarrocco, S., Malfatti, L., Baroncelli, R., & Vannacci, G., (2019). CRISPR-Cas for fungal genome editing: A new tool for the management of plant diseases. *Front. Plant Sci., 10*, 135.
4. Fróna, D., Szenderák, J., & Harangi-Rákos, M., (2019). The challenge of feeding the world. *Sustainability, 11*, 5816.
5. Mojica, F. J. M., Juez, G., & Rodriguez-Valera, F., (1993). Transcription at different salinities of *Haloferax mediterranei* sequences adjacent to partially modified PstI sites. *Mol. Microbiol., 9*, 613–621.
6. Lander, E. S., (2016). The heroes of CRISPR. *Cell, 164*, 18–28.
7. Abd-Elsalam, K. A., & Lim, K. T., (2021). Can CRISPRized crops save the global food supply? In: Abd-Elsalam, K. A., & Lim, K. T., (eds.), *CRISPR and RNAi Systems; Nanobiotechnology for Plant Protection* (pp. 1–14). Elsevier.
8. Scott, M. J., Gould, F., Lorenzen, M., Grubbs, N., Edwards, O., & O'Brochta, D., (2018). Agricultural production: Assessment of the potential use of Cas9-mediated gene drive systems for agricultural pest control. *J. Resp. Innov., 5*, S98–S120.
9. Shelake, R. M., Pramanik, D., & Kim, J. Y., (2019). Exploration of plant-microbe interactions for sustainable agriculture in CRISPR era. *Microorgan., 7*, 269.
10. Barrangou, R., & Notebaart, R. A., (2019). CRISPR-directed microbiome manipulation across the food supply chain. *TIM, 27*, 489–496.
11. Pan, M., & Barrangou, R., (2020). Combining omics technologies with CRISPR-based genome editing to study food microbes. *Curr. Opin. Biotechnol., 61*, 198–208.
12. Haq, Q. M. I., & Hussain, T., (2020). CRISPR/Cas9: A new revolutionary science in agricultural and horticulture. In: Hakeem, K. R., & Pirzadah, T. B., (eds.), *Nanobiotechnology in Agriculture: An Approach Towards Sustainability; Nanotechnology in the Life Sciences* (pp. 195–205). Springer International Publishing: Cham.
13. Wiesner-Hanks, T., & Nelson, R., (2016). Multiple disease resistance in plants. *Annu. Rev. Phytopathol., 54*, 229–252.
14. Chen, K., Wang, Y., Zhang, R., Zhang, H., & Gao, C., (2019). CRISPR/Cas genome editing and precision plant breeding in agriculture. *Annu. Rev. Plant Bio., 70*, 667–697.
15. Mushtaq, M., Sakina, A., Wani, S. H., Shikari, A. B., Tripathi, P., Zaid, A., Galla, A., Abdelrahman, M., Sharma, M., Singh, A. K., et al., (2019). Harnessing genome editing techniques to engineer disease resistance in plants. *Front. Plant Sci., 10.*
16. Tyagi, S., Mulla, S. I., Lee, K. J., Chae, J. C., & Shukla, P., (2018). VOCs-mediated hormonal signaling and crosstalk with plant growth promoting microbes. *Crit. Rev. Biotechnol., 38*, 1277–1296.
17. Vannier, N., Agler, M., & Hacquard, S., (2019). Microbiota-mediated disease resistance in plants. *PLOS Pathogens, 15*, e1007740.
18. Silva, M. S., Arraes, F. B. M., Campos, M. D. A., Grossi-de-Sa, M., Fernandez, D., Cândido, E. D. S., Cardoso, M. H., Franco, O. L., & Grossi-de-Sa, M. F., (2018). Review: Potential biotechnological assets related to plant immunity modulation applicable in engineering disease-resistant crops. *Plant Sci., 270*, 72–84.
19. Kamber, T., Pothier, J. F., Pelludat, C., Rezzonico, F., Duffy, B., & Smits, T. H. M., (2017). Role of the type VI secretion systems during disease interactions of *Erwinia amylovora* with its plant host. *BMC Genomics, 18*, 628.

20. Kachroo, A., Vincelli, P., & Kachroo, P., (2017). Signaling mechanisms underlying resistance responses: What have we learned, and how is it being applied? *Phytopathology®*, *107*, 1452–1461.
21. Lõhmus, A., Hafrén, A., & Mäkinen, K., (2017). Coat protein regulation by CK2, CPIP, HSP70, and CHIP Is required for potato virus A replication and coat protein accumulation. *J. Virol., 91.*
22. Musidlak, O., Buchwald, W., & Nawrot, R., (2014). Plant defense responses against viral and bacterial pathogen infections. Focus on RNA-binding proteins (RBPs). *Herba Pol., 60.*
23. Pritchard, L., & Birch, P. R. J., (2014). The zigzag model of plant–microbe interactions: Is it time to move on? *Mol. Plant PathoL., 15*, 865–870.
24. Kubicek, C. P., Starr, T. L., & Glass, N. L., (2014). Plant cell wall–degrading enzymes and their secretion in plant-pathogenic fungi. *AnnU. Rev. Phytopathol., 52*, 427–451.
25. Amine, G. M., Gacem, H., Boukerouis, D., & Wink, J., (2021). CRISPR–Cas systems as antimicrobial agents for agri-food pathogens. In: Abd-Elsalam, K. A., & Lim, K. T., (eds.), *CRISPR and RNAi Systems; Nanobiotechnology for Plant Protection* (pp. 361–386). Elsevier.
26. Kumar, S. P., & Shukla, P., (2015). Systems biology as an approach for deciphering microbial interactions. *Brief. Funct. Genomics, 14*, 166–168.
27. Jin, Z., (2017). The molecular mechanism of CRISPR/Cas9 system and its application in gene therapy of human diseases. *Microrev. Cell. Mol. Biol., 3.*
28. Tyagi, S., Kumar, R., Kumar, V., Won, S. Y., & Shukla, P., (2021). Engineering disease resistant plants through CRISPR-Cas9 technology. *GM Crops Food, 12*, 125–144.
29. Sattar, M. N., Iqbal, Z., Tahir, M. N., Shahid, M. S., Khurshid, M., Al-Khateeb, A. A., & Al-Khateeb, S. A., (2017). CRISPR/Cas9: A practical approach in date palm genome editing. *Front. Plant Sci., 8.*
30. Makarova, K. S., Zhang, F., & Koonin, E. V., (2017). SnapShot: Class 1 CRISPR-Cas systems. *Cell, 168*, 946–946.e941.
31. Makarova, K. S., Zhang, F., & Koonin, E. V., (2017). SnapShot: Class 2 CRISPR-Cas systems. *Cell, 168*, 328–328.e321.
32. Shmakov, S., Abudayyeh, O. O., Makarova, K. S., Wolf, Y. I., Gootenberg, J. S., Semenova, E., Minakhin, L., Joung, J., Konermann, S., Severinov, K., et al., (2015). Discovery and functional characterization of diverse class 2 CRISPR-Cas systems. *Mol. Cell, 60*, 385–397.
33. Ji, X., Wang, D., & Gao, C., (2019). CRISPR editing-mediated antiviral immunity: A versatile source of resistance to combat plant virus infections. *Sci. China Life Sci., 62*, 1246–1249.
34. Garcia-Doval, C., & Jinek, M., (2017). Molecular architectures and mechanisms of class 2 CRISPR-associated nucleases. *Curr. Opin. Struct. Biol., 47*, 157–166.
35. Bao, A., Burritt, D. J., Chen, H., Zhou, X., Cao, D., & Tran, L. S. P., (2019). The CRISPR/Cas9 system and its applications in crop genome editing. *Crit. Rev. Biotechnol., 39*, 321–336.
36. Ahmad, S., Wei, X., Sheng, Z., Hu, P., & Tang, S., (2020). CRISPR/Cas9 for development of disease resistance in plants: Recent progress, limitations, and future prospects. *Brief. Funct. Genomics, 19*, 26–39.
37. Liu, M., Rehman, S., Tang, X., Gu, K., Fan, Q., Chen, D., & Ma, W., (2019). Methodologies for improving HDR efficiency. *Front. Genet., 9.*

38. Hahn, F., & Nekrasov, V., (2019). CRISPR/Cas precision: Do we need to worry about off-targeting in plants? *Plant Cell Rep., 38*, 437–441.
39. Shan, Q., Wang, Y., Li, J., & Gao, C., (2014). Genome editing in rice and wheat using the CRISPR/Cas system. *Nat. Protoc., 9*, 2395–2410.
40. Brauer, E. K., Balcerzak, M., Rocheleau, H., Leung, W., Schernthaner, J., Subramaniam, R., & Ouellet, T., (2019). Genome editing of a deoxynivalenol-induced transcription factor confers resistance to *Fusarium graminearum* in Wheat. *MPMI, 33*, 553–560.
41. Zaynab, M., Sharif, Y., Fatima, M., Afzal, M. Z., Aslam, M. M., Raza, M. F., Anwar, M., Raza, M. A., Sajjad, N., Yang, X., et al., (2020). CRISPR/Cas9 to generate plant immunity against pathogen. *Microb. Pathog., 141*, 103996.
42. Zhang, P., Du, H., Wang, J., Pu, Y., Yang, C., Yan, R., Yang, H., Cheng, H., & Yu, D., (2020). Multiplex CRISPR/Cas9-mediated metabolic engineering increases soya bean isoflavone content and resistance to soya bean mosaic virus. *Plant Biotechnol. J., 18*, 1384–1395.
43. Borrelli, V. M. G., Brambilla, V., Rogowsky, P., Marocco, A., & Lanubile, A., (2018). The enhancement of plant disease resistance using CRISPR/Cas9 technology. *Front. Plant Sci., 9*.
44. Wang, Z., Tyler, B. M., & Liu, X., (2018). Protocol of *Phytophthora capsici* transformation using the CRISPR-Cas9 system. In: Ma, W., & Wolpert, T., (eds.), *Plant Pathogenic Fungi and Oomycetes: Methods and Protocols; Methods in Molecular Biology* (pp. 265–274). Springer: New York, NY.
45. Manghwar, H., Lindsey, K., Zhang, X., & Jin, S., (2019). CRISPR/Cas system: Recent advances and future prospects for genome editing. *Trend. Plant Sci., 24*, 1102–1125.
46. Fang, Y., & Tyler, B. M., (2016). Efficient disruption and replacement of an effector gene in the oomycete *Phytophthora sojae* using CRISPR/Cas9. *Mol. Plant Pathol., 17*, 127–139.
47. Miao, J., Chi, Y., Lin, D., Tyler, B. M., & Liu, X., (2018). Mutations in ORP1 conferring oxathiapiprolin resistance confirmed by genome editing using CRISPR/Cas9 in *Phytophthora capsici* and *P. sojae*. *Phytopathology, 108*, 1412–1419.
48. Huck, S., Bock, J., Girardello, J., Gauert, M., & Pul, Ü., (2019). Marker-free genome editing in *Ustilago trichophora* with the CRISPR-Cas9 technology. *RNA Biol., 16*, 397–403.
49. Liu, D., Chen, X., Liu, J., Ye, J., & Guo, Z., (2012). The rice ERF transcription factor OsERF922 negatively regulates resistance to *Magnaporthe oryzae* and salt tolerance. *J. Exp. Bot., 63*, 3899–3911.
50. Shan, Q., Wang, Y., Li, J., Zhang, Y., Chen, K., Liang, Z., Zhang, K., Liu, J., Xi, J. J., Qiu, J. L., et al., (2013). Targeted genome modification of crop plants using a CRISPR-Cas system. *Nat. Biotechnol., 31*, 686–688.
51. Nalam, V. J., Alam, S., Keereetaweep, J., Venables, B., Burdan, D., Lee, H., Trick, H. N., et al., (2015). Facilitation of *Fusarium graminearum* Infection by 9-lipoxygenases in *Arabidopsis* and wheat. *MPMI, 28*, 1142–1152.
52. Liang, Y., Han, Y., Wang, C., Jiang, C., & Xu, J. R., (2018). Targeted deletion of the USTA and UvSLT2 genes efficiently in *Ustilaginoidea virens* with the CRISPR-Cas9 system. *Front. Plant Sci., 9*.
53. Zheng, Z., Nonomura, T., Appiano, M., Pavan, S., Matsuda, Y., Toyoda, H., Wolters, A. M. A., Visser, R. G. F., & Bai, Y., (2013). Loss of function in Mlo orthologs reduces

susceptibility of pepper and tomato to powdery mildew disease caused by *Leveillula taurica*. *PLOS One*, *8*, e70723.
54. Thomazella, D. P. D. T., Brail, Q., Dahlbeck, D., & Staskawicz, B., (2016). CRISPR-Cas9 mediated mutagenesis of a DMR6 ortholog in tomato confers broad-spectrum disease resistance. *bioRxiv,* 064824.
55. Zhang, Y., Bai, Y., Wu, G., Zou, S., Chen, Y., Gao, C., & Tang, D., (2017). Simultaneous modification of three homoeologs of TaEDR1 by genome editing enhances powdery mildew resistance in wheat. *Plant J.*, *91*, 714–724.
56. Zheng, Z., Appiano, M., Pavan, S., Bracuto, V., Ricciardi, L., Visser, R. G. F., Wolters, A. M. A., & Bai, Y., (2016). Genome-wide study of the tomato SlMLO gene family and its functional characterization in response to the powdery mildew fungus *Oidium neolycopersici*. *Front. Plant Sci., 7*.
57. Nekrasov, V., Staskawicz, B., Weigel, D., Jones, J. D. G., & Kamoun, S., (2013). Targeted mutagenesis in the model plant *Nicotiana benthamiana* using Cas9 RNA-guided endonuclease. *Nat. Biotechnol., 31*, 691–693.
58. Li, F., Fan, G., Lu, C., Xiao, G., Zou, C., Kohel, R. J., Ma, Z., Shang, H., Ma, X., Wu, J., et al., (2015). Genome sequence of cultivated upland cotton (*Gossypium hirsutum* TM-1) provides insights into genome evolution. *Nat. Biotechnol., 33*, 524–530.
59. Zhang, Z., Ge, X., Luo, X., Wang, P., Fan, Q., Hu, G., Xiao, J., et al., (2018). Simultaneous editing of two copies of Gh14-3-3d confers enhanced transgene-clean plant defense against *Verticillium dahliae* in allotetraploid upland cotton. *Front. Plant Sci., 9*.
60. Mendes, R. J., Luz, J. P., Santos, C., & Tavares, F., (2021). CRISPR genotyping as complementary tool for epidemiological surveillance of *Erwinia amylovora* outbreaks. *PLOS One*, *16*, e0250280.
61. Gumtow, R., Wu, D., Uchida, J., & Tian, M., (2017). A *Phytophthora palmivora* extracellular cystatin-like protease inhibitor targets papain to contribute to virulence on papaya. *MPMI*, *31*, 363–373.
62. Fister, A. S., Landherr, L., Maximova, S. N., & Guiltinan, M. J., (2018). Transient expression of CRISPR/Cas9 machinery targeting TcNPR3 enhances defense response in *Theobroma cacao*. *Front. Plant Sci., 9*.
63. Sun, Q., Lin, L., Liu, D., Wu, D., Fang, Y., Wu, J., & Wang, Y., (2018). CRISPR/Cas9-mediated multiplex genome editing of the BnWRKY11 and BnWRKY70 genes in *Brassica napus* L. *Int. J. Mol. Sci., 19*, 2716.
64. Javed, M. R., Abbasi, A. Z., Akhtar, M. J., Ghafoor, S., Afzal, M. A., Majeed, Z., & Umer, B., (2021). Current trends and recent progress of genetic engineering in genus *Phytophthora* using CRISPR systems. In: Abd-Elsalam, K. A., & Lim, K. T., (eds.), *CRISPR and RNAi Systems, Nanobiotechnology for Plant Protection* (pp. 183–209). Elsevier.
65. Hahn, M., & Scalliet, G., (2020). One cut to change them all: CRISPR/Cas, a groundbreaking tool for genome editing in botrytis cinerea and other fungal plant pathogens. *Phytopathology®*, *111*, 474–477.
66. Wang, Y., Cheng, X., Shan, Q., Zhang, Y., Liu, J., Gao, C., & Qiu, J. L., (2014). Simultaneous editing of three homoeoalleles in hexaploid bread wheat confers heritable resistance to powdery mildew. *Nat. Biotechnol., 32*, 947–951.
67. Santillán, M. M. I., Bracuto, V., Koseoglou, E., Appiano, M., Jacobsen, E., Visser, R. G. F., Wolters, A. M. A., & Bai, Y., (2020). CRISPR/Cas9-targeted mutagenesis of the

tomato susceptibility gene PMR4 for resistance against powdery mildew. *BMC Plant Biol., 20*, 284.

68. Backer, R., Mahomed, W., Reeksting, B. J., Engelbrecht, J., Ibarra-Laclette, E., & Van, D. B. N., (2015). Phylogenetic and expression analysis of the NPR1-like gene family from *Persea americana* (Mill.). *Front. Plant Sci., 6*.
69. Kong, G., Wan, L., Deng, Y. Z., Yang, W., Li, W., Jiang, L., Situ, J., et al., (2019). Pectin acetylesterase PAE5 is associated with the virulence of plant pathogenic oomycete *Peronophythora litchii. Physiol. Mol. Plant Pathol., 106*, 16–22.
70. Hoogen, J. V. D., & Govers, F., (2018). Attempts to implement CRISPR/Cas9 for genome editing in the oomycete *Phytophthora infestans. bioRxiv*, 274829.
71. Wei, Y., Liu, W., Hu, W., Liu, G., Wu, C., Liu, W., Zeng, H., He, C., & Shi, H., (2017). Genome-wide analysis of autophagy-related genes in banana highlights MaATG8s in cell death and autophagy in immune response to *Fusarium* wilt. *Plant Cell Rep., 36*, 1237–1250.
72. Wang, W., Pan, Q., He, F., Akhunova, A., Chao, S., Trick, H., & Akhunov, E., (2018). Transgenerational CRISPR-Cas9 activity facilitates multiplex gene editing in allopolyploid wheat. *CRISPR J., 1*, 65–74.
73. Schaart, J. G., Van De, W. C. C. M., Lotz, L. A. P., & Smulders, M. J. M., (2016). Opportunities for products of new plant breeding techniques. *Trends Plant Sci., 21*, 438–449.
74. Wang, F., Wang, C., Liu, P., Lei, C., Hao, W., Gao, Y., Liu, Y. G., & Zhao, K., (2016). Enhanced rice blast resistance by CRISPR/Cas9-targeted mutagenesis of the ERF transcription factor gene OsERF922. *PLOS One, 11*, e0154027.
75. Xie, K., & Yang, Y., (2013). RNA-guided genome editing in plants using a CRISPR–Cas system. *Mol. Plant, 6*, 1975–1983.
76. Su, Y. C., Xu, L. P., Xue, B. T., Wu, Q. B., Guo, J. L., Wu, L. G., & Que, Y. X., (2013). Molecular cloning and characterization of two pathogenesis-related β-1,3-glucanase genes ScGluA1 and ScGluD1 from sugarcane infected by *Sporisorium scitamineum. Plant Cell Rep., 32*, 1503–1519.
77. Zhang, M., Liu, Q., Yang, X., Xu, J., Liu, G., Yao, X., Ren, R., Xu, J., & Lou, L., (2020). CRISPR/Cas9-mediated mutagenesis of Clpsk1 in watermelon to confer resistance to *Fusarium oxysporum* f.sp. niveum. *Plant Cell Rep, 39*, 589–595.
78. Idnurm, A., Urquhart, A. S., Vummadi, D. R., Chang, S., Van De, W. A. P., & López-Ruiz, F. J., (2017). Spontaneous and CRISPR/Cas9-induced mutation of the osmosensor histidine kinase of the canola pathogen *Leptosphaeria maculans. Fungal Biol. Biotechnol., 4*, 12.
79. Schuster, M., Schweizer, G., Reissmann, S., & Kahmann, R., (2016). Genome editing in *Ustilago maydis* using the CRISPR–Cas system. *Fungal Genet. Biol., 89*, 3–9.
80. Schuster, M., Schweizer, G., & Kahmann, R., (2018). Comparative analyses of secreted proteins in plant pathogenic smut fungi and related basidiomycetes. *Fungal Genet. Biol., 112*, 21–30.
81. Foster, A. J., Martin-Urdiroz, M., Yan, X., Wright, H. S., Soanes, D. M., & Talbot, N. J., (2018). CRISPR-Cas9 ribonucleoprotein-mediated co-editing and counterselection in the rice blast fungus. *Sci. Rep., 8*, 14355.
82. Langner, T., Kamoun, S., & Belhaj, K., (2018). CRISPR crops: Plant genome editing toward disease resistance. *Annu. Rev. Phytopathol., 56*, 479–512.

83. Acevedo-Garcia, J., Kusch, S., & Panstruga, R., (2014). Magical mystery tour: MLO proteins in plant immunity and beyond. *New Phytol., 204*, 273–281.
84. Zhang, T., Zheng, Q., Yi, X., An, H., Zhao, Y., Ma, S., & Zhou, G., (2018). Establishing RNA virus resistance in plants by harnessing CRISPR immune system. *Plant Biotechnol. J., 16*, 1415–1423.
85. Bashar, K. K., & Hanif, M. A. (2021). Crop gene editing against biotic stresses via CRISPR/Cas9 tools: a review. *Archives of Phytopathology and Plant Protection, 54*(15–16), 1159–1181.
86. Zeng, W., Melotto, M., & He, S. Y., (2010). Plant stomata: A checkpoint of host immunity and pathogen virulence. *Curr. Opin. Biotechnol., 21*, 599–603.
87. Eş, I., Gavahian, M., Marti-Quijal, F. J., Lorenzo, J. M., Mousavi, K. A., Tsatsanis, C., Kampranis, S. C., & Barba, F. J., (2019). The application of the CRISPR-Cas9 genome editing machinery in food and agricultural science: Current status, future perspectives, and associated challenges. *Biotechnol. Adv., 37*, 410–421.
88. Zafar, K., Khan, M. Z., Amin, I., Mukhtar, Z., Yasmin, S., Arif, M., Ejaz, K., & Mansoor, S., (2020). Precise CRISPR-Cas9 mediated genome editing in super basmati rice for resistance against bacterial blight by targeting the major susceptibility gene. *Front. Plant Sci., 11*.
89. Jiang, W., Zhou, H., Bi, H., Fromm, M., Yang, B., & Weeks, D. P., (2013). Demonstration of CRISPR/Cas9/sgRNA-mediated targeted gene modification in *Arabidopsis*, tobacco, sorghum and rice. *Nucleic Acids Res., 41*, e188–e188.
90. Xu, Z., Xu, X., Gong, Q., Li, Z., Li, Y., Wang, S., Yang, Y., Ma, W., Liu, L., Zhu, B., et al., (2019). Engineering broad-spectrum bacterial blight resistance by simultaneously disrupting variable TALE-binding elements of multiple susceptibility genes in rice. *Mol. Plant, 12*, 1434–1446.
91. Zhou, J., Peng, Z., Long, J., Sosso, D., Liu, B., Eom, J. S., Huang, S., Liu, S., Cruz, C. V., Frommer, W. B., et al., (2015). Gene targeting by the TAL effector PthXo2 reveals cryptic resistance gene for bacterial blight of rice. *Plant J., 82*, 632–643.
92. Jin, S., Zong, Y., Gao, Q., Zhu, Z., Wang, Y., Qin, P., Liang, C., Wang, D., Qiu, J. L., Zhang, F., et al., (2019). Cytosine, but not adenine, base editors induce genome-wide off-target mutations in rice. *Sci., 364*, 292–295.
93. Malnoy, M., Viola, R., Jung, M. H., Koo, O. J., Kim, S., Kim, J. S., Velasco, R., & Nagamangala, K. C., (2016). DNA-free genetically edited grapevine and apple protoplast using CRISPR/Cas9 ribonucleoproteins. *Front. Plant Sci., 7*.
94. Peng, A., Chen, S., Lei, T., Xu, L., He, Y., Wu, L., Yao, L., & Zou, X., (2017). Engineering canker-resistant plants through CRISPR/Cas9-targeted editing of the susceptibility gene CsLOB1 promoter in citrus. *Plant Biotechnol. J., 15*, 1509–1519.
95. Jia, H., Zhang, Y., Orbović, V., Xu, J., White, F. F., Jones, J. B., & Wang, N., (2017). Genome editing of the disease susceptibility gene CsLOB1 in citrus confers resistance to citrus canker. *Plant Biotechnol. J., 15*, 817–823.
96. Ortigosa, A., Gimenez-Ibanez, S., Leonhardt, N., & Solano, R., (2019). Design of a bacterial speck resistant tomato by CRISPR/Cas9-mediated editing of SlJAZ2. *Plant Biotechnol. J., 17*, 665–673.
97. Oliva, R., Ji, C., Atienza-Grande, G., Huguet-Tapia, J. C., Perez-Quintero, A., Li, T., Eom, J. S., Li, C., Nguyen, H., Liu, B., et al., (2019). Broad-spectrum resistance to bacterial blight in rice using genome editing. *Nat. Biotechnol., 37*, 1344–1350.

98. Kim, Y. A., Moon, H., & Park, C. J., (2019). CRISPR/Cas9-targeted mutagenesis of Os8N3 in rice to confer resistance to *Xanthomonas oryzae* pv. oryzae. *Rice, 12*, 67.
99. Li, S., Shen, L., Hu, P., Liu, Q., Zhu, X., Qian, Q., Wang, K., & Wang, Y., (2019). Developing disease-resistant thermosensitive male sterile rice by multiplex gene editing. *Journal of Integrative Plant Biol., 61*, 1201–1205.
100. Sastry, K. S., & Zitter, T. A., (2014). Management of virus and viroid diseases of crops in the tropics. In: Sastry, K. S., & Zitter, T. A., (eds.), *Plant Virus and Viroid Diseases in the Tropics:\: Epidemiology and Management* (Vol. 2, pp. 149–480.). Springer Netherlands: Dordrecht.
101. Ali, Z., Abulfaraj, A., Idris, A., Ali, S., Tashkandi, M., & Mahfouz, M. M., (2015). CRISPR/Cas9-mediated viral interference in plants. *Genome Biol., 16*, 238.
102. Li, F., Liu, W., & Zhou, X., (2019). Pivoting plant immunity from theory to the field. *Sci. China Life Sci., 62*.
103. Zhang, Y., Malzahn, A. A., Sretenovic, S., & Qi, Y., (2019). The emerging and uncultivated potential of CRISPR technology in plant science. *Nat. Plants, 5*, 778–794.
104. Iqbal, Z., Sattar, M. N., & Shafiq, M., (2016). CRISPR/Cas9: A tool to circumscribe cotton leaf curl disease. *Front. Plant Sci., 7*.
105. Shahid, M. S., Sattar, M. N., Iqbal, Z., Raza, A., & Al-Sadi, A. M., (2020). Next-generation sequencing and the CRISPR-Cas nexus: A molecular plant virology perspective. *Front. Microbiol., 11*, 3456.
106. Chen, W., Qian, Y., Wu, X., Sun, Y., Wu, X., & Cheng, X., (2014). Inhibiting replication of begomoviruses using artificial zinc finger nucleases that target viral-conserved nucleotide motif. *Virus Genes, 48*, 494–501.
107. Cheng, X., Li, F., Cai, J., Chen, W., Zhao, N., Sun, Y., Guo, Y., Yang, X., & Wu, X., (2015). Artificial TALE as a convenient protein platform for engineering broad-spectrum resistance to begomoviruses. *Viruses, 7*, 4772–4782.
108. Baltes, N. J., Hummel, A. W., Konecna, E., Cegan, R., Bruns, A. N., Bisaro, D. M., & Voytas, D. F., (2015). Conferring resistance to geminiviruses with the CRISPR–Cas prokaryotic immune system. *Nat. Plants, 1*, 1–4.
109. Ji, X., Zhang, H., Zhang, Y., Wang, Y., & Gao, C., (2015). Establishing a CRISPR–Cas-like immune system conferring DNA virus resistance in plants. *Nat. Plants, 1*, 1–4.
110. Tashkandi, M., Ali, Z., Aljedaani, F., Shami, A., & Mahfouz, M. M., (2018). Engineering resistance against tomato yellow leaf curl virus via the CRISPR/Cas9 system in tomato. *Plant Signal. Behav., 13*, e1525996.
111. Ali, Z., Ali, S., Tashkandi, M., Zaidi, S. S. E. A., & Mahfouz, M. M., (2016). CRISPR/Cas9-mediated immunity to geminiviruses: Differential interference and evasion. *Sci. Rep., 6*, 26912.
112. Yin, K., Han, T., Xie, K., Zhao, J., Song, J., & Liu, Y., (2019). Engineer complete resistance to cotton leaf curl Multan virus by the CRISPR/Cas9 system in *Nicotiana benthamiana*. *Phytopathol. Res., 1*, 9.
113. Liu, H., Soyars, C. L., Li, J., Fei, Q., He, G., Peterson, B. A., Meyers, B. C., et al., (2018). CRISPR/Cas9-mediated resistance to cauliflower mosaic virus. *Plant Direct., 2*, e00047.
114. Khan, S., Mahmood, M., Rahman, S., Rizvi, F., & Ahmad, A., (2020). Evaluation of the CRISPR/Cas9 system for the development of resistance against cotton leaf curl virus in model plants. *Plant Prot. Sci., 56*(2020), 154–162.

115. Kis, A., Hamar, É., Tholt, G., Bán, R., & Havelda, Z., (2019). Creating highly efficient resistance against wheat dwarf virus in barley by employing CRISPR/Cas9 system. *Plant Biotechnol. J.*, *17*, 1004–1006.
116. Tripathi, J. N., Ntui, V. O., Ron, M., Muiruri, S. K., Britt, A., & Tripathi, L., (2019). CRISPR/Cas9 editing of endogenous banana streak virus in the B genome of *Musa* spp. overcomes a major challenge in banana breeding. *Commun. Biol.*, *2*, 1–11.
117. Uniyal, A. P., Yadav, S. K., & Kumar, V., (2019). The CRISPR–Cas9, genome editing approach: A promising tool for drafting defense strategy against begomoviruses including cotton leaf curl viruses. *J. Plant Biochem. Biotechnol.*, *28*, 121–132.
118. Rao, X. Q., Wu, Z., Zhou, L. I. N., Sun, J. I. E., & Li, H. P., (2017). Genetic diversity of banana bunchy top virus isolates from China. *Acta Virol.*, *61*, 217–222.
119. Price, A. A., Sampson, T. R., Ratner, H. K., Grakoui, A., & Weiss, D. S., (2015). Cas9-mediated targeting of viral RNA in eukaryotic cells. *PNAS*, *112*, 6164–6169.
120. Abudayyeh, O. O., Gootenberg, J. S., Essletzbichler, P., Han, S., Joung, J., Belanto, J. J., Verdine, V., Cox, D. B. T., Kellner, M. J., Regev, A., et al., (2017). RNA targeting with CRISPR–Cas13. *Nature*, *550*, 280–284.
121. Abudayyeh, O. O., Gootenberg, J. S., Konermann, S., Joung, J., Slaymaker, I. M., Cox, D. B. T., Shmakov, S., Makarova, K. S., Semenova, E., Minakhin, L., et al., (2016). C2c2 is a single-component programmable RNA-guided RNA-targeting CRISPR effector. *Sci.*
122. Aman, R., Ali, Z., Butt, H., Mahas, A., Aljedaani, F., Khan, M. Z., Ding, S., & Mahfouz, M., (2018). RNA virus interference via CRISPR/Cas13a system in plants. *Genome Biol.*, *19*, 1.
123. Aman, R., Mahas, A., Butt, H., Ali, Z., Aljedaani, F., & Mahfouz, M., (2018). Engineering RNA virus interference via the CRISPR/Cas13 machinery in *Arabidopsis*. *Viruses*, *10*, 732.
124. Zhang, T., Zhao, Y., Ye, J., Cao, X., Xu, C., Chen, B., An, H., Jiao, Y., Zhang, F., Yang, X., et al., (2019). Establishing CRISPR/Cas13a immune system conferring RNA virus resistance in both dicot and monocot plants. *Plant Biotechnol. J.*, *17*, 1185–1187.
125. Hameed, A., Shan-e-Ali, Z. S., Sattar, M. N., Iqbal, Z., & Tahir, M. N., (2019). CRISPR technology to combat plant RNA viruses: A theoretical model for potato virus Y (PVY) resistance. *Microb. Pathog.*, *133*, 103551.
126. Hadidi, A., Flores, R., Candresse, T., & Barba, M., (2016). Next-generation sequencing and genome editing in plant virology. *Front. Microbiol.*, *7*.
127. Ghorbani, F. P., Farsi, M., Seifi, A., & Mirshamsi, K. A., (2020). Virus-induced CRISPR-Cas9 system improved resistance against tomato yellow leaf curl virus. *Mol. Biol. Rep.*, *47*, 3369–3337.
128. Mehta, D., Stürchler, A., Anjanappa, R. B., Zaidi, S. S. E. A., Hirsch-Hoffmann, M., Gruissem, W., & Vanderschuren, H., (2019). Linking CRISPR-Cas9 interference in cassava to the evolution of editing-resistant geminiviruses. *Genome Biol.*, *20*, 80.
129. Roy, A., Zhai, Y., Ortiz, J., Neff, M., Mandal, B., Mukherjee, S. K., & Pappu, H. R., (2019). Multiplexed editing of a begomovirus genome restricts escape mutant formation and disease development. *PLOS One*, *14*, e0223765.
130. Khan, M. Z., Haider, S., Mansoor, S., & Amin, I., (2019). Targeting plant ssDNA viruses with engineered miniature CRISPR-Cas14a. *Trends Biotechnol.*, *37*, 800–804.
131. Aquino-Jarquin, G., (2019). CRISPR-Cas14 is now part of the artillery for gene editing and molecular diagnostic. *NBM*, *18*, 428–431.

132. Harrington, L. B., Burstein, D., Chen, J. S., Paez-Espino, D., Ma, E., Witte, I. P., Cofsky, J. C., et al., (2018). Programmed DNA destruction by miniature CRISPR-Cas14 enzymes. *Sci., 362*, 839–842.
133. Chandrasekaran, J., Brumin, M., Wolf, D., Leibman, D., Klap, C., Pearlsman, M., Sherman, A., et al., (2016). Development of broad virus resistance in non-transgenic cucumber using CRISPR/Cas9 technology. *Mol. Plant Pathol., 17*, 1140–1153.
134. Pyott, D. E., Sheehan, E., & Molnar, A., (2016). Engineering of CRISPR/Cas9-mediated potyvirus resistance in transgene-free *Arabidopsis* plants. *Mol. Plant Pathol., 17*, 1276–1288.
135. Macovei, A., Sevilla, N. R., Cantos, C., Jonson, G. B., Slamet-Loedin, I., Čermák, T., Voytas, D. F., et al., (2018). Novel alleles of rice eIF4G generated by CRISPR/Cas9-targeted mutagenesis confer resistance to rice tungro spherical virus. *Plant Biotechnol. J., 16*, 1918–1927.
136. Mahas, A., Aman, R., & Mahfouz, M., (2019). CRISPR-Cas13d mediates robust RNA virus interference in plants. *Genome Biol, 20*, 263.
137. Zhan, X., Zhang, F., Zhong, Z., Chen, R., Wang, Y., Chang, L., Bock, R., Nie, B., & Zhang, J., (2019). Generation of virus-resistant potato plants by RNA genome targeting. *Plant Biotechnol. J., 17*, 1814–1822.
138. Chaudhary, K., (2018). CRISPR/Cas13a targeting of RNA virus in plants. *Plant Cell Rep., 37*, 1707–1712.
139. Abudayyeh, O. O., Gootenberg, J. S., Kellner, M. J., & Zhang, F., (2019). Nucleic acid detection of plant genes using CRISPR-Cas13. *CRISPR J., 2*, 165–171.
140. Hashimoto, M., Neriya, Y., Yamaji, Y., & Namba, S., (2016). Recessive resistance to plant viruses: Potential resistance genes beyond translation initiation factors. *Front. Microbiol., 7*.
141. Cao, Y., Zhou, H., Zhou, X., & Li, F., (2020). Control of plant viruses by CRISPR/Cas system-mediated adaptive immunity. *Front. Microbiol., 11*.
142. Varanda, C. M. R., Félix, M. D. R., Campos, M. D., Patanita, M., & Materatski, P., (2021). Plant viruses: From targets to tools for CRISPR. *Viruses, 13*, 141.
143. Li, F., Yang, X., Bisaro, D. M., & Zhou, X., (2018). The βC1 protein of geminivirus–betasatellite complexes: A target and repressor of host defenses. *Mol. Plant, 11*, 1424–1426.
144. Wang, A., (2015). Dissecting the molecular network of virus-plant interactions: The complex roles of host factors. *Annu. Rev. Phytopathol., 53*, 45–66.
145. Yamanaka, T., Ohta, T., Takahashi, M., Meshi, T., Schmidt, R., Dean, C., Naito, S., & Ishikawa, M., (2000). TOM1, an *Arabidopsis* gene required for efficient multiplication of a tobamovirus, encodes a putative transmembrane protein. *PNAS, 97*, 10107–10112.
146. Lozano-Durán, R., Rosas-Díaz, T., Luna, A. P., & Bejarano, E. R., (2011). Identification of host genes involved in geminivirus infection using a reverse genetics approach. *PLOS ONE, 6*, e22383.
147. Sanfaçon, H., (2015). Plant translation factors and virus resistance. *Viruses, 7*, 3392–3419.
148. Pramanik, D., Shelake, R. M., Park, J., Kim, M. J., Hwang, I., Park, Y., & Kim, J. Y., (2021). CRISPR/Cas9-mediated generation of pathogen-resistant tomato against tomato yellow leaf curl virus and powdery mildew. *Int. J. Mol. Sci., 22*, 1878.
149. Gomez, M. A., Lin, Z. D., Moll, T., Chauhan, R. D., Hayden, L., Renninger, K., Beyene, G., Taylor, N. J., Carrington, J. C., Staskawicz, B. J., et al., (2019). Simultaneous

CRISPR/Cas9-mediated editing of cassava eIF4E isoforms nCBP-1 and nCBP-2 reduces cassava brown streak disease symptom severity and incidence. *Plant Biotechnol. J., 17*, 421–434.

150. Bastet, A., Zafirov, D., Giovinazzo, N., Guyon-Debast, A., Nogué, F., Robaglia, C., & Gallois, J. L., (2019). Mimicking natural polymorphism in eIF4E by CRISPR-Cas9 base editing is associated with resistance to potyviruses. *Plant Biotechnol. J., 17*, 1736–1750.
151. Helm, M., Qi, M., Sarkar, S., Yu, H., Whitham, S. A., & Innes, R. W., (2019). Engineering a decoy substrate in soybean to enable recognition of the soybean mosaic virus NIa protease. *MPMI, 32*, 760–769.
152. Bastet, A., Lederer, B., Giovinazzo, N., Arnoux, X., German-Retana, S., Reinbold, C., Brault, V., Garcia, D., Djennane, S., Gersch, S., et al., (2018). Trans-species synthetic gene design allows resistance pyramiding and broad-spectrum engineering of virus resistance in plants. *Plant Biotechnol. J., 16*, 1569–1581.
153. Khan, Z., Saboor, T., Ashfaq, M., Saddique, A., & Khanum, P., (2021). Disease resistance in crops through CRISPR/Cas. In: Ahmad, A., Khan, S. H., & Khan, Z., (EDS.), *CRISPR Crops: The Future of Food Security* (pp. 151–175). Springer: Singapore.
154. Duke, S. O., (1995). *Herbicide-Resistant Crops: Agricultural, Economic, Environmental, Regulatory, and Technological Aspects* (p. 438). CRC Press.
155. Huang, Y., Chen, Y., Zeng, B., Wang, Y., James, A. A., Gurr, G. M., Yang, G., et al., (2016). CRISPR/Cas9 mediated knockout of the abdominal-A homeotic gene in the global pest, diamondback moth (*Plutella xylostella*). *Insect Biochem. Mol. Biol., 75*, 98–106.
156. Ali, M. A., Abbas, A., Azeem, F., Javed, N., & Bohlmann, H., (2015). Plant-nematode interactions: From genomics to metabolomics. *Int. J. Agric. Biol., 17*, 1071–1082.
157. Ali, M. A., Azeem, F., Abbas, A., Joyia, F. A., Li, H., & Dababat, A. A., (2017). Transgenic strategies for enhancement of nematode resistance in plants. *Front. Plant Sci., 8*.
158. Decraemer, W., & Hunt, D. J. (2006). Structure and classification. In *Plant Nematology*; Perry, R. N., & Moens, M., (Eds.). Wallingford, Oxfordshire: CAB International, 3–32.
159. Murray, G. M., & Brennan, J. P., (2009). Estimating disease losses to the Australian wheat industry. *Australas. Plant Pathol., 38*, 558–570.
160. Smiley, R. W., Dababat, A. A., Iqbal, S., Jones, M. G. K., Maafi, Z. T., Peng, D., Subbotin, S. A., & Waeyenberge, L., (2017). Cereal cyst nematodes: A complex and destructive group of *heterodera* species. *Plant Dis., 101*, 1692–1720.
161. Dickinson, D. J., & Goldstein, B., (2016). CRISPR-based methods for *Caenorhabditis* elegans genome engineering. *Genetics, 202*, 885–901.
162. Paix, A., Folkmann, A., & Seydoux, G., (2017). Precision genome editing using CRISPR-Cas9 and linear repair templates in *C. elegans*. *Methods, 121–122*, 86–93.
163. Kang, J., (2016). *Application of CRISPR/Cas9-Mediated Genome Editing for Studying Soybean Resistance to Soybean Cyst Nematode*. Thesis, University of Missouri–Columbia.
164. Mayer, K. F. X., Waugh, R., Langridge, P., Close, T. J., Wise, R. P., Graner, A., Matsumoto, T., Sato, K., Schulman, A., Muehlbauer, G. J., et al., (2012). A physical, genetic, and functional sequence assembly of the barley genome. *Nature, 491*, 711–716.
165. Van, S. C. C. N., & Takken, F. L. W., (2014). Susceptibility genes 101: How to be a good host. *Annu. Rev. Phytopathol., 52*, 551–581.

166. Montecillo, J. A. V., Chu, L. L., & Bae, H., (2020). CRISPR-Cas9 system for plant genome editing: Current approaches and emerging developments. *Agronomy, 10*, 1033, doi: 10.3390.
167. Nadakuduti, S. S., Buell, C. R., Voytas, D. F., Starker, C. G., & Douches, D. S., (2018). Genome editing for crop improvement – applications in clonally propagated polyploids with a focus on potato (*Solanum tuberosum* L.). *Front. Plant Sci., 9*.
168. Ellens, K. W., Levac, D., Pearson, C., Savoie, A., Strand, N., Louter, J., & Tibelius, C., (2019). Canadian regulatory aspects of gene editing technologies. *Transgenic Res., 28*, 165–168.
169. Callaway, E., (2018). CRISPR plants now subject to tough GM laws in European Union. *Nature, 560*, 16, 17.
170. Hickey, L. T., Hafeez, A. N., Robinson, H., Jackson, S. A., Leal-Bertioli, S. C. M., Tester, M., Gao, C., et al., (2019). Breeding crops to feed 10 billion. *Nat. Biotechnol., 37*, 744–754.

CHAPTER 9

CRISPR-CAS-MEDIATED GENOME EDITING IN MAIZE – PRESENT, PAST, AND FUTURE

RAHIL SHAHZAD[1‡], SHAKRA JAMIL[1‡], GHALIB T G ALSUBAIE[2], SHAKEEL AHMAD[2], AMINA NISAR[3], SAJID UR RAHMAN[1], IQRAR AHMAD[4], and RANA MUHAMMAD ATIF[5,6,7]

[1]*Agricultural Biotechnology Research Institute, AARI, Faisalabad, Pakistan*

[2]*Seed Centre and Plant Genetic Resources Bank, Ministry of Environment, Water and Agriculture, Riyadh, Saudi Arabia.*

[3]*Department of Plant Breeding and Genetics, University of Agriculture, Faisalabad, Punjab, Pakistan*

[4]*Centre for Advanced Studies in Agriculture and Food Security, University of Agriculture, Faisalabad, Pakistan.*

[5]*Department of Plant Sciences, University of California Davis, California, United States*

[6]*Department of Plant Breeding and Genetics, University of Agriculture, Faisalabad 38000, Pakistan*

[7]*Precision Agriculture and Analytics Lab, Centre for Advanced Studies in Agriculture and Food Security, National Centre in Big Data and Cloud Computing, University of Agriculture, Faisalabad, Pakistan*

[‡]equal contribution

ABSTRACT

Maize is the third leading cereal crop after wheat and rice and occupies central position among crops for achieving the global food security. Various

CRISPR-Based Gene Editing: A Solution for Global Food Security. Shakeel Ahmad, Asif Ali Khan, and Muhammad Arshad Javed (Eds.)

technological breakthrough was achieved for improvement of maize crop, i.e., hybrid seed system, genomic assisted selection and transgenics developments. However, these technologies failed to keep pace with the ever-increasing global population and changing climate to ensure food security. Climate change is further hampering the yield of different crops by exposing them to various biotic and abiotic stresses with more severity. CRISPR-Cas technology is emerging as a promising tool for crop improvement at rapid pace to ensure food security. It offers precise modification of various genes with high specificity for development of climate smart and nutritionally enriched crops to overcome hunger and malnutrition. In this chapter, we have compared different genome editing (GE) tools with respect to their pros and cons. Further we have explored the current role and future prospects of the CRISPR-Cas technology for genetic improvement of the maize crop for climate resilience and to ensure global food security.

9.1 INTRODUCTION

Maize is 3rd leading cereal crop after wheat and rice with respect to area under cultivation and fulfillment of dietary requirements of masses. = Global trade for maize exceeds up to $52 billion dollars approximately. It is predicted that maize production will boost to as high as 1.315 B metric tons by 2030 (www.statista.com). As far as situation of maize crop in Pakistan is concerned it is currently grown on an area of 1,418,000 hectares with an annual production of 8.465 million tons. It contributes 3.4% as value added in Agriculture and 0.6% to national GDP. It is grown as a multipurpose crop in Pakistan and is used largely as food, feed, and fodder purpose. As compared to previous year maize production was increased by 7.4%. The huge surge in local and international maize production is associated with multiple factors with 50% share of genetic improvement for tolerance to various biotic and abiotic factors and climate resilience [1].

Many technologies were developed for increasing the productivity of the maize crop. Hybrid seed technology was introduced in maize, a century ago and still practiced with zeal and zest for increasing the productivity. This involves the development and advancement of the heterotic pools with good combining abilities and masking the effect of the deleterious alleles. Hybrid maize has provided additional 30–40 mounds of yield per acre as compared to open pollinated varieties [2–4]. Later on, marker assisted breeding (MAB) has paved way for crop improvement for various aspects. MAB was largely used for biotic and abiotic stress tolerance and nutritional quality enhancement of crops. MAB was used for assessment of genetic diversity of genotypes which helped breeders and geneticists in assigning inbred lines to various heterotic groups for designing of effective crossing plan for development of climate resilient maize hybrids [5–7]. MAB started with the set of dispersed markers covering a certain portion of the genome and now has advanced to the high-density SNPs [8, 9]. The advancement in the bioinformatics tools and phenomenal increase in the depth and breadth of genome sequencing data has spawned the era of genomic selection and paved way for cost effective genetic improvement of maize crop.

In the 20th genetic engineering emerged as a novel plant breeding tool. For the first time foreign genes were inserted in crops using bioengineering approaches. The era of transgenic breeding started with the development of Bt. cotton for control of control of cotton bollworms which later exploited in other crops for incorporation of herbicide tolerance, insect pest resistance and quality enhancement of crops [10, 11]. Despite these advancements in crop improvement techniques, food production is not sufficient to meet the growing dietary requirements of ever-increasing population which is expected to reach rise 8.5 billion by the end of 2030 [12, 13]. Further the regulatory issues associated with GM crops due to presence of foreign gene were hampering their potential due to delayed commercialization [14].

Keeping in view of the above said challenges and hindrance to achievement of sustainable development goal 2 (SDG2) which is related to zero hunger, there was an urgent need of innovative tool for accelerated crop improvement which could have transgene free nature. Early in the 21st century, CRISPR-Cas (Clustered Regularly Interspersed Short Palindromic Repeats-CRISPR Associated) technologies revolutionized genetic research and enhanced our ability to precisely manipulate genetic information. CRISPR-Cas system has opened new horizons for crop improvement by creating novel genetic variations that does not exist in nature [15]. CRISPR platforms are offering insertion, deletions, substitution, inversion, and translocation in a particular gene with high precision [16]. The latest breakthrough in CRISPR technology, i.e., base editing and prime editing have offered single base modifications in DNA and RNA, respectively [15, 17].

The main focus of this chapter revolves around exploring role of CRISPR-Cas system for improvement of maize crop for climate resilience and nutritional quality enhancement. Further, different development related to maize GE were also summarized which has enabled scientists to make use of this technology in maize improvement. We have also provided an overview of different GE tools, compared their efficiencies and provided an insight about their pros and cons. The latest developments in GE systems, i.e., base editing and prime editing were also discussed in detail. At the end we have discussed the different regulatory frameworks for commercialization of genome edited crops. What are main hindrances which are restricting the potential of GE technologies? We have compared the regulatory framework of different school of thoughts, i.e., process-based regulation and end product-based regulation and their effectiveness. We have also provided some future prospects related to metabolic pathways engineering for effective use of GE technologies to achieve zero hunger.

9.2 OVERVIEW OF DIFFERENT GENOME EDITING (GE) TOOLS

The advent of GE tools and their significance for generation of transgene free plants has attracted the plant scientists and consumers as well. The latest breakthrough in the technology, i.e., CRISPR-Cas9, base editing and prime editing are paving way for fast-track crop improvement in cost effective manner. These GE tools allow cutting and re-joining of the genetic material at a specific site allowing repairing of the faulty genes by insertions, deletions, and substitution. We have shed light on different GE tools and their mechanism of action. The comparison of different GE tools is listed in subsections (Figure 9.1 and Table 9.1).

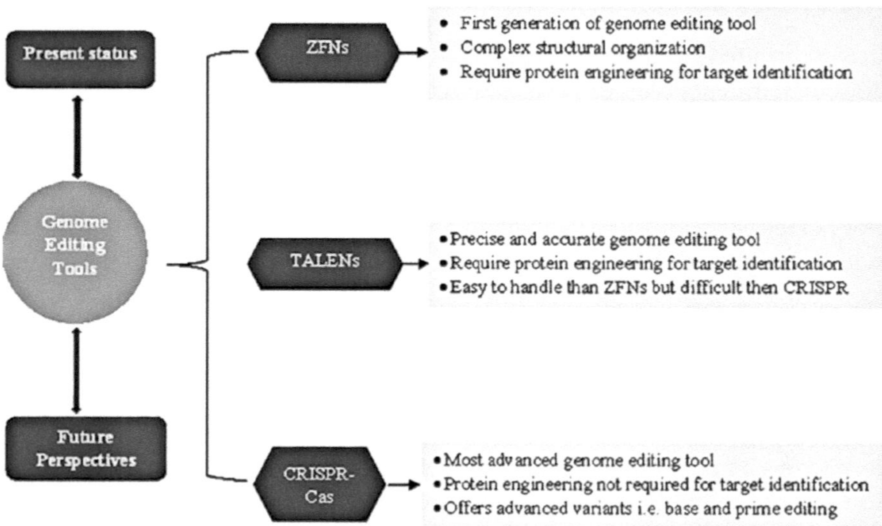

FIGURE 9.1 Comparison of different genome editing tools with respect to their merits and demerits.

9.2.1 ZINC FINGER NUCLEASES (ZFNS)

Zinc finger nucleases (ZFNs) are the first generation of GE tools which uses engineered nucleases developed after characterization of function of the ZFNs domain (Cys2-His2 (C2H2) zinc finger). The Zinc finger domain is usually comprised of 30 amino acids which are folded in *ββα configuration*. The genetic analysis has exhibited that Zinc finger domain interact with DNA by inserting its alpha helix and each protein have the ability to recognize three tandem repeats in the DNA [18]. These ZFNs introduces double stranded breaks (DSBs) at target region and error-prone non-homologous end joining (NHEJ) repair is responsible for particular insertions/deletions at ZFN cleavage site. The normal ZNFs monomer is comprised of two different domains, i.e., C2H2 domain at the N terminus whereas C terminus contains *Fok*I DNA cleavage domain. Dimerization of *Fok*I domain is necessary for activity of ZFNs [19].

Fok1 is primarily used in cleavage domain, having non-specific DNA cleavage activity of C terminal and DNA binding domain at N terminal. Furthermore, DNA binding domain consists of Zinc Finger Domains (3–6 in numbers) and each domain has ability to identify target sequence comprising of 3 bp. Two zinc finger arrays (ZFAs) (each of 18–24 bp) bind with target sequences in opposite direction and both are separated by 5–8 base pairs. This spacing is essential characteristics of ZFNs system as Fok1 dimerizes and creates DSB by recognizing it.

ZFNs as GE tools were firstly used in Tobacco and *Arabidopsis*. After that, these were also reported in rice, maize, wheat, and other plants. Recent study has reported the efficient ZFNs-mediated GE in allohexaploid wheat to improve herbicide (imidazolinone) tolerance by targeting the coded sequence of acetohydroxyacid synthase (*AHAS*). As a result, transgenic allohexaploid herbicide tolerant wheat plants were developed with a mutation frequency of 2.9% [21]. Shukla et al. [22] reported targeted modification of inositol phosphate for herbicide tolerance in maize through cleavage of inositol-1,3,5,6-pentakisphosphate kinase 1 (*IPK1*), one of the genes of phytic acid biosynthesis pathway. Jung et al. [23] developed transgenic rice lines via alterations of *SSIVa* gene, that is involved in starch biosynthesis pathway. However, there are some limitations of ZFNs, i.e., these have very complex structural organization and rest are explained in Table 9.1.

9.2.2 TRANSCRIPTION ACTIVATOR-LIKE EFFECTORS NUCLEASES (TALENS)

TALENs are comprised of TALE array combined with *Fokl* endonucleases. These effectors are categorized as Type II effector proteins, which

TABLE 9.1 Comparison of Different Genome Editing Tools for Their Efficiency and Structural Requirements

Point of Comparison	ZFNs	TALENs	CRISPR-Cas9	Base Editing	Prime Editing
Components	Zn finger domain and FokI nuclease domain.	TALE domain and FokI nuclease domain.	gRNA and Cas9.	nCas9 or dCas9 and deaminases.	pegRNA and nCas9.
Catalytic domain	Restriction endonucleases *Fok1*.	Restriction endonucleases *Fok1*.	No catalytic domain.	nCas9 or dCas9.	dCas9
Structural proteins	Dimeric	Dimeric	Monomeric	Monomeric	Monomeric
Protein engineering	Required	Required	Not required	Not required	Not required
Target site	24–36 bases	24–59 bases	20–22 bases	Single base	1–33 bases
gRNA designing	Not required	Not required	Easy designing	Easy designing	Easy designing
Cloning	Pre-requisite	Pre-requisite	Not compulsory	Not compulsory	Not compulsory
Multiplexing	Difficult	Difficult	Possible	Possible	Possible
Mode of action	DSBs	DSBs	DSBs or single stranded nicks.	No breaks required.	No breaks required
Mutation frequency	High	High	Low	Low	Low
Off-target effects	Low	Low	Low	Low	Low
Generation of large-scale libraries.	Impossible	Technically difficult	Easy	Easy	Easy

Source: The table has been modified from previously published review article [20] with some modifications.

are found in plant pathogenic bacteria *Xanthomonas*, which infects cotton and rice plants [24]. Structurally TALENs contains a central DNA binding domain (consisting of 33–35 amino acids tandem repeats; each tandem repeat is specific for single nucleotide present in target sequence), highly conserved acidic activation domain (AD) for transcription and nuclear localization signals (NLS) at C terminal. While N terminal contains translocation and secretion signal [25]. Structural arrangement of TALENs is easier than Zinc-Finger Nucleases (ZFNs). But it's mode of action is similar to ZFNs. Moreover, Type II secretion mechanism is applied for transferring two TALENs into a cell and subsequently to nucleus for binding to target DNA in opposite direction 12–30 base pairs apart. Then, TALENs stimulates host transcription machinery for transcription initiation. After binding of TALENs' to target site, *FokI* dimerizes to induce DSB at spacer region [26].

Precise GE through TALENs was explored in different crops, i.e., *Arabidopsis,* soybean, barley, maize, and rice for improvement of qualitative as well as quantitative traits and understands the functions of unknown genes. TALENs' backbone *N287C230* show low mutation rate (0–6.6%), however, its C-terminal domain have high (up to 25%) mutation frequency [27]. TALENs-mediated mutagenesis in fatty acid desaturase 2 gene improved soybean oil quality [28], in rice *Xanthomonas oryzae* resistance was incorporated by disrupting *OsSWEET14* gene. When it comes to comparison of TALENs and ZFNs. These are better than ZFNs due to precision, accuracy, and high success rates. Regardless of benefits, these also have some limitations, i.e., these are unable to edit methylated target sites [29].

9.2.3 CLUSTERED REGULARLY INTERSPACED SHORT PALINDROMIC REPEATS AND ASSOCIATED PROTEIN CAS9 (CRISPR/CAS9)

CRISPR-Cas system was discovered and characterized in 1990s. CRISPR/Cas system are categorized into 2 main classes, 6 types and 19 subtypes depending upon structural arrangements of different parts and composition of effector nuclease's [30]. The effector of class 1 encompasses number of proteins for multiple functions, forming a complex while the effector of class 2 contains only one protein with multi-domains [31]. The three main components of CRISPR/Cas system are expression, interference, and adaption. At the first stage, CRISPR array contains similar sequences to target sequences (protospacers), these are transcribed into pre-CRISPR RNA (pre-crRNA). Pre-crRNAs are able to interact with the trans-activating crRNAs (tracrRNA). As crRNA/tracrRNA complex is developed, it adheres to Cas9 protein and RNase cut the long pre-crRNAs into crRNA/tracrRNA complexes (gRNA). At interference stage, Cas9 complex is guided by crRNA/tracrRNA to target the sequence and gRNA bind with targeted sequence after identification of protospacer adjacent motif (PAM) sequence. Nuclease and helicase domain of Cas9 cut the target sequence and DSBs are repaired by HDR/NHEJ. As a results desired mutants are obtained and are heritable also [32].

CRISPR/Cas9 system was applied in multiple crops including cereals, pulses, legumes, vegetables, oilseeds, fodders, and fiber crops, few examples are described below for an overview of the potential applications of the technology. Ethylene responsive factor 3 (*TaERF3*) and wheat dehydration responsive element binding protein 2 (*TaDREB2*) were targeted for development of abiotic stress responsive wheat varieties [33]. It was exploited in maize with special focus on albino marker gene, *Zmzb7*. The agrobacterium mediated transformation was performed in maize embryos with 31% mutational frequency in T_0 lines [34]. CRISPR has advantage over other GE tools being precise, efficient, economical, and time saving technology. It has been widely adopted for development of abiotic and biotic stress responsive crops with higher yield and increased nutritional value [15, 35].

9.2.4 BASE EDITING

The ever-expanding knowledge about genome wide association mapping has developed association of agronomically important traits with one or more SNPs followed by candidate gene tagging. The identification of SNPs is paving way for accelerated crop improvement by genomics assisted breeding programs. However, their integration in different breeding programs is almost impossible by means of conventional breeding techniques. The advanced GE tools, i.e., base editing comes here for rescue as it offers the advantage of single base editing at target site without requirement of DBS and donor repair templates. The cytosine base editors (CBE) and adenine base editors (ABE) are the two popularly used base editors [36]. In CBE, the catalytically dead Cas9 (dCas9) or Cas9 nickase (nCas9) is fused with cytidine deaminase for deamination of cytosine (C) in a narrow window of non-target strand. The deamination of C converts it to Uracil (U) which is read as T during DNA replication process hence resulting in C.G. to T.A. base substitution [37]. In case of ABE, dCas9 or nCas9 are fused with adenosine deaminase to catalyze deamination of adenine (A). The deamination of A converts it to Inosine (I) in a narrow window which is identified as guanine (G) during DNA replication process as a result A.T. to G.C. base substitution occurs [16, 38].

Base editing is one step advanced offering RNA editing using RNA editors for precise RNA editing. The scientists are now able to edit RNA using REPAIR (RNA-editing for programmable A to I (G) replacement). The REPAIR is comprised of RNA-targeting CRISPR effector Cas13 and catalytic domain of $ADAR_2$ which leads towards editing of A base to I which is read as G by RNA machinery of the cell [39]. Similarly, C to U conversion in RNA is also possible by utilizing the RESCUE (RNA-editing for a specific C to U exchange) technology. The RESCUE technology utilizes dead Cas13 (dCas13) fused with cytidine deaminase which catalyzes substitution of C to U in RNA transcript [40].

9.2.5 PRIME EDITING

The availability of the CBE and ABE have offered base transition however still no tool available for base transversion. The advent of prime editing in 2019 which allowed all 12 kind of base substitution, precise insertion of up to 44 base pair and deletion of about 80 bp and combination of these edits simultaneously have opened up new horizons for crop improvement. The machinery for prime editing is comprised of nCas9 (H840A) along with reverse transcriptase and prime editing guide RNA (pegRNA). The reverse transcriptase template contains genetic information of target site. The prime binding site pairs with nCas9 (H840A) nicked ssDNA and induce reverse transcription and incorporate reverse transcriptase information into the genome. The reverse transcription is followed by equilibrium between edited 3′ end and non-edited 5′, ligation and repair result in induction of desired mutation at target site. The prime editors are not constrained by PAM sequence hence these offers relatively wide range of base substitution, deletions, and insertions (+1 to +33, counting from the first base 3′ of the pegRNA-induced nick) [41].

9.3 THE DEVELOPMENT OF CRISPR-CAS SYSTEM IN MAIZE

The recent developments in the field of genomics and availability of large-scale sequencing platforms have equipped maize geneticists with jumbo genetic datasets for crop improvement. Further subsections discuss the advancements which have paved way for use of CRISPR-Cas9 in maize improvement.

9.3.1 *CRISPR-SYSTEM IN MAIZE*

The CRISPR-Cas9 was first exploited in maize back in 2014 when Xing et al. [42] developed toolkit for efficient GE in maize and *Arabidopsis* simultaneously. This study has laid the foundation for CRISPR based GE in maize. Later on, another study conducted by Svitashev et al. [43] developed a detailed methodology for targeted GE using CRISPR/Cas system. They used codon optimized *Streptococcus pyogenes* Cas9 endonuclease along with single guide RNA (gRNA) for five different regions, i.e., *acetolactate synthase* genes (*ALS1* and *ALS2*), male fertility genes (*Ms26* and *Ms45*) and *liguleless 1* (*LIG1*) gene in immature maize embryos through biolistic transformation. All the five genes were mutated and desired gene edits were obtained in different experiments exhibiting successful exploitation of CRISPR-Cas9 for GE in maize.

Up till now the efficiency and inheritance pattern of mutations induced by CRISPR-Cas9 was not studied in maize. Keeping in view of these facts [44] studied the genetic efficiency and inheritance pattern of maize mutants obtained through CRISPR-Cas9. They used type II CRISPR-Cas9 system for mutagenesis at desired target site along with codon-optimized Cas9 protein and non-coding guide RNA obtained by using maize U6 snRNA promoter. The results showed targeted mutagenesis at above 90 loci with average cleavage efficiency of 10.67% in genome as confirmed through protoplast assay. The mutations induced in germ cells were inherited in progeny, i.e., phytoene synthase gene (*PSY1*) mutants were stably inherited. Further off-target assay using computational assay did not identify any off-target effects in genome and transcriptome assay also confirmed no difference between transcriptome of Cas9 edited and non-edited wild type plants except target genes. This study confirmed the safe use of the CRISPR-Cas9 system for mutagenesis in complex genome, i.e., of maize. Further developments about use of CRISPR-Cas9 system in maize are illustrated in Table 9.2.

9.3.2 ADVANCEMENT IN REAGENT DELIVERY METHODS IN MAIZE

The above said GE method generate desired mutants at target site with good precision but also lead towards integration of foreign DNA and selectable marker gene alongside desired mutations which is highly undesirable. These integrations put the mutant plants under strict GM regulations before commercialization which is a very tedious and cost ineffective approach. To avoid these unnecessary regulations GE crops needs to be transgene and marker free. Development of marker free transgenic maize plants was first exploited in maize in 2016 by Svitashev et al. [45]. They delivered Cas9 and gRNA as ribonucleoproteins (RNPs) in immature embryos using particle bombardment approach. They further elaborated the use of transiently expressed Cas9 and gRNA as *in vitro* transcribed RNA molecule for delivery of DNA free gene editing machinery in maize protoplast. The process delivered transgene free gene edited crops with high mutation frequency and low off-target effects.

In another study, Koiso et al. [46] developed transfection method for delivery of CRISPR-Cas system in maize egg cells and zygotes. They used polyethylene glycol calcium (PEG-Ca^{2+}) transfection-mediated transformation for development of transgene free mutants in maize. The transformation efficiency of PEG-Ca^{2+} was 80% showing stable genetic transformation and importantly zygotes were developed into healthy mass of cells indicating high success rate of this method for reagent delivery. Furthermore, efficiency of Cas9 and Cas12a proteins were compared for targeted mutagenesis in maize. The maize *glossy2* gene was used in this study owing to potential target site for both proteins. Constructs carrying SpCas9-guide RNA (gRNA) and LbCas12a-CRISPR RNA (crRNA) were introduced in immature embryos of maize inbred line (B104). The on-target mutant analysis depicted that approximately 90–100% of Cas9 edited plants harbored indel mutations and 63–77% of them were bi-allelic and homozygous whereas Cas12 edited plants showed mutation efficiency of 0–66%. The off-target genome wide analysis predicted total of 18 and 67 off-target for Cas9 and Cas12a, respectively. These results highlighted significance of Cas9 for gene editing in maize with good efficiency as compared to Cas12a. Sideways, these findings also suggested that Cas12a system needs further exploration for efficient utilization in maize GE [47]. In another study it was observed that although CRISPR-Cas9 has mutation high frequency in the T_1 generation however, CRISPR/Cpf_1 generated more types of new mutations in T_2.

Further, it was observed that editing efficiency of CRISPR/Cpf_1 correlates directly with Cpf_1 expression level [48].

For development of transgenic free gene editing system in maize, Cas9 and gRNA targeting *MATL-1* locus was stably transferred in NP2222 (non-haploid inducer lines). The biallelic mutants were self-pollinated to obtain homozygous plants for frame shift mutation in MATL locus and Cas9. These mutants were crossed with an in inbred line (B14-v1) expressing *virescent* (*v1*) marker which impart pale leaf phenotype for identification of maternal haploids. Later on mutants' study revealed that 03 out of 95 haploids lack Cas9 as well as wild type *ZmMATL* target site. Further sequencing and flow cytometry-based studies confirmed that all these three haploids were mutants for *ZmMATL* locus hence this study has identified a novel method for gene editing using haploid inducer line as medium for delivery of reagent and obtaining of transgene free mutant plants [49]. Reporter-assisted selection of genome edited plants was less explored in maize; a research group has recently reported this phenomenon in maize endosperm and embryo. They have established seed fluorescence reporter (SFR)-assisted GE platform in maize and mutants' plants are selected based on the fluorescence emittance from embryo, endosperm, or both [50]. Further examples explaining the reagent delivery methods are summarized in Table 9.2.

9.3.3 ADVANCEMENTS IN BREEDING TECHNOLOGIES OF MAIZE

CRISPR-Cas9 and its different variants have paved way for accelerated breeding of crops by providing various tools for efficient crop improvement. The following section discuss in detail about technological advancements in maize breeding through CRISPR-Cas system which have paved way for accelerated maize breeding. The accelerated introduction of desired/targeted mutations in maize genome for crop improvement was the first development in this domain. An RNA guided endonucleases system (RGEN) was used as an *in vivo* desired target mutator (DTM) in maize for reduction of linkage drag in breeding procedure. They used *LIGULELESS1* (*LG1*) locus as a proof of concept and recorded 51.5%–91.2% mutation frequency in T_0 plants. The T_1 plants with stable DTM expression were crossed with six diverse maize inbred lines and 11.79%–28.71% of population were mutants for DTM effect. The phenotypic analysis of mutant plants in F_2 generation exhibited 50% reduced leaf angle and mutations were heritable. The planting experimentation confirmed that DTM effect allowed more planting density and greater yield potential per unit of land. The authors highlighted use of stably expressed RGEN as an *in vivo* DTM to induce desired mutants in maize at rapid pace ignoring linkage drag effect [51].

CRISPR-Cas9 was used for generation of male sterile lines in maize for large scale hybrid seed production. The CRISPR-Cas9 vector targeting the *MS8* locus was designed and transformed in maize. The resultant plant did not show any mutations in T_0 generation however, mutants were observed in F_1 and F_2 plants for *MS8* locus. The mutations were heritable, and transgene free plants were obtained by selection among F_2 plants for desired mutants without transgene. The mutated MS locus may be transferred to other breeding programs through crossing [52]. CRISPR-Cas9 is also used for stacking of beneficial alleles in maize in the form of complex trait locus (CTL). A CTL is comprised of multiple preselected sites positioned within a small well-characterized chromosomal region where trait genes are inserted. The scientist's generated individual lines carrying site specific insertion landing pad (SSILP) targeted to a specific site and capable of receiving transgene through recombinase-mediated cassette exchange. These selected sites supported consistence gene expression and SSILP based insertion. These findings suggested that genetic information for two traits lying at a distance on a CTL may be combined using genetic recombination approach. The use of CRISPR for rearrangements and development of CTL is a major step towards development of climate resilient maize hybrids [53].

Quantitative variation was created in maize for different yield related traits by modification of promotor of CLAVATA–WUSCHEL (*CLE)* genes by CRISPR aided mutagenesis and generation of weak promotor alleles of *CLE* genes. The resultant mutants exhibited plenty of variation for different yield related traits which were later improved through selection [54]. The CRISPR application in maize is not limited to the above said factors. It is also used for generation of large-scale mutant libraries, development of multiple genomes editing set up, gene detection and cytogenetics as illustrated in Table 9.2.

9.4 CLIMATE SMART MAIZE DEVELOPMENT THROUGH CRISPR-CAS

The changing climate is giving birth to various biotic [94, 95] and abiotic stresses and reducing crop yield [13]. Maize is also susceptible to various biotic and abiotic stresses as well as there is need for quality enhancement of crops to address malnutrition. Following section discuss in detail about recent developments in maize crop improvement using CRISPR-Cas9 (Figure 9.2).

TABLE 9.2 The Sequential Utilization of the CRISPR-Cas System in Maize for Gene Editing and Latest Technological Advancements for its Use in Gene Editing, Reagent Delivery Methods, and Cytogenetic Studies

Aspect Studied	Detailed Description	CRISPR-Cas System	References
Development of genome editing platform in maize.	Toolkit for CRISPR application in maize improvement.	CRISPR-Cas9	[42]
	Technology development for CRISPR-Cas9 application in mutation induction and gene editing.	CRISPR-Cas9	[43]
	Strategy development for GE in maize.	CRISPR-Cas9	[44]
	Toolkit for CRISPR application in maize improvement.	CRISPR-Cas9	[55]
	Base editing platform in maize.	Base editors	[56]
	Overexpression of the Cas9 using *ZmDMC1* promoter along with *U3* promoter for increased mutation frequency.	CRISPR-Cas9	[57]
	Safety assessment of Cas9 activity with reference to off-targeting.	CRISPR-Cas9	[58]
	Strategy development for GE in maize.	CRISPR-Cas9	[59]
	Comparison of single and multiple genomes editing in maize.	CRISPR-Cas9	[60]
	Efficiency testing of ternary vector system that combines morphogenic regulator and CRISPR/Cas9 modules.	CRISPR-Cas9	[61]
	Development of transgene free genome editing system in maize.	CRISPR-Cas9	[62]
	Toolkit for functional characterization of maize genes.	CRISPR-Cas9	[63]
	Exploitation of seed fluorescence reporter system in maize.	CRISPR-Cas9	[50]
	Development of visual maize Editing toolbox (ViMeBox) for easy identification of mutant plants.	CRISPR-Cas9	[64]
Innovative reagent delivery methods in maize.	Genome editing using RNPs.	CRISPR-Cas9	[45]
	Transfection method for delivery of reagents in maize embryos and zygotes.	CRISPR-Cas9	[46]
	Comparison of efficiency of different GE platforms.	CRISPR-Cas9 and CRISPR-Cas12a.	[47, 48]
	Use of haploid inducer lines for targeted genome editing in maize.	CRISPR-Cas9	[65]
	Temperature dependent efficiency testing of Cas12a.	CRISPR-Cas12a	[66]

TABLE 9.2 *(Continued)*

Aspect Studied	Detailed Description	CRISPR-Cas System	References
	Reagent delivery method through foxtail mosaic virus (FoMV) as vector in maize.	CRISPR-Cas9	[67]
	Reagent delivery method through barley stripe mosaic virus (BSMV).	CRISPR-Cas9	[68]
	Development of selectable marker free transformation system with double mutation frequency.	CRISPR-Cas9	[69]
	pRGEB32-BAR vector allows simultaneous delivery of multiple gRNAs.	CRISPR-Cas9	[70]
Application in plant breeding.	Generation of new alleles.	CRISPR-Cas9	[71]
	Generation of transgene free CMS lines for hybrid breeding at commercial scale.	CRISPR-Cas9	[52]
	Haploid induction in maize.	CRISPR-Cas9	[72]
	Haploid inducer lines as a medium for targeted GE in maize.	CRISPR-Cas9	[73]
	Development of H_2O_2 based method for detection of transgene free CRISPR-Cas9 derived plants.	CRISPR-Cas9	[74]
	Development of novel CMS lines system targeting *ZmMS26* for hybrid breeding at commercial scale.	CRISPR-Cas9	[75]
	Gene pyramiding through CRISPR-Cas9.	CRISPR-Cas9	[53]
	Large chromosomal inversions.	CRISPR-Cas9	[76]
	Promotor editing of *CLE* genes for yield maximization in maize.	CRISPR-Cas9	[54]
	Creation of aromatic maize through modification of *ZmBADH2a* and *ZmBADH2b* genes.	CRISPR-Cas9	[77]
	Generation of CENH3 null mutant in maize for haploid induction.	CRISPR-Cas9	[78]
Functional characterization of genes.	*ZmCOIa* and *ZmJAZ15* confirming resistance to Gibberella stalk rot.	CRISPR-Cas9	[79]
	ZmCLCg and *ZmPMP3* genes for salt tolerance in maize.	CRISPR-Cas9	[80]
	ZmMYB and its paralog *ZmMYB2* for genic male sterility.	CRISPR-Cas9	[81]
	Knl1 gene for controlling cell division.	CRISPR-Cas9	[82]
	ZmNL4 gene for controlling leaf size.	CRISPR-Cas9	[83]
	MS2 gene for anther development.	CRISPR-Cas9	[84]
	ZmYSL2 and sh4 genes in seed development.	CRISPR-Cas9	[85]

TABLE 9.2 *(Continued)*

Aspect Studied	Detailed Description	CRISPR-Cas System	References
	ZmWUS1 gene for regulation of meristem size.	CRISPR-Cas9	[86]
	ZmCDPK7 gene for heat tolerance.	CRISPR-Cas9	[87]
Other applications.	Generation of mutant libraries.	CRISPR-Cas9	[88]
	Development of MGE platform in maize.	CRISPR-Cas9	[89]
	Development of gene identification kit for GMO specifically.	CRISPR-Cas12a	[90]
	Cytogenetic studies through CRISPR/Cas9-based RNA-guided endonuclease-*in situ* labeling (RGEN-ISL).	CRISPR-Cas9	[91]
	Development of CRISPR-FISH method which is helpful in cytogenetic studies of the maize.	CRISPR-Cas9	[92]
	Functional analysis of genes.	CRISPR-Cas9	[93]

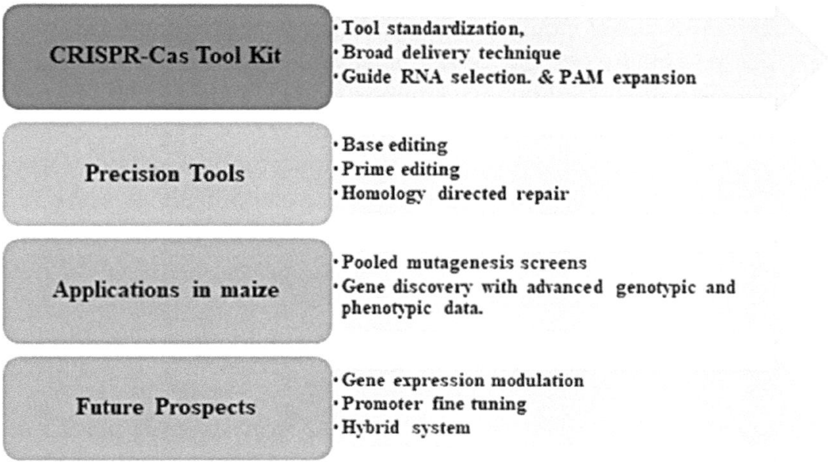

FIGURE 9.2 Crop improvement in maize using CRISPR-Cas system.

The technological development for exploitation of CRISPR-Cas system in maize is already discussed in the above section. This section covers the successful application of CRISPR-Cas system in development of climate smart and nutritionally enhanced crops. CRISPR-Cas9 was exploited for integration of drought tolerance in maize by targeting *ARGOS8* gene which reduces ethylene sensitivity and improves grain yield. The resultant plants showed an increase of five bushels per acre in yield as compared to corresponding wild type plants [96]. Further expression of *ZmCCT9* gene was repressed by targeted knocking out of through CRISPR-Cas9 to develop day neutral plants for their adaptation to higher altitude with long day lengths [97].

Maize is vulnerable to lodging due to high plant height and weak plant stature. These characteristics feature leads towards lodging and yield losses. *Gibberellin-oxidase20-3* (*GA20ox3*) gene was knocked down to suppress the production of GA precursors (GA15, GA24, GA9, GA44, and GA20). The resultant plant showed decreased plant height and increased lodging tolerance. These results were further verified through supplementation of the GA to the mutant plants which revived the wild type plants [98]. Apart from these well characterized studies, potential exists for improvement of maize crop by targeting various S genes (Which promote susceptibility). *ZmVPP1* is an S gene which promote drought stress in its wild form however this gene may be modified by insertion of the 366 bp segment in the promotor region possessing three MYB cis elements which converts drought sensitive plants to drought tolerant plants. CRISPR-Cas system may be used to integrate this modification for development of drought tolerant maize plants [99].

Hence potential exists for development of climate resilient maize crop through gene editing for biotic and abiotic stress tolerance as highlighted in Table 9.2 under the section functional characterization of genes. The advent of advance GE tools, i.e., base editing and prime editing has provided opportunity for single base pair modification and even modifications in the RNA is also possible. These developments are paving way for fast-track GE in maize for climate resilience.

9.5 REGULATORY ASPECT AND RISK ASSESSMENT OF GE CROPS

The regulation of genome edited crops is a controversial matter across various geographies. Scientists from some countries are of the view point that GE crops should be not regulated and treated equivalent to product evolved from traditional breeding techniques, if the end product is transgene free. However, some other believes that genome edited crops should be regulated following the SOPs developed for genetically modified (GM) crops. So collectively two schools of thoughts persist about regulation of GE crops, i.e., process-based regulation and end product-based regulation [100].

The supporters of the process-based regulation are of firm believe, if the transgenic technology is used at any stage of crop development, the plants to be commercialized should be regulated as GMO irrespective of fact that end product is transgene free. However, supporters of end product-based regulations states that if end product is transgene free plants to be commercialized should be released for commercial cultivation without passing through strict GMO regulations after proper safety assessments. The first model is implemented in Europe, China, India, New Zealand, and Australia which states that the products of new breeding techniques (NBTs) will be regulated on the base of process used for their development. The second relatively less stringent model of regulation is focused on the US, Argentina, and Canada which is focused on novelty of trait into question on a case-to-case basis irrespective of technology used for its development. This model has provision for registration of GE crops as non-GMOs if the end-product is free from any foreign DNA [101].

The process-based regulation of GE crops across various countries is hampering biotechnological advancement such as EU because multinational research focused companies are shifting their business to countries which have end product-based regulation, i.e., US [102]. Currently, various EU countries, i.e., Sweden and France have raised voices against process based regulatory framework [103]. Recently, French Agriculture Minister Julien Denormandie, in his interview with Reuters at the Agriculture Ministry in Paris, France stated that we see the crops developed through GE different than GMOs in contrast to EU decision to put these crops under strict GMO regulation (https://www.reuters.com/article/france-agriculture-gmo/france-backs-non-gmo-regulation-for-crop-gene-editing-in-eu-idINL8N2JT4A3). These regulatory developments are indicating that future of GE crops is brighter, and these will play a key role in achieving global food security [100].

9.6 CONCLUSION

CRISPR-Cas has increased opportunity for crop improvement by offering various genetic tools since its inception. It offers platform for precise modification of genes to understand their role. Further, it promises to increase pace of germplasm improvement in cost and time efficient manner. Functions of many maize genes have been explored through whole genome sequencing and employing various bioinformatics and transcriptomics tools. CRISPR-Cas technology offers two-way advantages, i.e., knowing the functions of unknown genes through precise mutagenesis and crop improvement through modification of S genes and converting them to beneficial alleles. CRISPR-Cas has opened up new avenues for maize breeders and geneticists for development of climate resilient and nutritional enriched crop varieties. Although, CRISPR has made various technological breakthrough however there is much more to achieve yet. The multiplex genome editing (MGE) tools have developed in maize and other crops for crop improvement. Now there is need to utilize these MGE for metabolic engineering by targeting various genes involved in different metabolic pathways. Further there is need to develop tissue culture free gene editing system to avoid laborious tissue culture process and gain high success rate.

The main bone of contention in a way of another green revolution is controversial regulatory aspect of genome edited crops. The end product-based regulation is relatively good one which and environmentally safe as well. Therefore, it is emphasized that end product-based regulation of GE crops should be practiced throughout the world for harvesting of maximum benefits from the technology. The GE crops free from transgene should be regulated like products of conventional mutagenesis and should reach to market in minimum possible time.

KEYWORDS

- **CRISPR-Cas**
- **genome editing tools**
- **global food security**
- **multiplex genome editing**
- **mutagenesis**
- **transcription activator-like effectors nucleases**
- **zinc finger nucleases**

REFERENCES

1. Nuccio, M. L., Claeys, H., & Heyndrickx, K. S., (2021). CRISPR-Cas technology in corn: A new key to unlock genetic knowledge and create novel products. *Molecular Breeding, 41*, 1–19.
2. Aslam, M., Sohail, Q., Maqbool, M., Ahmad, S., & Shahzad, R., (2017). Combining ability analysis for yield traits in diallel crosses of maize. *Journal of Animal and Plant Sciences, 27*, 136–143.
3. Yousaf, M. I., Hussain, K., Hussain, S., Shahzad, R., Ghani, A., Arshad, M., Mumtaz, A., & Akhter, N., (2017). Morphometric and phenological characterization of maize (*Zea mays* L.) germplasm under heat stress. *International Journal of Biology and Biotechnology, 14*, 271–278.
4. Yousaf, M. I., Bhatti, M. H., Ghani, A., Shehzad, A., Hussain, A., Shahzad, R., Hafeez, M. A., et al., (2021). Variation among maize (*Zea Mays* L.) hybrids in response to heat stress: Hybrid selection criteria *Turkish Journal of Field Crops, 26*, 8–17.
5. Sultana, R., Jamil, S., Aslam, M., Shahzad, R., Fatima, R., Maqbool, M. A., & Iqbal, M. Z., (2019). Overview of quality protein maize and molecular breeding approaches for its development. *International Journal of Biosciences, 14*, 533–542.
6. Jamil, S., Shahzad, R., Yasmeen, E., Rahman, S. U., Younas, M., & Iqbal, M. Z., (2020c). DNA fingerprinting of Pakistani maize hybrids and parental lines using simple sequence repeat markers. *Pak. J. Bot., 52*, 2133–2145.
7. Iqbal, M. Z., Jamil, S., Shahzad, R., Bilal, K., Qaisar, R., Nisar, A., Kanwal, S., & Bhatti, M. K., (2021b). DNA fingerprinting of crops and its applications in the field of plant breeding. *J. Agric. Res., 59*, 13–28.
8. Andorf, C., Beavis, W. D., Hufford, M., Smith, S., Suza, W. P., Wang, K., Woodhouse, M., et al., (2019). Technological advances in maize breeding: Past, present, and future. *Theoretical and Applied Genetics, 132*, 817–849.
9. Iqbal, M. Z., Jamil, S., Shahzad, R., Bilal, K., Qaisar, R., Nisar, A., Kanwal, S., & Bhatti, M. K., (2021a). DNA fingerprinting of crops and its applications in the field of plant breeding *J. Agric. Res., 59*, 13–28.

10. Jamil, S., Arshad, S., Kanwal, S., Razzaq, H., & Shahzad, R., (2019). Impact of transgenic crops on global food security: A review. *Journal of Agricultural Research, 57*(4), 245–258.
11. Jamil, S., Shahzad, R., Rahman, S. U., Iqbal, M. Z., Yaseen, M., Ahmad, S., & Fatima, R., (2020b). The level of Cry1Ac endotoxin and its efficacy against *H. armigera* in Bt cotton at large scale in Pakistan. *GM Crops & Food, 12*, 1–17.
12. Shahzad, R., Jamil, S., Alsubaie, Ghalib, T. G., Ahmad, S., Nisar, A., Rahman, S. U., Ahmad, I., & Atif, R. M. (2021a). Omics approaches for improving abiotic stress tolerance in rice: Recent advances and future prospects. In: Aftab, T., & Hakeem, K. R., (eds.), *Frontiers in Plant-Soil Interaction; Molecular Insights into Plant Adaptation,* (pp. 199–220). 125 London Wall, London EC2Y 5AS, United Kingdom: Elsevier.
13. Shahzad, R., Jamil, S., Ahmad, S., Nisar, A., Amina, Z., Saleem, S., Iqbal, M. Z., et al., (2021b). Harnessing the potential of plant transcription factors in developing climate resilient crops to improve global food security: Current and future perspectives. *Saudi Journal of Biological Sciences*.
14. Ahmad, A., Ghouri, M. Z., Munawar, N., Ismail, M., Ashraf, S., & Aftab, S. O., (2021a). Regulatory, ethical, and social aspects of CRISPR crops. in *CRISPR Crops* (pp. 261–287). Springer.
15. Ahmad, S., Wei, X., Sheng, Z., Hu, P., & Tang, S., (2020). CRISPR/Cas9 for development of disease resistance in plants: Recent progress, limitations, and future prospects. *Briefings in Functional Genomics, 19*, 26–39.
16. Monsur, M. B., Shao, G., Lv, Y., Ahmad, S., Wei, X., Hu, P., & Tang, S., (2020). Base editing: The ever expanding clustered regularly interspaced short palindromic repeats (CRISPR) tool kit for precise genome editing in plants. *Genes, 11*, 466.
17. Zafar, S. A., Zaidi, S. S. E. A., Gaba, Y., Singla-Pareek, S. L., Dhankher, O. P., Li, X., Mansoor, S., & Pareek, A., (2020). Engineering abiotic stress tolerance via CRISPR/Cas-mediated genome editing. *Journal of Experimental Botany, 71*, 470–479.
18. Paschon, D. E., Lussier, S., Wangzor, T., Xia, D. F., Li, P. W., Hinkley, S. J., Scarlott, N. A., et al., (2019). Diversifying the structure of zinc finger nucleases for high-precision genome editing. *Nature Communications, 10*, 1–12.
19. Fiaz, S., Ahmad, S., Noor, M. A., Wang, X., Younas, A., Riaz, A., Riaz, A., & Ali, F., (2019). Applications of the CRISPR/Cas9 system for rice grain quality improvement: Perspectives and opportunities. *International Journal of Molecular Sciences, 20*, 888.
20. Kamburova, V. S., Nikitina, E. V., Shermatov, S. E., Buriev, Z. T., Kumpatla, S. P., Emani, C., & Abdurakhmonov, I. Y., (2017). Genome editing in plants: An overview of tools and applications. *International Journal of Agronomy, 2017*.
21. Ran, Y., Patron, N., Kay, P., Wong, D., Buchanan, M., Cao, Y. Y., Sawbridge, T., et al., (2018). Zinc finger nuclease-mediated precision genome editing of an endogenous gene in hexaploid bread wheat (*Triticum aestivum*) using a DNA repair template. *Plant Biotechnology Journal, 16*, 2088–2101.
22. Shukla, V. K., Doyon, Y., Miller, J. C., Dekelver, R. C., Moehle, E. A., Worden, S. E., Mitchell, J. C., et al., (2009). Precise genome modification in the crop species *Zea mays* using zinc-finger nucleases. *Nature, 459*, 437–441.
23. Jung, Y. J., Nogoy, F. M., Lee, S. K., Cho, Y. G., & Kang, K. K., (2018). Application of ZFN for site directed mutagenesis of rice SSIVa gene. *Biotechnology and Bioprocess Engineering, 23*, 108–115.

24. Boch, J., Scholze, H., Schornack, S., Landgraf, A., Hahn, S., Kay, S., Lahaye, T., Nickstadt, A., & Bonas, U., (2009). Breaking the code of DNA binding specificity of TAL-type III effectors. *Science, 326*, 1509–1512.
25. Doyle, E. L., Booher, N. J., Standage, D. S., Voytas, D. F., Brendel, V. P., Vandyk, J. K., & Bogdanove, A. J., (2012). TAL effector-nucleotide targeter (TALE-NT) 2.0: Tools for TAL effector design and target prediction. *Nucleic Acids Research, 40*, W117–W122.
26. Li, T., Huang, S., Zhao, X., Wright, D. A., Carpenter, S., Spalding, M. H., Weeks, D. P., & Yang, B., (2011). Modularly assembled designer TAL effector nucleases for targeted gene knockout and gene replacement in eukaryotes. *Nucleic Acids Research, 39*, 6315–6325.
27. Zhang, Z., Mao, Y., Ha, S., Liu, W., Botella, J. R., & Zhu, J. K., (2016). A multiplex CRISPR/Cas9 platform for fast and efficient editing of multiple genes in *Arabidopsis. Plant Cell Reports, 35*, 1519–1533.
28. Haun, W., Coffman, A., Clasen, B. M., Demorest, Z. L., Lowy, A., Ray, E., Retterath, A., et al., (2014). Improved soybean oil quality by targeted mutagenesis of the fatty acid desaturase 2 gene family. *Plant Biotechnology Journal, 12*, 934–940.
29. Ansari, W. A., Chandanshive, S. U., Bhatt, V., Nadaf, A. B., Vats, S., Katara, J. L., Sonah, H., & Deshmukh, R., (2020). Genome editing in cereals: Approaches, applications, and challenges. *International Journal of Molecular Sciences, 21*, 4040.
30. Shmakov, S., Smargon, A., Scott, D., Cox, D., Pyzocha, N., Yan, W., Abudayyeh, O. O., et al., (2017). Diversity and evolution of class 2 CRISPR–Cas systems. *Nature Reviews Microbiology, 15*, 169–182.
31. Makarova, K. S., Wolf, Y. I., Alkhnbashi, O. S., Costa, F., Shah, S. A., Saunders, S. J., Barrangou, R., et al., (2015). An updated evolutionary classification of CRISPR–Cas systems. *Nature Reviews Microbiology, 13*, 722–736.
32. Jackson, S. A., Mckenzie, R. E., Fagerlund, R. D., Kieper, S. N., Fineran, P. C., & Brouns, S. J., (2017). CRISPR-Cas: Adapting to change. *Science, 356.*
33. Kim, D., Alptekin, B., & Budak, H., (2018). CRISPR/Cas9 genome editing in wheat. *Functional & Integrative Genomics, 18*, 31–41.
34. Feng, C., Yuan, J., Wang, R., Liu, Y., Birchler, J. A., & Han, F., (2016). Efficient targeted genome modification in maize using CRISPR/Cas9 system. *Journal of Genetics and Genomics, 43*, 37–43.
35. Haque, E., Taniguchi, H., Hassan, M., Bhowmik, P., Karim, M. R., Śmiech, M., Zhao, K., et al., (2018). Application of CRISPR/Cas9 genome editing technology for the improvement of crops cultivated in tropical climates: Recent progress, prospects, and challenges. *Frontiers in Plant Science, 9*, 617.
36. Bharat, S. S., Li, S., Li, J., Yan, L., & Xia, L., (2019). Base editing in plants: Current status and challenges. *The Crop Journal.*
37. Komor, A. C., Kim, Y. B., Packer, M. S., Zuris, J. A., & Liu, D. R., (2016). Programmable editing of a target base in genomic DNA without double-stranded DNA cleavage. *Nature, 533*, 420–424.
38. Nishida, K., Arazoe, T., Yachie, N., Banno, S., Kakimoto, M., Tabata, M., Mochizuki, M., et al., (2016). Targeted nucleotide editing using hybrid prokaryotic and vertebrate adaptive immune systems. *Science, 353.*
39. Cox, D. B., Gootenberg, J. S., Abudayyeh, O. O., Franklin, B., Kellner, M. J., Joung, J., & Zhang, F., (2017). RNA editing with CRISPR-Cas13. *Science, 358*, 1019–1027.

40. Abudayyeh, O. O., Gootenberg, J. S., Franklin, B., Koob, J., Kellner, M. J., Ladha, A., Joung, J., et al., (2019). A cytosine deaminase for programmable single-base RNA editing. *Science, 365*, 382–386.
41. Anzalone, A. V., Randolph, P. B., Davis, J. R., Sousa, A. A., Koblan, L. W., Levy, J. M., Chen, P. J., et al., (2019). Search-and-replace genome editing without double-strand breaks or donor DNA. *Nature, 576*, 149–157.
42. Xing, H. L., Dong, L., Wang, Z. P., Zhang, H. Y., Han, C. Y., Liu, B., Wang, X. C., & Chen, Q. J., (2014). A CRISPR/Cas9 toolkit for multiplex genome editing in plants. *BMC Plant Biology, 14*, 1–12.
43. Svitashev, S., Young, J. K., Schwartz, C., Gao, H., Falco, S. C., & Cigan, A. M., (2015). Targeted mutagenesis, precise gene editing, and site-specific gene insertion in maize using Cas9 and guide RNA. *Plant Physiology, 169*, 931–945.
44. Zhu, J., Song, N., Sun, S., Yang, W., Zhao, H., Song, W., & Lai, J., (2016). Efficiency and inheritance of targeted mutagenesis in maize using CRISPR-Cas9. *Journal of Genetics and Genomics, 43*, 25–36.
45. Svitashev, S., Schwartz, C., Lenderts, B., Young, J. K., & Cigan, A. M., (2016). Genome editing in maize directed by CRISPR–Cas9 ribonucleoprotein complexes. *Nature Communications, 7*, 1–7.
46. Koiso, N., Toda, E., Ichikawa, M., Kato, N., & Okamoto, T., (2017). Development of gene expression system in egg cells and zygotes isolated from rice and maize. *Plant Direct, 1*, e00010.
47. Lee, K., Zhang, Y., Kleinstiver, B. P., Guo, J. A., Aryee, M. J., Miller, J., Malzahn, A., et al., (2019a). Activities and specificities of CRISPR/Cas9 and Cas12a nucleases for targeted mutagenesis in maize. *Plant Biotechnology Journal, 17*, 362–372.
48. Gong, C., Huang, S., Song, R., & Qi, W., (2021). Comparative study between the CRISPR/Cpf1 (Cas12a) and CRISPR/Cas9 systems for multiplex gene editing in maize. *Agriculture, 11*, 429.
49. Kelliher, T., Starr, D., Su, X., Tang, G., Chen, Z., Carter, J., Wittich, P. E., et al., (2019). One-step genome editing of elite crop germplasm during haploid induction. *Nature Biotechnology, 37*, 287–292.
50. Yan, Y., Zhu, J., Qi, X., Cheng, B., Liu, C., & Xie, C., (2021). Establishment of an efficient seed fluorescence reporter-assisted CRISPR/Cas9 gene editing in maize. *Journal of Integrative Plant Biology*.
51. Li, C., Liu, C., Qi, X., Wu, Y., Fei, X., Mao, L., Cheng, B., et al., (2017a). RNA-guided Cas9 as an *in vivo* desired-target mutator in maize. *Plant Biotechnology Journal, 15*, 1566–1576.
52. Chen, R., Xu, Q., Liu, Y., Zhang, J., Ren, D., Wang, G., & Liu, Y., (2018). Generation of transgene-free maize male sterile lines using the CRISPR/Cas9 system. *Frontiers in Plant Science, 9*, 1180.
53. Gao, H., Mutti, J., Young, J. K., Yang, M., Schroder, M., Lenderts, B., Wang, L., et al., (2020). Complex trait loci in maize enabled by CRISPR-Cas9 mediated gene insertion. *Frontiers in Plant Science, 11*, 535.
54. Liu, L., Gallagher, J., Arevalo, E. D., Chen, R., Skopelitis, T., Wu, Q., Bartlett, M., & Jackson, D., (2021). Enhancing grain-yield-related traits by CRISPR–Cas9 promoter editing of maize CLE genes. *Nature Plants, 7*, 287–294.
55. Char, S. N., Neelakandan, A. K., Nahampun, H., Frame, B., Main, M., Spalding, M. H., Becraft, P. W., et al., (2017). An *Agrobacterium*-delivered CRISPR/Cas9 system

for high-frequency targeted mutagenesis in maize. *Plant Biotechnology Journal, 15*, 257–268.

56. Zong, Y., Wang, Y., Li, C., Zhang, R., Chen, K., Ran, Y., Qiu, J. L., et al., (2017). Precise base editing in rice, wheat, and maize with a Cas9-cytidine deaminase fusion. *Nature Biotechnology, 35*, 438.
57. Feng, C., Su, H., Bai, H., Wang, R., Liu, Y., Guo, X., Liu, C., et al., (2018). High-efficiency genome editing using a dmc1 promoter-controlled CRISPR/Cas9 system in maize. *Plant Biotechnology Journal, 16*, 1848–1857.
58. Young, J., Zastrow-Hayes, G., Deschamps, S., Svitashev, S., Zaremba, M., Acharya, A., Paulraj, S., et al., (2019). CRISPR-Cas9 editing in maize: Systematic evaluation of off-target activity and its relevance in crop improvement. *Scientific Reports, 9*, 1–11.
59. Lee, K., Zhu, H., Yang, B., & Wang, K., (2019b). An *Agrobacterium*-mediated CRISPR/Cas9 platform for genome editing in maize. In *Plant Genome Editing with CRISPR Systems* pp. 121–143). Springer.
60. Doll, N. M., Gilles, L. M., Gérentes, M. F., Richard, C., Just, J., Fierlej, Y., Borrelli, V. M., et al., (2019). Single and multiple gene knockouts by CRISPR–Cas9 in maize. *Plant Cell Reports, 38*, 487–501.
61. Zhang, Q., Zhang, Y., Lu, M. H., Chai, Y. P., Jiang, Y. Y., Zhou, Y., Wang, X. C., & Chen, Q. J., (2019). A novel ternary vector system united with morphogenic genes enhances CRISPR/Cas delivery in maize. *Plant Physiology, 181*, 1441–1448.
62. Barone, P., Wu, E., Lenderts, B., Anand, A., Gordon-Kamm, W., Svitashev, S., & Kumar, S., (2020). Efficient gene targeting in maize using inducible CRISPR-Cas9 and marker-free donor template. *Molecular Plant, 13*, 1219–1227.
63. Gentzel, I. N., Park, C. H., Bellizzi, M., Xiao, G., Gadhave, K. R., Murphree, C., Yang, Q., et al., (2020). A CRISPR/dCas9 toolkit for functional analysis of maize genes. *Plant Methods, 16*, 1–9.
64. Xu, J., Yin, Y., Jian, L., Han, B., Chen, Z., Yan, J., & Liu, X., (2021a). Seeing is believing: A visualization toolbox to enhance selection efficiency in maize genome editing. *Plant Biotechnology Journal, 19*, 872.
65. Kelliher, T., Starr, D., Richbourg, L., Chintamanani, S., Delzer, B., Nuccio, M. L., Green, J., et al., (2017). MATRILINEAL, a sperm-specific phospholipase, triggers maize haploid induction. *Nature, 542*, 105–109.
66. Malzahn, A. A., Tang, X., Lee, K., Ren, Q., Sretenovic, S., Zhang, Y., Chen, H., et al., (2019). Application of CRISPR-Cas12a temperature sensitivity for improved genome editing in rice, maize, and *Arabidopsis*. *BMC Biology, 17*, 1–14.
67. Mei, Y., Beernink, B. M., Ellison, E. E., Konečná, E., Neelakandan, A. K., Voytas, D. F., & Whitham, S. A., (2019). Protein expression and gene editing in monocots using foxtail mosaic virus vectors. *Plant Direct, 3*, e00181.
68. Hu, J., Li, S., Li, Z., Li, H., Song, W., Zhao, H., Lai, J., et al., (2019). A barley stripe mosaic virus-based guide RNA delivery system for targeted mutagenesis in wheat and maize. *Molecular Plant Pathology, 20*, 1463–1474.
69. Peterson, D., Barone, P., Lenderts, B., Schwartz, C., Feigenbutz, L., St. Clair, G., Jones, S., & Svitashev, S., (2021). Advances in *Agrobacterium* transformation and vector design result in high frequency targeted gene insertion in maize. *Plant Biotechnology Journal.*
70. Hunter, C. T., (2021). CRISPR/Cas9 targeted mutagenesis for functional genetics in maize. *Plants, 10*, 723.

71. Li, J., Sun, Y., Du, J., Zhao, Y., & Xia, L., (2017b). Generation of targeted point mutations in rice by a modified CRISPR/Cas9 system. *Molecular Plant, 10*, 526–529.
72. Dong, L., Li, L., Liu, C., Liu, C., Geng, S., Li, X., Huang, C., et al., (2018). Genome editing and double-fluorescence proteins enable robust maternal haploid induction and identification in maize. *Molecular Plant, 11*, 1214–1217.
73. Wang, B., Zhu, L., Zhao, B., Zhao, Y., Xie, Y., Zheng, Z., Li, Y., et al., (2019). Development of a haploid-inducer mediated genome editing system for accelerating maize breeding. *Molecular Plant, 12,* 597–602.
74. Wu, T. M., Huang, J. Z., Oung, H. M., Hsu, Y. T., Tsai, Y. C., & Hong, C. Y., (2019). H_2O_2-based method for rapid detection of transgene-free rice plants from segregating CRISPR/Cas9 genome-edited progenies. *International Journal of Molecular Sciences, 20*, 3885.
75. Qi, X., Zhang, C., Zhu, J., Liu, C., Huang, C., Li, X., & Xie, C., (2020). Genome editing enables next-generation hybrid seed production technology. *Molecular Plant, 13*, 1262–1269.
76. Schwartz, C., Lenderts, B., Feigenbutz, L., Barone, P., Llaca, V., Fengler, K., & Svitashev, S., (2020). CRISPR–Cas9-mediated 75.5-Mb inversion in maize. *Nature Plants, 6*, 1427–1431.
77. Wang, Y., Liu, X., Zheng, X., Wang, W., Yin, X., Liu, H., Ma, C., et al., (2021b). Creation of aromatic maize by CRISPR/Cas. *Journal of Integrative Plant Biology.*
78. Wang, N., Gent, J. I., & Dawe, R. K., (2021a). Haploid induction by a maize cenh3 null mutant. *Science Advances, 7*, eabe2299.
79. Ma, L., Sun, Y., Ruan, X., Huang, P. C., Wang, S., Li, S., Zhou, Y., et al., (2021). Genome-wide characterization of jasmonates signaling components reveals the essential role of ZmCOI1a-ZmJAZ15 action module in regulating maize immunity to *Gibberella* stalk rot. *International Journal of Molecular Sciences, 22*, 870.
80. Luo, M., Zhang, Y., Li, J., Zhang, P., Chen, K., Song, W., Wang, X., et al., (2021). Molecular dissection of maize seedling salt tolerance using a genome-wide association analysis method. *Plant Biotechnology Journal.*
81. Li, Z., Zhu, T., Liu, S., Jiang, Y., Liu, H., Zhang, Y., Xie, K., et al., (2021). Genome-wide analyses on transcription factors and their potential microRNA regulators involved in maize male fertility. *The Crop Journal.*
82. Su, H., Liu, Y., Wang, C., Liu, Y., Feng, C., Sun, Y., Yuan, J., et al., (2021). Knl1 participates in spindle assembly checkpoint signaling in maize. *Proceedings of the National Academy of Sciences, 118*.
83. Gao, L., Yang, G., Li, Y., Sun, Y., Xu, R., Chen, Y., Wang, Z., et al., (2021). A kelch-repeat superfamily gene, ZmNL4, controls leaf width in maize (*Zea mays* L.). *The Plant Journal.*
84. Xu, Q., Yang, L., Kang, D., Ren, Z., & Liu, Y., (2021b). Maize MS2 encodes an ATP-binding cassette transporter that is essential for another development. *The Crop Journal.*
85. He, Y., Yang, Q., Yang, J., Wang, Y. F., Sun, X., Wang, S., Qi, W., et al., (2021). shrunken4 is a mutant allele of ZmYSL2 that affects aleurone development and starch synthesis in maize. *Genetics.*
86. Chen, Z., Li, W., Gaines, C., Buck, A., Galli, M., & Gallavotti, A., (2021). Structural variation at the maize WUSCHEL1 locus alters stem cell organization in inflorescences. *Nature Communications, 12*, 1–12.

87. Zhao, Y., Du, H., Wang, Y., Wang, H., Yang, S., Li, C., Chen, N., et al., (2021). The calcium-dependent protein kinase ZmCDPK7 functions in heat-stress tolerance in maize. *Journal of Integrative Plant Biology, 63*, 510–527.
88. Liu, J., Fernie, A. R., & Yan, J., (2020). The past, present, and future of maize improvement: Domestication, genomics, and functional genomic routes toward crop enhancement. *Plant Communications, 1*, 100010.
89. Qi, W., Zhu, T., Tian, Z., Li, C., Zhang, W., & Song, R., (2016). High-efficiency CRISPR/Cas9 multiplex gene editing using the glycine tRNA-processing system-based strategy in maize. *BMC Biotechnology, 16*, 1–8.
90. Wu, H., Qian, C., Wu, C., Wang, Z., Wang, D., Ye, Z., Ping, J., et al., (2020). End-point dual specific detection of nucleic acids using CRISPR/Cas12a based portable biosensor. *Biosensors and Bioelectronics, 157*, 112153.
91. Němečková, A., Wäsch, C., Schubert, V., Ishii, T., Hřibová, E., & Houben, A., (2019). CRISPR/Cas9-based RGEN-ISL allows the simultaneous and specific visualization of proteins, DNA repeats, and sites of DNA replication. *Cytogenetic and Genome Research, 159*, 48–53.
92. Potlapalli, B. P., Schubert, V., Metje-Sprink, J., Liehr, T., & Houben, A., (2020). Application of Tris-HCl allows the specific labeling of regularly prepared chromosomes by CRISPR-FISH. *Cytogenetic and Genome Research, 160*, 156–165.
93. Lei, H. Y., Zhao, Q. S., Bai, F. L., Song, H. F., & Wang, Z. J., (2021). Identification of developing-related gene ZmCen using CRISPR/Cas9 in maize. *China Biotechnology, 40*, 49–57.
94. Jamil, S., Shahzad, R., Ahmad, S., Fatima, R., Zahid, R., Anwar, M., Iqbal, M. Z., & Wang, X., (2020a). Role of genetics, genomics and breeding approaches to combat stripe rust of wheat. *Frontiers in Nutrition, 7*, 173.
95. Jamil, S., Shahzad, R., Rahman, S. U., Iqbal, M. Z., Yaseen, M., Ahmad, S., & Fatima, R., (2021). The level of Cry1Ac endotoxin and its efficacy against *H. armigera* in Bt cotton at large scale in Pakistan. *GM Crops & Food, 12*, 1–17.
96. Shi, J., Gao, H., Wang, H., Lafitte, H. R., Archibald, R. L., Yang, M., Hakimi, S. M., et al., (2017). ARGOS 8 variants generated by CRISPR-Cas9 improve maize grain yield under field drought stress conditions. *Plant Biotechnology Journal, 15*, 207–216.
97. Huang, C., Sun, H., Xu, D., Chen, Q., Liang, Y., Wang, X., Xu, G., et al., (2018). ZmCCT9 enhances maize adaptation to higher latitudes. *Proceedings of the National Academy of Sciences, 115*, E334–E341.
98. Zhang, J., Zhang, X., Chen, R., Yang, L., Fan, K., Liu, Y., Wang, G., Ren, Z., & Liu, Y., (2020). Generation of transgene-free semidwarf maize plants by gene editing of gibberellin-oxidase20-3 using CRISPR/Cas9. *Frontiers in Plant Science, 11*, 1048.
99. Liu, X., Zhang, C., Wang, X., Liu, Q., Yuan, D., Pan, G., Sun, S. S., & Tu, J., (2016). Development of high-lysine rice via endosperm-specific expression of a foreign lysine rich protein gene. *BMC Plant Biology, 16*, 147.
100. Ahmad, S., Shahzad, R., Jamil, S., Tabassum, J., Chaudhary, M. A. M., Atif, R. M., Iqbal, M. M., et al., (2021b). Regulatory aspects, risk assessment, and toxicity associated with RNAi and CRISPR methods. In *CRISPR and RNAi Systems* (pp. 687–721). Elsevier.
101. Schiemann, J., Robienski, J., Schleissing, S., Spök, A., Sprink, T., & Wilhelm, R. A., (2020). Plant genome editing–policies and governance. *Frontiers in Plant Science, 11*.

102. Anyshchenko, A., & Yarnold, J., (2020). From 'mad cow' crisis to synthetic biology: Challenges to EU regulation of GMOs beyond the European context. *International Environmental Agreements: Politics, Law, and Economics,* 1–14.
103. Eriksson, D., De Andrade, E., Bohanec, B., Chatzopolou, S., Defez, R., Eriksson, N. L., Van, D. M. P., et al., (2018). Why the European Union needs a national GMO opt-in mechanism. *Nature Biotechnology, 36*, 18, 19.

CHAPTER 10

HARNESSING THE POTENTIAL OF GENOMIC EDITION TOOLS IN SOYBEAN IMPROVEMENT: PROGRESS AND PROSPECTS

MUHAMMAD AFZAL,[1] SALEM SAFER ALGHAMDI,[1] HUSSEIN H. MIGDADI,[1,2] HIRA NAWAZ,[3] MUHAMMAD ALTAF KHAN,[1] and EHAB ELHARTY[1]

[1]*College of Food and Agriculture Sciences, Plant Production Department, King Saud University, Riyadh, Saudi Arabia*

[2]*National Agricultural Research Center, Baqa, Jordan*

[3]*Plant Pathology, Faculty of Agriculture, University of Agriculture, Faisalabad, Pakistan*

ABSTRACT

Soybean (*Glycine max*) is an economically valuable, leguminous crop that provides food to humans and animal feed rich in protein and oils. Genetic advancement in Soybean must be accelerated to gratify the constantly increased demand for soybean products and the changing climate. This increased demand for plant protein would require new, environment-friendly, and sustainable agricultural practices. Conventional approaches have greatly increased crop production, yield, and quality, but new approaches to developing legumes are required to meet the global food needs. The increasingly evident technologies in the field of genomic edition (ZFNs and TALENs), as well as the clustered regularly interspaced short palindromic repeats/CRISPR-associated proteins (CRISPR/Cas), have shown wide potential

CRISPR-Based Gene Editing: A Solution for Global Food Security. Shakeel Ahmad, Asif Ali Khan, and Muhammad Arshad Javed (Eds.)

applications for targeted genes functioning in research and improvement of essential agronomic parameters in several crops in recent years and CRISPR/Cas system has also shown wide potential applications for genome research, as well as improvement in agricultural traits of many crops. We summarize the importance, challenges, current application, and potential perspectives of the soybean genome editing (GE) tools that can be used to help for the development of the soybean.

10.1 INTRODUCTION

Today in the world almost a billion people suffer from chronic malnutrition, while the loss in biodiversity and the growing uncertainties of climate change degrade our farming systems at the same time [1]. With the global population expected to exceed 9 billion by 2050, today's agriculture will face huge challenges, demanding higher yields, better quality plants, and less input [2]. The U.N. sustainable development goals adopted in 2015 include eliminating hunger and ensuring global food security by 2030. For the world's population to be fed and hunger to be eradicated by 2030, food production must increase and be more consistent [3]. Although conventional breeding needs a labor force and usually takes several years for progress toward commercial varying crosses from the initial stages of screening phenotypes and genotypes. The transfer of genes, (transgenes) or gene elements of known function to the elite crops is the product of genetically modified crops (GM) with beneficial traits.

Although GM crops are assured to maintain global food security, their use is largely affected by non-founded safety and health issues. Governments have created significant cost barriers to the rapid and widespread adoption of new GM traits by regulatory frameworks that aim to protect human and environmental biosafety [92]. Consequently, there are only a few cultivated plant species with the advantages of GM traits. The editing of the genome is defined as a collection of advanced molecular biology techniques, which enable accurate, effective, and targeted changes on the specific locus [4]. The mechanism of genomic editing (GE) technology involves the use of artificial sequence-specific nuclease (SSNs) for the introduction of double-stranded breaks (SSB) within the target site of the genome, which is subsequently repaired through non-homologous end joining (NHEJs) [5].

Currently, important techniques like zinc finger nucleases (ZFNs), transcription activator-like effectors nuclease, and clustered regularly interspaced short palindromic repeats (CRISPR) [6, 7]. Despite their initial development, ZFN's and TALEN's are intricate and expensive so its extensive use is limited. The simplicity of the CRISPR systems has become famous in life/allied studies. The significant technique is CRISPR/Cas9 [93]. The toolbox for genome modification has been expanded with the selection of Cas9 orthologs and designed variants that represent the CRISPR/Cas9 system [8]. The dead nickase Cas9 and Cas9 were widely used for the basic editing, regulation of gene expression, editing epigenome, cell image, and further field of application [9]. Fortunately, genome editing (GE) techniques, such as CRISPR/Cas9, have transformed crop development by permitting robust and accurate genome alterations [10].

Soybean (*Glycine Max* L.) is an important crop for humans and animals as used as food it is a rich source of oil and protein. The use of conventional breeding approaches simply is inadequate to meet the growing demand for soy products and challenges to the agriculture environment. Traits that involve production, quality, and stress-tolerance, or resistance need a quick and accurate introduction of efficient breeding strategies to grow better species [11]. The emerging GE technology recently demonstrated its great value through its efficiency, accuracy, and user-friendly in the fundamental research and application of many crops in genetic modification [12]. Furthermore, through gene-editing tools, many new varieties of crops have been generated that are more resistant to abiotic stresses. The choice of a gene for targeted gene editing is extremely important. GET is mostly used to target the negative regulators of the desired feature [13]. Applications of these technologies in soybean are comparatively less in comparison with rice and other field crops [14]. All these technologies use typical SSNs for recognizing and generating dual-strand breaks (DSBs). DSBs can be either fixed by non-homologous termination joining (NHEJ), which can produce genetic knockouts or genetic recombination (HR), and which may cause gene replacement and insertion. The plant's endogenous reparatory systems can delete the DSB [15]. A wide range of plants have used many gene-knockout mutants, gene substitutes, inserted mutants, and many of them have been proven to be useful for crop improvements. Other methods for improving crop plants, i.e., CRISPR/Cas12 and base editing have been used in the future for recombinase and primary editing tools that will allow precise and efficient editing [16]. Base editors can create accurate point mutations by performing a permanent base conversion at an exact location, with very low levels. Various nucleobase deaminases have been fused with Cas9, Cas13, or Cas12a (Cpf1) proteins are used to create different plant base editors [17].

CRISPR and RNAi, two modern biotechniques, have played a key role in agricultural development, medicine, and biotechnology. These systems are successful because they regulate the final product and have minimal toxicities. GMOs (genetically modified organisms) crops, for example, are generally subjected to thorough risk evaluations, and toxicity tests [18]. The risk of altering the genome through genome-editing technology is much lower than that associated with GM crops because most editing only alters a few nucleotides and produces changes that do not differ in all-natural populations [19]. However, the toxicity in RNAi and CRISPR base editing is less in crops [18]. There is no way to differentiate between a 'naturally occurring' mutation and a gene editor once the genomic-editing agents have been separated. Thus, it must be possible to improve crops rapidly and precisely by introducing GE into modern breeding programs.

This is partly because it recalcitrates the processing and regeneration of Agrobacterium, which seriously hindered recent biotechnology, includes genomic edition, to improve soybean [20, 21]. Furthermore, the development of a diploid forms an allotetraploid also limits the investigation of soybeans with gene disorder. The present review summarizes the prospects of soybean genomic edition, importance, current application, challenge, and based on recent advance in genome-editing technology, to provide theoretical reference for better use to enhance the fundamental study of soybean.

10.2 CHALLENGES FOR FOOD SECURITY

The growing global population poses a threat to food security objectives that are further complicated by climate change and restricted arable land because of desertification, salinity, human intervention, and disease. Conventional food production methods are inadequate because of abiotic stresses [22]. To protect plants from changing climate hazards and abiotic threats, breeders have targeted genes that are relevant to these specific characteristics and favored plants that are better able to resist abiotic stress conditions. Researchers also have targeted genes that have been exposed to abiotic stress (i.e., salt, drought, and toxic metals). The growth of pathogens, an accelerated threat from weather changes, can contribute to globalization and trade. Moreover, it also includes biotic resistance in the form of fungi, bacteria, viruses, and nematodes were also threats for a global production reduction [23]. Despite the efforts of several government policies to promote the growth of vegetables through subsidies, there have been decreases in the production of legumes, including common beans. Soya is the only legume with a surface area of 3.45 million ha in 2012 that grows more in Europe than all other legumes in 2012. However, legumes are growing in America, Oceania, and cultivations are traded worldwide [24]. Selective breeding of insects and resistance to diseases in some plants has progressed. However, legumes showed lower yield gains than major grain crops because of few capital crops, long selection cycles, failure to capitalize on harvest index, investment in sprouting, and high genotype X environment (G x E). There is a serious need for persistent feature adjustments to stressful conditions and compatibility with other crops to increasing yield and quality. The use of genetic resource use in particular integrated approaches, e.g., marker-assisted selection, new breeding technologies (NBT), and feature engineering, i.e., CRISPR/Cas9 can address these needs through genetic yield gains and other complex properties [25]. These techniques can be used also to optimize the nutritional content of crops through optimizing the profiles of human vitamins and minerals. Genomic techniques have allowed the alteration of key metabolic pathways and, as a result, increased vitamin, and iron in addition to decreasing compounds, i.e., phytic acids and amino acid acrylamide [26]. Several fortified crops, such as wheat, maize, and rice (golden rice) have been developed to solve nutrient deficit problems, i.e., β-carotene [27].

10.3 DEVELOPMENTS IN GENOMIC EDITION

10.3.1 ZFNS AND TALENS

The ZFN is a basic GE tool which was first developed. First-generation GE uses the TALENs and ZFNs, procedures that are time-consuming and labor-intensive. The zinc finger protein (ZFP) array and a nuclease FokI are comprised of a customized C2H2 module [28]. The ZFP array can recognize a particular DNA sequence as a DNA binding domain while the FokI nuclease reduces the target DNA site to DSB. A single C_2H_2 ZFP contains 30 amino acids. A stable ββα structure folds into zinc ion chelating, with the C_2H_2 residue conserved [29]. Each ZFP is linked to a triple nucleotide in the DNA sequence by the α-helix side chain. The triplet specificity can be changed by using different ZFP's, and a ZFP array can be developed to identify a 9 to 12 bp long sequence by linking 3–4 ZFPs to a tandem [28]. A ZFP array is fused in to form a ZFN module with a FokI nuclease. As FokI, only by building a dimer can exercise its endonuclease activity [30]. The basic model and illustration of ZFNs and TALENs were presented in Figure 10.1.

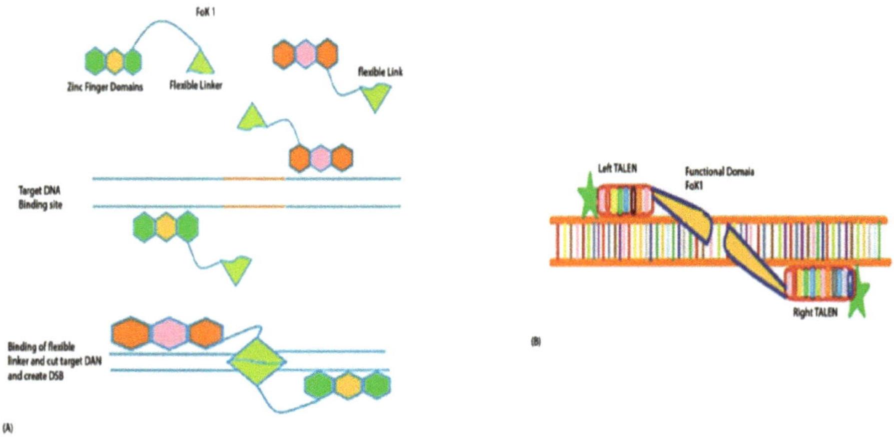

FIGURE 10.*1* (A) Illustration of the basic structure of zinc finger nuclease (ZFN). DNA-binding regions are joined to the catalytic domain of the nonspecific Fok1 endonuclease to generate ZFNs. DNA can be bound by ZFN, which can be predetermined in terms of sequence. This requires dimerization, therefore the tailored ZFNs must work in pairs to function. To activate Fok1 dimerization, the zinc finger-binding domain arranges two Fok1 units in the proper orientation over the target sequence. (B) Illustration of the basic structure of TALENs. There are repetitive domains in the DNA-binding domain (DBD). There are 34 amino acid residues in each repeat domain, and each repeat domain recognizes one nucleotide. They both (left and right TALENs) have a wild-type FokI domain, which makes them homodimeric. TALENs contain a mutated FoK1 domain.

The inverted DSB of the genome is required to produce two ZFN modules (with a 5–7 bp gap) was present [31]. ZFN's for plants and animals were also castoff for GE 10 years ago [32]. The TALEN system was built with an 18–20 bp nucleotide such as ZFN. TALEN requires a 14–20 bp distance FOOK1 dimerization [33]. However, due to their base-pair size, TALENs are shown to be highly specific to the required areas, compared with ZFNs. In addition, due to the initial thymidine foundations, their wide size, and repeating nature, the TALENs are easy to design and assemble. Since its development, ZFNs are used in many plants to target genes and effective genetic mutations have been created in maize [34], Arabidopsis Thaliana [35], soybean [36], and Tobacco [37]. Moreover, the successful applications of TALENs were conducted in plants like *Brachypodium* [38], tobacco [39], rice [40], Arabidopsis [41] and Soybean [42] was reported. However, because of the unlimited difficulties and high cost of this editing, the large applications of ZFNs are restricted in plants. Additionally, new, and convenient GE have been developed and more varieties have been made. Some important progress with mega-nucleases system was presented in Table 10.1.

TABLE 10.1 The Traits Improved Using Meganucleases System in Soybean

Editing Type	Specific Gene	Repair Type	Traits Improved	References
TALENs	FAD2-1A, FAD2-1B	NHEJ	High oleic acid contents.	[42]
TALENs	FAD2-1A, FAD2-1B, FAD3A	NHEJ	High oleic, low linoleic contents.	[43]
CRISPR/Cas9	ALS	HR	Herbicide resistance	[44]
RGEN	–	Biolistic agrobacterium	Oil quality	[94]
ZFNs	–	Agrobacterium	Gene mutation	[95]

10.3.2 CRISPR/CAS SYSTEM

CRISPR/Cas is a prokaryotic system that is adaptive to the recognition and destruction of phages and other pathogens in bacteria [45]. Applications in GE and the re-regulation of desirable genes in multiple crop plants are currently being studied in the GE tool CRISPR/Cas9 [38]. The construction of the Cas effectors modules has been divided into two classes. Class 1 contains multi-sub-unit Cas protein complexes of type-I, type-III, and type-IV. Class 2 includes type-II, type-V, and type-VI, and only one larger Cas proteins are available. Both classes require CRISPR RNA (crRNA) to lead Cas nuclease to its target place to cut targeted sequences [46]. The Cas9 system represents type II, composed of 3 different parts, i.e., crRNA, tracrRNA, and Cas9. The illustration of the basic CRISPR/Cas9 systems is presented in Figure 10.2. CRISPR, first discovered in 1987 from the Escherichia Coli genome, consists of 29 repeated nucleotide fragments in length, along with 32 nucleotides interspersed variable fragments. In 2012, cr-RNA and tracrRNA were integrated into a single guide (sg-RNA) and the targeted division of the Cas9/sg-RNA system was efficiently implemented [47]. The target specification can only be attained by altering the combination of the sg-RNA sequence combined with the targeted DNA in this simplified system that makes it low-cost and easy to use to ZFNs and TALENs. In addition,

Cas9, a screened marker, and several sg-RNA expressions cases can then assemble in a vector to produce flexible tools for multiple genomes editing [47].

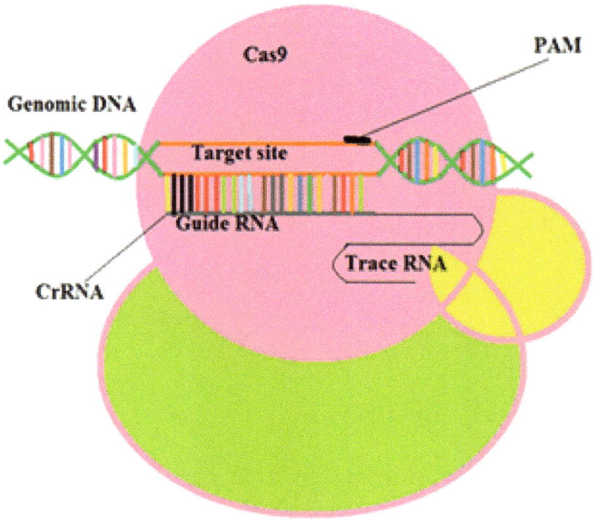

FIGURE 10.2 Illustrative diagram of Cas9 and Cpf1 activities. The target specificity is given by the 17–20 nt located at the 5′-ends of the sgRNA sequence. Cas9 has 5′-NGG-3′ PAM sequence and Cpf1 has 5′-TTTV-3′ PAM sequence. DNA ends after nuclease activity. Cas9 leads to blunt-end and Cpf1 to staggered cut. Polycistronic tRNA-gRNA (PTG) gene construct and targeting activity for Cas9. PTG is composed of tRNA-gRNA repeats and is upregulated by ZmU6 promoter or TaU3 promoter according to the experimental design as different terminator regions (T) are adopted.

For the CRISPR/Cas9 system, two conditions are required to perform on targeted sequences, one being site-specific binding between a 50-end 20-bp sg-RNA sequences and target site and the other being existing protospacer neighboring motif (PAM) downstream [47]. There are two methods of editing the CRISPR/Cas9 based genome in which the sg-RNA/Cas9 component is transported in the cell. One of these methods is stable express sg-RNA or Cas9, DNA by developing plasmid CRISPR/Cas9 other is transforming the gene of interest (GOI) by particle bombardment method into the plant genome [48]. The other is the CRISPR/Cas9 transitional expression system in which ribonucleoproteins (RNPs) or *in vitro* (IVT's) are pre-assembled by particle bombardment to edit DNA-free genomes in cells [49]. The stable expression system was used commonly in plant GE, thanks to the simple research designs and repetitive operations. Moreover, it is likely to do edited genomic, transgenic allowed offspring mutant by isolations of targeted DNA edition and sg-RNA/Cas9, DNA addition sites. Since these mutants do not differ from those derived from natural mutagen, these are safe to use in the species of required crops [50]. CRISPR/Cas9 transient expression system, which is an editing system for DNA-free genomes, totally avoids the hazard of integration of external DNA fragments into the genome that helps in large measures to relieve monitoring concern and can generate editing mutants. Preliminary studies have shown that repair and partition replicons are part of the role of the CRISPR locus in cellular DNA. In 2005, the system was suggested to be included as part of an adaptive prokaryotic immune system and sequences placed between identical repeats from the invading phage and plasmid genome were reported [51, 52]. Later, important improvement of agronomical traits, changed patterns of gene expression, obtaining transgenes-free crops, and the setting up of CRISPR mutation libraries have been achieved [50]. With time, the CRISPR/Cas9 system has flourished around the globe and is used for various kinds of vegetables like tobacco, Arabidopsis, olive rape, rice, soybean, maize, potato wheat, and mushroom [50].

The efficacy of the CRISPR/Cas-12-a system for plant genomic edition has been demonstrated in recent reports. Lb-Cas-12a and Fn-Cas-12a have been employed effectively for inherited rice and tobacco indel mutations [53]. Nuclease and Cas12a-RNP are also used for DNA-free desired genes mutagen or reposition in soybeans, tobacco, and more recent progress in the transient expression system of RNPs [54] and *Oryza sativa* [55].

10.4 ADVANCEMENTS IN THE GENOMIC EDITION TECHNIQUES

Today, in genomic edition research and applications, Cas9, Cas12a, and their catalytically impaired variants accompany each other but PAM is still limited to the number of destination sites. Consequently, scientists also explored more nucleases to add more flexibility to the toolbox for genomic edition. These nucleases have been obtained in two ways. The first is the use of Cas protein orthologous. These include Cms1 [56], StCas, NmCas9, and SaCas9 [8], Cas-12b [57], Cas13a [58], and Cas13b [59]. The CRISPR orthologous have various PAM sequences, desired length, and substrate specificities that can broaden the target range of genomic edition. The application of evolutionary Cases variants that have modifying PAM specificities is an alternate way to discourse the targeted ranged limitations. Those derived CRISPR/Cas variable systems offer the option of selecting genome-wide editing targets. Some other important variants were SaCas9, SpCas9-NG [96], VRER/Cas9 [60], x-Cas9 [97], Lb-Cas12a-PR [61] have been successfully used for the plant's genomes.

10.5 DEVELOPMENTS OF SOYBEAN WITH GENOMIC EDITION

Since soybean genome sequencing is completed, 46,000–56,000 annotated genes have been characterized as a major challenge and research priority for soybean researchers [20]. Transgenic technology is a useful method in functional genomics research and crop genetic improvement, but it is still associated with some limitations in the application in soybean. Over the last few decades, the processing and bombardment of agro-bacterium-mediated soya plants have been used commonly. In general, Agrobacterium-mediated transformation is better if its simple protocol, low cost, a single or a small number of insertions, and relatively rare rearrangements are considered [98]. Transgenic analysis of the gene function often involves integrating the foreign DNA sequence into the plant genome, which results in overexpression or silencing of the target gene [62]. Also, potential risks generate during transformation, such as unintended insertion of genes, endogenous disorder, and unpredictable gene expression [63]. As a result, transgenic plants often have potential bio-security problems and regulatory restrictions on GMOs. Besides, when a diploid was derived from palaeotetraploid, soya has a highly duplicated genome with approximately 75% predicted genes that maintain several exemplary copies. This leads to a high degree of genetic redundancy [21]. On the other hand, it is often difficult to correlate the routine random mutagenic (physical, chemical, or T-DNA) method, as one homolog can be compensated for by redundant homologous copies [64]. Similarly, RNA-based technology on silence often silences the entire gene family and a single gene copy is difficult to silence [65], and the phenotype may be unstable because of the gene product's incomplete depletion [66]. Some important genes and genes functions used for soybean improvement were presented in Table 10.2.

TABLE 10.2 CRISPR/Cas9-based Genomic Edition in Soybean

Crop Name	Gene Name	Function	References
Soybean	Glutamine synthase	Deletion (2 bp)	[67]
Soybean	CH120	Deletion (1–32)	[67]
Soybean	GFP	Loss of florescence	[68]
Soybean	Glyma07g14530	Deletion, DNP, and replace.	[68]
Soybean	miR1514, miR1509	Greater than 95% indel frequency in 4 miR1514 and 3 out of 4 miR1509.	[68]
Medicago truncatula	β-glucuronidase (GUS)	GUS staining loss.	[67]
Soybean	GmPDS11, GmPDS18	Single target efficiency of AtU6-26 promotor for Albino.	[68]
Soybean	GmDc13a	Transmission failed in the mutation of T1 generation.	[69]
Soybean	GmDrb2a and GmDrb2b	Drought stress sensitivity, dark green coloration.	[69]
Medicago truncatula	MtHen1	Seed mottling imbibition.	[69]
Soybean	Drb2a and Drb2b genes	Drought resistance and oil contents.	[70]

To speed up the researcher and breeding of soybean genes functioning, precise, and efficient genetic engineering techniques must be established. At least in part, existing available technology for genome edition has provided the opportunity to address the challenges above. First, genomic edition technology can mutate not only a single gene but can also edit several interest genes at the same time by using a single transformation, thus making the soybean and other polyploid plants ideal for searching. Second, an important feature of GE technology is that the integration sequences into plant genome are a component of genomic edition and not foreign genes of interest. These genomic edition components can be removed after editing the target gene, creating transgene-free mutants, which are safer for race use and are easier than market access by GMO rules [71]. Finally, given the poor efficiency of genetic transformation, direct genomic edition without transformation and tissue culture is one possible solution. A variety with relatively high process efficiency is normally edited to improve one or more features, and then the edited plant with editing components may be used as a donor. Ma et al. [99] suggested that elite lines recalcitrant for the transformation of the edited target gene or genomic edition components themselves can be integrated through hybridization with contributor plant and subsequently repeated crossing [72].

Agrobacterium based transformation method is an easy and quick tool for research in soybean gene functions, and transgenic hair roots can only be produced for a short period (a couple of weeks) [73]. That's why the mega-nuclease system (ZFN's and TALEN's) and CRISPR/Cas9 were first used in the soybean, researchers favored the use of a hair-root transformation system to quickly assess the effectiveness of these genome-editing tools before the laborious transformation of whole soybeans. Thus, in 2011, the first example of genomic edition for soya was achieved with ZFNs technology in heritage mutations in two *DICER-LIKE* homologous genes, *DCL4a* and *DCL4b* [74]. TALEN technology was then used in soybean to transform the two genes *FAD2-1A* and *FAD2-1B* into a high variety of oleic acid soybean. Besides, the technology was used in soybean for the first time [42]. The CRISPR/Cas9 system mutation effectiveness of endogenous or exogenous soybean genes in hairy roots has been successfully evaluated by scientists, setting an example in using the CRISPR/Cas technique to test soybean genes [75, 76]. Almost 76% of targeted mutagenesis was achieved via NHEJ, which also resulted in the obtaining by Cas9/gRNA induced NHEJ of mutated acetol gate-based soybean (*ALS1*) genes with targeted gene integration at the target site via HR [44]. The strong capability of these GE tools helps to make a valued feature in the short-term by making specific gene modification was shown by all those initial achievements.

Besides, some researchers have worked to evaluate the gene functions and improve agronomic properties by applying technology to genome edition and entire plant transformations in addition to improvements of the genome edition system to facilitate better application in soybean research. To analyze their soya bean mi-RNA function in the pathway, Curtin used ZFN's based mutagenesis technology to create single and double gene mutant *DCL-1a* and *DCL-1b* [64]. Cai reported homozygous mutant in FLOWERING LOCUS-T2-a, delayed soybean blooming obtained by the CRISPR/Cas9 mediated mutagenesis (*GmFt2-a*) [77]. They also mutated *GmFT5-a* by the CRISPR/Cas9 system, then by crossing with ft2a mutant, they obtained *ft2a–ft5a* double mutant, which flowered approximately 31 days later than wild plant [78].

Although genomic edition technology is now the prevailing choice of CRISPR/Cas9, it is less widely available in highly specific genome regions due to the large limitation of NGG PAM.

Since the ZFNs and TALEN target range is limitless, they represent important instruments in addition to CRISPR/Cas9. The combined use of these systems will support soybean functionality and reproduction. Curtin used CRISPR/Cas9 and TALENs for instance to mutate a gene collection that encrypts core machinery proteins used in *soya*-and *Medicago truncatula* for small RNA processing [79]. Along with earlier described mutants formed by ZFN's, the resulted mutant plants constituted a range of mutant's resources for further study of minor RNAs in the leguminous plant [64]. High-performance CRISPR/Cas9 mutagenesis was used to build a broad mutant genome library or collection of mutants linked to convinced function in pooled function screen that provided rich mammalian cells, rice, and tomato resources [100]. The successful CRISPR library depends, however, heavily on the effective genetic transformation system. CRISPR's soybean library is not only an enormous problem but also very valuable in supporting researchers and breeders. A first CRISPR soybean library has recently been built aiming at over 100 candidate genes. An improved process was also produced using several key stages to show that CRISPR/Cas9 technology is practicable and effective to carry out large scales mutagenic in soybean [20]. A study was reported by Mao et al. [80] in rice suggested that to create long-grain erect-panicle geng/japonica rice germplasm, altering grain shape genes was a successful strategy. Furthermore, it was also suggested that GS3 and TGW6 alleles can be used directly for breeding long-grain shape geng/japonica rice while modifying GW8 resulted in great appearance quality but low yield.

10.6 PROSPECTS

With its strong site-oriented modification capability, genomic edition technology has been replacing outdated mutagenesis methods are supposed to transform breeding fundamentally, but it still has a few challenges to deal with in soybean breeding. The complexity of the soy genome makes functional gene analysis difficult. The absence of plenty of genetic resources is the main limiting factor for improving agronomic characteristics using genetic management techniques including genomic edition. Since previous soya genomic edition applications focused mostly on modifying several known gene, more translated genome resource is a prerequisite for the use of soya reproductive genomic edition technology. For efficient genetic screen function loss, the CRISPR-based transformed library can be used. A changed genomic library in soybean is extremely difficult to build, but it is possible to create a small library related to certain functions. The genome-editing process is still very long-term but intensive for those cultivars which are intractable to genetic change cannot produce gene-editing tools, even if crosses them through with another Genomic edition variety and then with repeated markers that support back-crossing. Technological breakthroughs have resulted in recent efforts to overcome these obstacles. To progress the easiness and competence of transformations, morphogenic regulators such as Baby Boom and Wuschel were shown [81]. Replica based on Geminivirus has acted as an operative supplier of gene-focusing SSNs (including ZFNs, TALENs, and CRISPR/Cas9) to plants, leading to an increase in the gene copy number [101]. Plasmid coated carbon dot was also shown to facilitate spray-on gene edition in wheat quickly [82]. All this could be a locus to encourage the application of genome soybean editing. A haploid inductor line that expresses initially Cas9/sg-RNA through transformation or crossing is obtained by the technical route; pollens from the haploid inducer line are then transferred as a male parent to a stigma of all inbred lines; the genome of the female parents is edited in the transient zygotic state by CRISPR/Cas9, and the male parent is subject to post-fertilization genome [72]. HI-Edit-IMGE method allows a single cross to be directly edited for monocot and dicot species, thereby preventing genetic transformations and repetitive backcrossing, which can speed up breeding. Moreover, their use has been further expanded through the establishment of synthetic haploid initiation systems and the wide crosses viability. Applications of these two methods HI-Edit/IMGE along with CRISPR/Cas9 technology accelerate clonal seed production from hybrids [83]. The process of breeding soybean is greatly accelerated if this breeding method is implemented in soybean.

Finally, we can conclude, meganuclease system, CRISPR/Cas, and their variants represent a rich toolkit to edit the genome that promotes future soybean and other plant research on genomic and genetic functionality. Currently, soybean is a predominant choice because of its effectiveness, complex edition, and higher performance mutagens abilities as well as the development of CRISPR/Cas9 technology. But soybean genomic edition will have more flexible choices with advanced genomic edition systems. While we may have to make great efforts to make use of these instruments, because genomic editing and the economic importance of soya are very potential, we are looking forward to meeting these challenges soon.

10.7 CONCLUDING REMARKS

The growing population has created a global need for better crop yields. Gene editing by programmable nucleases can be used quickly to introduce positive features that increase crop yields. However, both ZFN and TALEN have had their successes because of the simplicity and effectiveness of CRISPR/Cas and their flexibility, which has been discovered most recently and has had a major impact. However, mega-nucleases system (ZFNs, TALENS) has had their successes because of the simplicity and effectiveness of CRISPR/Cas and their flexibility, which has been discovered most recently and has had a major impact. As the ability of this technology is investigated and developed, it is likely to play a role as restrictive enzymes in molecular biology. CRISPR/Cas technology methods for transforming facilities limit the speed and ease of research and development. In recent times these methods have not altered or improved, and research has great implications for a modern, easy method of transforming plants like viruses or nanoparticles or for efficient transformation of RNPs through the cell wall. Pressure must be put on regulators to assist in this and to secure the approval of GM crops, which will allow more research, development, and use of these technologies to feed the growing population of the world.

KEYWORDS

- **CRISPR/Cas9**
- **crop improvement**
- **crop yields**
- **genome editing**
- **meganuclease**
- **restrictive enzymes**
- **soybean**

REFERENCES

1. Foley, J. A., Ramankutty, N., Brauman, K. A., Cassidy, E. S., Gerber, J. S., Johnston, M., et al., (2011). Solutions for a cultivated planet. *Nature, 478*, 337–342.
2. Tilman, D., Balzer, C., Hill, J., & Befort, B. L., (2011). Global food demand and the sustainable intensification of agriculture. *Proc. Natl. Acad. Sci. U S A., 108*, 20260–20264.

3. Ahmad, S., Tang, L., Shahzad, R., Mawia, A. M., Rao, G. S., Jamil, S., & Wei, X., (2021a). CRISPR-based crop improvements: A way forward to achieve zero hunger. *Journal of Agricultural and Food Chemistry*.
4. Chen, K., & Gao, C., (2013). Targeted genome modification technologies and their applications in crop improvements. *Plant Cell Rep., 33*, 575–583.
5. Pacher, M., & Puchta, H., (2017). *From Classical Mutagenesis to Nuclease-Based Breeding–Directing Natural DNA Repair for a Natural End-Product*. Wiley Online Library.
6. Zetsche, B., Gootenberg, J. S., Abudayyeh, O. O., Slaymaker, I. M., Makarova, K. S., Essletzbichler, P., Volz, S. E., Joung, J., et al., (2015). Cpf1 is a single RNA-guided endonuclease of a class 2 CRISPR-Cas system. *Cell, 163*, 759–771.
7. Cong, Á., Ran, F., Cox, D., Lin, S., Barretto, R., Habib, N., Hsu, P., et al., (2013). Multiplex genome engineering using CRISPR/Cas systems. *Science, 339*, 819–823.
8. Cebrian-Serrano, A., & Davies, B., (2017). CRISPR-Cas orthologues and variants: Optimizing the repertoire, specificity, and delivery of genome engineering tools. *Mammalian Genome, 28*, 247–261.
9. Moradpour, M., & Abdulah, S. N. A., (2020). CRISPR/dCas9 platforms in plants: Strategies and applications beyond genome editing. *Plant Biotechnol. J., 18*, 32–44. https://doi.org/10.1111/pbi.13232.
10. Ahmad, S., Wei, X., Sheng, Z., Hu, P., & Tang, S., (2020). CRISPR/Cas9 for development of disease resistance in plants: Recent progress, limitations, and future prospects. *Briefings in Functional Genomics, 19*(1), 26–39.
11. Li, B., Rui, H., Li, Y., Wang, Q., Alariqi, M., Qin, L., Sun, L., Ding, X., Wang, F., & Zou, J., (2019). Robust CRISPR/Cpf1 (Cas12a)-mediated genome editing in allotetraploid cotton (*Gossypium hirsutum*). *Plant Biotechnology Journal, 17*, 1862–1864.
12. Bao, A., Burritt, D. J., Chen, H., Zhou, X., Cao, D., & Tran, L. S. P., (2019). The CRISPR/Cas9 system and its applications in crop genome editing. *Critical Reviews in Biotechnology, 39*, 321–336.
13. Ahmad, S., Sheng, Z., Jalal, R. S., Tabassum, J., Ahmed, F. K., Hu, S., & Hu, P., (2021b). CRISPR–Cas technology towards improvement of abiotic stress tolerance in plants. In: *CRISPR and RNAi Systems* (pp. 755–772): Elsevier.
14. Du, H., et al., (2016). Efficient targeted mutagenesis in soybean by TALENs and CRISPR/Cas9 *Journal of Biotechnology, 217*, 90–97.
15. Symington, L. S., & Gautier, J., (2011). Double-strand break end resection and repair pathway choice. *Annu. Rev. Genet., 45*, 247–271.
16. Tabassum, J., Ahmad, S., Hussain, B., Mawia, A. M., Zeb, A., & Ju, L., (2021). Applications and potential of genome-editing systems in rice improvement: Current and future perspectives. *Agronomy, 11*(7), 1359.
17. Monsur, M. B., Shao, G., Lv, Y., Ahmad, S., Wei, X., Hu, P., & Tang, S., (2020). Base editing: The ever expanding clustered regularly interspaced short palindromic repeats (CRISPR) tool kit for precise genome editing in plants. *Genes, 11*(4), 466.
18. Ahmad, S., Shahzad, R., Jamil, S., Tabassum, J., Chaudhary, M. A. M., Atif, R. M., & Sheng, Z., (2021c). Regulatory aspects, risk assessment, and toxicity associated with RNAi and CRISPR methods. In: *CRISPR and RNAi Systems* (pp. 687–721). Elsevier.
19. Voytas, D. F., & Gao, C., (2014). Precision genome engineering and agriculture: Opportunities and regulatory challenges. *PLoS Biol., 12*, e1001877.

20. Bai, M. Y., Yuan, J. H., Kuang, H. Q., Gong, P. P., Li, S. N., Zhang, Z. H., Liu, B., Sun, J. F., Yang, M. X., Yang, L., et al., (2019). Generation of a multiplex mutagenesis population via pooled CRISPR-Cas9 in soya bean. *Plant Biotechnol. J.,* e13239 https://doi.org/10.1111/pbi.13239.
21. Schmutz, J., Cannon, S. B., Schlueter, J., Ma, J. X., Mitros, T., Nelson, W., Hyten, D. L., Song, Q. J., Thelen, J. J., Cheng, J. L., et al., (2010). Genome sequence of the paleopolyploid soybean. *Nature, 463*, 178–183. https://doi.org/10.1038/nature08670.
22. Jones, J. D., et al., (2014). Elevating crop disease resistance with cloned genes. *Philosophical Transactions of the Royal Society B: Biological Sciences, 369*, 20130087.
23. Butt, H., Jamil, M., Wang, J. Y., Al-Babili, S., & Mahfouz, M., (2018). Engineering plant architecture via CRISPR/Cas9-mediated alteration of strigolactone biosynthesis *BMC Plant Biology, 18*, 1–9.
24. Rubiales, D., & Mikić, A., (2015). *Introduction: Legumes in Sustainable Agriculture, 34*, 1–3.
25. Annicchiarico, P., Barrett, B., Brummer, E. C., Julier, B., & Marshall, A. H., (2015). Achievements and challenges in improving temperate perennial forage legumes. *Critical Reviews in Plant Sciences, 34*, 327–380.
26. Jiang, M., Liu, Y., Liu, Y., Tan, Y., Huang, J., & Shu, Q., (2019). Mutation of inositol 1, 3, 4-trisphosphate 5/6-kinase6 impairs plant growth and phytic acid synthesis in rice. *Plants, 8*, 114.
27. Mugode, L., et al., (2014). Carotenoid retention of biofortified provitamin A maize (*Zea mays* L.) after Zambian traditional methods of milling, cooking, and storage. *Journal of Agricultural and Food Chemistry, 62*, 6317–6325.
28. Urnov, F. D., Miller, J. C., Lee, Y. L., Beausejour, C. M., Rock, J. M., Augustus, S., Jamieson, A. C., et al., (2005). Highly efficient endogenous human gene correction using designed zinc-finger nucleases. *Nature, 435*, 646–651.
29. Pavletich, N., & Pabo, C., (1991). Zinc structure of a recognition: Complex Zif268-DNA. *Adv. Sci., 252*, 809–817.
30. Bitinaite, J., Wah, D. A., Aggarwal, A. K., & Schildkraut, I., (1998). FokI dimerization is required for DNA cleavage. *Proceedings of the National Academy of Sciences,* 95, 10570–10575.
31. Wright, D. A., Thibodeau-Beganny, S., Sander, J. D., Winfrey, R. J., Hirsh, A. S., Eichtinger, M., Fu, F. L., Porteus, M. H., Dobbs, D., Voytas, D. F., et al., (2006). Standardized reagents and protocols for engineering zinc finger nucleases by modular assembly. *Nat. Protoc., 1*, 1637–1652. https://doi.org/10.1038/nprot.2006.259.
32. Govindan, G., & Ramalingam, S., (2016). Programmable site-specific nucleases for targeted genome engineering in higher eukaryotes. *Journal of Cellular Physiology, 231*, 2380–2392.
33. Stephens, J., & Barakate, A., (2017). *Gene Editing Technologies–ZFNs, TALENs, and CRISPR/Cas9,* 157–161.
34. Shukla, V. K., Doyon, Y., Miller, J. C., DeKelver, R. C., Moehle, E. A., Worden, S. E., Mitchell, J. C., Arnold, N. L., Gopalan, S., Meng, X. D., et al., (2009). Precise genome modification in the crop species *Zea mays* using zinc-finger nucleases. *Nature, 459*, 437–441. https://doi.org/10.1038/nature07992.
35. Lloyd, A., Plaisier, C. L., Carroll, D., & Drews, G. N., (2005). Targeted mutagenesis using zinc finger nucleases in *Arabidopsis*. *Proc. Natl. Acad. Sci. U. S. A., 102*, 2232–2237.https://doi.org/10.1073/pnas.0409339102.

36. Curtin, S. J., Zhang, F., Sander, J. D., Haun, W. J., Starker, C., Baltes, N. J., Reyon, D., et al., (2011). Targeted mutagenesis of duplicated genes in soybean with zinc-finger nucleases. *Plant Physiology, 156*, 466–473.
37. Townsend, J. A., Wright, D. A., Winfrey, R. J., Fu, F., Maeder, M. L., Joung, J. K., & Voytas, D. F., (2009). High-frequency modification of plant genes using engineered zinc-finger nucleases. *Nature, 459*, 442–445.
38. Yang, J., (2019). Recent publications view all. *International Journal of Geographical Information Science, 10*, 55.
39. Zhang, Y., et al., (2013). Transcription activator-like effector nucleases enable efficient plant genome engineering. *Plant Physiology, 161*, 20–27.
40. Li, T., Liu, B., Spalding, M. H., Weeks, D. P., & Yang, B., (2012). High-efficiency TALEN-based gene editing produces disease-resistant rice. *Nature Biotechnology, 30*, 390.
41. Cermak, T., et al., (2011). Efficient design and assembly of custom TALEN and other TAL effector-based constructs for DNA targeting. *Nucleic Acids Research, 39*.
42. Haun, W., Coffman, A., Clasen, B. M., Demorest, Z. L., Lowy, A., Ray, E., Retterath, A., Stoddard, T., Juillerat, A., Cedrone, F., et al., (2014). Improved soybean oil quality by targeted mutagenesis of the fatty acid desaturase 2 gene family. *Plant Biotechnol. J., 12*, 934–940. https://doi.org/10.1111/pbi.12201.
43. Demorest, Z. L., Coffman, A., Baltes, N. J., Stoddard, T. J., Clasen, B. M., Luo, S., Retterath, A., et al., (2016). Direct stacking of sequence-specific nuclease-induced mutations to produce high oleic and low linolenic soybean oil. *BMC Plant Biology, 16*, 1–8.
44. Li, Z., Liu, Z. B., Xing, A., Moon, B. P., Koellhoffer, J. P., Huang, L., Ward, R. T., et al., (2015). Cas9-guide RNA directed genome editing in soybean. *Plant Physiology, 169*, 960–970.
45. Jansen, R., Van, E. J. D. A., Gaastra, W., & Schouls, L. M., (2002). Identification of genes that are associated with DNA repeats in prokaryotes. *Molecular Microbiology, 43*.
46. Koonin, E. V., Makarova, K. S., & Zhang, F., (2017). Diversity, classification, and evolution of CRISPR-Cas systems. *Current Opinion in Microbiology, 37*, 67–78.
47. Jinek, M., Chylinski, K., Fonfara, I., Hauer, M., Jennifer, D., & Emmanuelle, C., (2012). A programmable dual-RNA-guided DNA endonuclease in adaptive bacterial immunity. *Science*, 816–821.
48. Ma, X., Zhang, Q., Zhu, Q., Liu, W., Chen, Y., Qiu, R., Wang, B., et al., (2015). A robust CRISPR/Cas9 system for convenient, high-efficiency multiplex genome editing in monocot and dicot plants. *Molecular Plant, 8*, 1274–1284.
49. Zhang, Y., Liang, Z., Zong, Y., Wang, Y., Liu, J., Chen, K., Qiu, J. L., & Gao, C., (2016). Efficient and transgene-free genome editing in wheat through transient expression of CRISPR/Cas9 DNA or RNA. *Nature Communications, 7*, 1–8.
50. Bao, A. L., Burritt, D. J., Chen, H. F., Zhou, X. N., Cao, D., & Tran, L. S. P., (2019a). The CRISPR/Cas9 system and its applications in crop genome editing. *Crit. Rev. Biotechnol., 39*, 321–336. https://doi.org/10.1080/07388551.2018.1554621.
51. Mojica, F. J., García-Martínez, J., & Soria, E., (2005). Intervening sequences of regularly spaced prokaryotic repeats derive from foreign genetic elements. *Journal of Molecular Evolution, 60*, 174–182.

52. Pourcel, C., Salvignol, G., & Vergnaud, G., (2005). CRISPR elements in *Yersinia pestis* acquire new repeats by preferential uptake of bacteriophage DNA and provide additional tools for evolutionary studies. *Microbiology, 151*, 653–663.
53. Xu, R., Qin, R., Li, H., Li, D., Li, L., Wei, P., & Yang, J., (2017). Generation of targeted mutant rice using a CRISPR-Cpf1 system. *Plant Biotechnology Journal, 15*, 713–717.
54. Kim, H., Kim, S. T., Ryu, J., Kang, B. C., Kim, J. S., & Kim, S. G., (2017). CRISPR/Cpf1-mediated DNA-free plant genome editing. *Nature Communications, 8*, 1–7.
55. Li, J., Luo, J., Xu, M., Li, S., Zhang, J., Li, H., Yan, L., Zhao, Y., & Xia, L., (2019). Plant genome editing using xCas9 with expanded PAM compatibility. *J. Genet. Genomics, 46*, 277–280.
56. Begemann, M. B., Gray, B. N., January, E., Gordon, G. C., He, Y., Liu, H., Wu, X., et al., (2017). Precise insertion and guided editing of higher plant genomes using Cpf1 CRISPR nucleases. *Scientific Reports, 7*, 1–6.
57. Teng, F., Cui, T., Feng, G., Guo, L., Xu, K., Gao, Q., Li, T., et al., (2018). Repurposing CRISPR-Cas12b for mammalian genome engineering. *Cell Discovery, 4*, 1–15.
58. Abudayyeh, O. O., Gootenberg, J. S., Essletzbichler, P., Han, S., Joung, J., Belanto, J. J., Verdine, V., et al., (2017). RNA targeting with CRISPR–Cas13. *Nature, 550*, 280–284.
59. Cox, D. B., Gootenberg, J. S., Abudayyeh, O. O., Franklin, B., Kellner, M. J., Joung, J., & Zhang, F., (2017). RNA editing with CRISPR-Cas13. *Science, 358*, 1019–1027.
60. Hu, X., Wang, C., Fu, Y., Liu, Q., Jiao, X., & Wang, K., (2016). Expanding the range of CRISPR/Cas9 genome editing in rice. *Molecular Plant, 9*, 943–945.
61. Zhong, Z., Zhang, Y., You, Q., Tang, X., Ren, Q., Liu, S., Yang, L., et al., (2018). Plant genome editing using FnCpf1 and LbCpf1 nucleases at redefined and altered PAM sites. *Molecular Plant, 11*, 999–1002.
62. Rajeev, K. S., Anunanthini, P., & Ramalingam, S., (2015). Epigenetic silencing in transgenic plants. *Frontiers in Plant Science, 6*, 693.
63. Heinemann, J. A., Agapito-Tenfen, S. Z., & Carman, J. A., (2013). A comparative evaluation of the regulation of GM crops or products containing dsRNA and suggested improvements to risk assessments. *Environment International, 55*, 43–55.
64. Curtin, S. J., Michno, J. M., Campbell, B. W., Gil-Humanes, J., Mathioni, S. M., Hammond, R., Gutierrez-Gonzalez, J. J., et al., (2016). MicroRNA maturation and microRNA target gene expression regulation are severely disrupted in soybean dicer-like1 double mutants. *G3: Genes, Genomes, Genetics, 6*, 423–433.
65. Guo, D., Shibuya, R., Akiba, C., Saji, S., Kondo, T., & Nakamura, J., (2016). Active sites of nitrogen-doped carbon materials for oxygen reduction reaction clarified using model catalysts. *Science, 351*, 361–365.
66. Rusecka, J., Kaliszewska, M., Bartnik, E., & Tońska, K., (2018). Nuclear genes involved in mitochondrial diseases caused by instability of mitochondrial DNA. *Journal of Applied Genetics, 59*, 43–57.
67. Jacobs, T. B., LaFayette, P. R., Schmitz, R. J., & Parrott, W. A., (2015). Targeted genome modifications in soybean with CRISPR/Cas9. *BMC Biotechnology, 15*, 16.
68. Du, H., Zeng, X., Zhao, M., Cui, X., Wang, Q., Yang, H., Cheng, H., & Yu, D., (2016). Efficient targeted mutagenesis in soybean by TALENs and CRISPR/Cas9. *Journal of Biotechnology, 217*, 90–97.
69. Ji, J., Zhang, C., Sun, Z., Wang, L., Duanmu, D., & Fan, Q., (2019). Genome editing in cowpea *Vigna unguiculata* using CRISPR-Cas9. *International Journal of Molecular Sciences, 20*, 2471.

70. Gao, R., Feyissa, B. A., Croft, M., & Hannoufa, A., (2018). Gene editing by CRISPR/Cas9 in the obligatory outcrossing *Medicago sativa*. *Planta, 247*, 1043–1050.
71. Bao, A., Zhang, C., Huang, Y., Chen, H., Zhou, X., & Cao, D., (2020). Genome editing technology and application in soybean improvement. *Oil Crop Science*, *5*, 31–40.
72. Wang, B., Zhu, L., Zhao, B., Zhao, Y., Xie, Y., Zheng, Z., Li, Y., et al., (2019). Development of a haploid-inducer mediated genome editing system for accelerating maize breeding. *Molecular Plant, 12*, 597–602.
73. Chen, L., Cai, Y., Liu, X., Yao, W., Guo, C., Sun, S., Wu, C., et al., (2018). Improvement of soybean *Agrobacterium*-mediated transformation efficiency by adding glutamine and asparagine into the culture media. *International Journal of Molecular Sciences, 19*, 3039.
74. Piatek, A., Ali, Z., Baazim, H., Li, L., Abulfaraj, A., Al-Shareef, S., Aouida, M., & Mahfouz, M. M., (2015). RNA-guided transcriptional regulation in planta via synthetic dC as9-based transcription factors. *Plant Biotechnology Journal, 13*, 578–589.
75. Michno, J. M., Wang, X., Liu, J., Curtin, S. J., Kono, T. J., & Stupar, R. M., (2015). CRISPR/Cas mutagenesis of soybean and *Medicago truncatula* using a new web-tool and a modified Cas9 enzyme. *GM Crops & Food, 6*, 243–252.
76. Sun, X., Hu, Z., Chen, R., Jiang, Q., Song, G., Zhang, H., & Xi, Y., (2015). Targeted mutagenesis in soybean using the CRISPR-Cas9 system. *Scientific Reports, 5*, 1–10.
77. Cai, Y., Chen, L., Liu, X., Guo, C., Sun, S., Wu, C., Jiang, B., et al., (2018). CRISPR/Cas9-mediated targeted mutagenesis of GmFT2a delays flowering time in soya bean. *Plant Biotechnology Journal, 16*, 176–185.
78. Cai, Y., Wang, L., Chen, L., Wu, T., Liu, L., Sun, S., Wu, C., et al., (2020). Mutagenesis of GmFT2a and GmFT5a mediated by CRISPR/Cas9 contributes for expanding the regional adaptability of soybean. *Plant Biotechnology Journal, 18*, 298–309.
79. Curtin, S. J., et al., (2018). Crispr/Cas9 and Talen s generate heritable mutations for genes involved in small RNA processing of glycine max and *Medicago truncatula. Plant Biotechnology Journal, 16,* 1125–1137.
80. Mao, T., Zhu, M., Sheng, Z., Shao, G., Jiao, G., Mawia, A. M., & Wei, X., (2021). Effects of grain shape genes editing on appearance quality of erect-panicle Geng/Japonica rice. *Rice, 14*, 74. https://doi.org/10.1186/s12284-021-00517-5.
81. Mookkan, M., Nelson-Vasilchik, K., Hague, J., Zhang, Z. J., & Kausch, A. P., (2017). Selectable marker independent transformation of recalcitrant maize inbred B73 and sorghum P898012 mediated by morphogenic regulators BABY BOOM and WUSCHEL2. *Plant Cell Reports, 36*, 1477–1491.
82. Doyle, C., Higginbottom, K., Swift, T. A., Winfield, M., Bellas, C., Benito-Alifonso, D., Fletcher, T., et al., (2019). A simple method for spray-on gene editing in planta. *BioRxiv*, 805036.
83. Khanday, I., Skinner, D., Yang, B., Mercier, R., & Sundaresan, V., (2019). A male-expressed rice embryogenic trigger redirected for asexual propagation through seeds. *Nature, 565*, 91–95.
84. Gao, C., (2015). Genome editing in crops: From bench to field. *Natl. Sci. Rev., 2*, 13–15.
85. Heinemann, M., Eifert, B., & Heiliger, C., (2013). Band structure and phase stability of the copper oxides Cu_2O, CuO, and Cu_4O_3. *Physical Review B, 87,* 115111.
86. Hua, K., Tao, X., Han, P., Wang, R., & Zhu, J. K., (2019). Genome engineering in rice using Cas9 variants that recognize NG PAM sequences. *Molecular Plant, 12*, 1003–1014.

87. Li, B., Rui, H. P., Li, Y. J., Wang, Q. Q., Alariqi, M., Qin, L., Sun, L., Ding, X., Wang, F. Q., Zou, J. W., et al., (2019a). Robust CRISPR/Cpf1 (Cas12a)-mediated genome editing in allotetraploid cotton (*Gossypium hirsutum*). *Plant Biotechnol. J., 17*, 1862–1864. https://doi.org/10.1111/pbi.13147.
88. Li, S., Cong, Y., Liu, Y., Wang, T., Shuai, Q., Chen, N., Gai, J., & Li, Y., (2017). Optimization of *Agrobacterium*-mediated transformation in soybean. *Frontiers in Plant Science, 8*, 246.
89. Li, S., Li, J., He, Y., Xu, M., Zhang, J., Du, W., Zhao, Y., & Xia, L., (2019). Precise gene replacement in rice by RNA transcript-templated homologous recombination. *Nature Biotechnology, 37*, 445–450.
90. Meng, X., Hu, X., Liu, Q., Song, X., Gao, C., Li, J., & Wang, K., (2018). Robust genome editing of CRISPR-Cas9 at NAG PAMs in rice. *Science China Life Sciences, 61*, 122–125.
91. Petolino, J. F., (2015). Genome editing in plants via designed zinc finger nucleases. *In Vitro Cellular & Developmental Biology-Plant, 51*, 1–8.
92. Prado, J. R., Segers, G., Voelker, T., Carson, D., Dobert, R., Phillips, J., et al., (2014). Genetically engineered crops: From idea to product. *Annu Rev Plant Biol., 65*, 769–790.
93. Chen, K., & Gao, C. (2014). Targeted genome modification technologies and their applications in crop improvements. *Plant Cell Reports, 33*, 575–583.
94. Jiang, W., Zhou, H., Bi, H., Fromm, M., Yang, B., & Weeks, D. P. (2013). Demonstration of CRISPR/Cas9/sgRNA-mediated targeted gene modification in Arabidopsis, tobacco, sorghum and rice. *Nucleic Acids Research, 41*(20), e188–e188.
95. Curtin, S. J., Anderson, J. E., Starker, C. G., Baltes, N. J., Mani, D., Voytas, D. F., & Stupar, R. M. (2013). Targeted mutagenesis for functional analysis of gene duplication in legumes. *Legume Genomics: Methods and Protocols,* 25–42.
96. Hua, K., Tao, X., Han, P., Wang, R., & Zhu, J.-K. (2019). Genome engineering in rice using Cas9 variants that recognize NG PAM sequences. *Molecular Plant, 12*(7), 1003–1014.
97. Li, J.-Y., Luo, J.-M., Xu, M.-L., Li, S.-Y., Zhang, J.-H., Li, H.-Y., Yan, L., Zhao, Y.-D., & Xia, L.-Q. (2019). *Plant Genome Editing Using xCas9 with Expanded PAM Compatibility*.
98. Li, J., Sun, Y., Du, J., Zhao, Y., & Xia, L. (2017). Generation of targeted point mutations in rice by a modified CRISPR/Cas9 system. *Molecular Plant, 10*(3), 526–529.
99. Ma, J.-F., Yang, F., Mahida, S. N., Zhao, L., Chen, X., Zhang, M. L., Sun, Z., Yao, Y., Zhang, Y.-X., & Zheng, G.-Y. (2016). TBX5 mutations contribute to early-onset atrial fibrillation in Chinese and Caucasians. *Cardiovascular Research, 109*(3), 442–450.
100. Meng, X., Yu, H., Zhang, Y., Zhuang, F., Song, X., Gao, S., Gao, C., & Li, J. (2017). Construction of a genome-wide mutant library in rice using CRISPR/Cas9. *Molecular Plant, 10*(9), 1238–1241.
101. Wang, M., Mao, Y., Lu, Y., Tao, X., & Zhu, J.-k. (2017). Multiplex gene editing in rice using the CRISPR-Cpf1 system. *Molecular Plant, 10*(7), 1011–1013.

CHAPTER 11

CRISPR-CAS SYSTEM: A NEW TOOLBOX FOR DEVELOPING TRANSGENE-FREE MUTANT IN PLANTS

UMAR AKRAM,[1,5] AWAIS FAROOQ,[2] SUTAR SUHAS BHARAT,[3] RABIA KALSOOM,[4] FURQAN AHMAD,[5] SHOAIB UR REHMAN,[5] MUHAMMAD SHABAN,[6] WAQAS MALIK,[6] and ZULFIQAR ALI[5]

[1]*Biotechnology Research Institute, Chinese Academy of Agricultural Sciences, Beijing, P.R. China*

[2]*Department of Plant Breeding and Genetics, University of Agriculture, Faisalabad, Pakistan*

[3]*Institute of Crop Sciences, Chinese Academy of Agricultural Sciences, Beijing, P.R. China*

[4]*School of Biological Science and Technology, Beijing Forestry University, Beijing, P.R. China*

[5]*Institute of Plant Breeding and Biotechnology, MNS University of Agriculture, Multan, Pakistan*

[6]*Genomics Lab, Department of Plant Breeding and Genetics, Bahauddin Zakariya University, Multan, Pakistan*

ABSTRACT

CRISPR/Cas technology has improved molecular biology for highly specific and accurate modification of genomes. The rapid elimination of the

CRISPR-Based Gene Editing: A Solution for Global Food Security. Shakeel Ahmad, Asif Ali Khan, and Muhammad Arshad Javed (Eds.)

transgene after successful genome editing (GE) has been a challenge in crops plant biotechnology because of their extended life cycle and occurrence of multiploidy, not only to minimize off-target mutations and to eliminate transgene position effects, but also to ensure the delivery gene editing machinery free plants to the consumers. In sexually reproducing crop plants genetic segregation could play a key role in removal of CRISPR transgene. Elimination of transgene through marker-assisted tracking decreases both the time and labor required for identification of the ideal transgene-free plants. The sequential activation of edited genes and suicide genes can be utilized to program self-elimination of transgene that could greatly accelerate the detection of transgene-free and gene-edited plants. Non-transgenic techniques like nanotechnology, transient expression without DNA integration and ribonucleoprotein (RNP) transfection can also be used to generate transgene-free plants. Marker gene-based selection is sometime misleading for CRISPR/Cas mediated genome edited plants because plants lacking transgene will not have resistance and thus will not survive. For this type of plants two or more generations must be screened to evaluate the presence of the transgene.

11.1 INTRODUCTION

Transgene free genome modification growing as a modern and rapidly expanding vogue in plant science research because of its clear edge. Indeed, transgene free genome editing (GE) set the stage to targeted gene editing without disrupting non-targeted genome [1], also uplifted the potential to generate genetically customized organisms [2]. As a transformation tools the marker DNA free modification has also faced technological issues. Regardless of various applications of transformation technology, the transformation of RNA and protein into the plant system is most underdeveloped compared to animal system, and merely biolistic and transfection of protoplast possibly used. However, efforts to reprocess other transformation methods and tools are enacted due to the significant prospective of transgene free genome modification.

Formerly, genome engineering had emerged in the 1990s since the rise of effective tools for transformation in plant. Firstly, in an experiment, homologous recombination (HR)-mediated allele replacement was carried out by using heterologous DNA. Subsequently, Zink finger nuclease (ZFN) and transcription activator like effector nuclease (TALEN) came into existence as a genome engineering tools (Boettcher M,) Because of the complex design these technologies remained unpopular and unexplored.

The finding of bacterial adaptive immune system which is based upon RNA-directed endonuclease unlocked a new horizon for precise DNA editing. Further, the development of guide RNA that targets specific DNA sequence and a Cas endonuclease results in an advancement in the field of genetic engineering. DNA editing became easy in almost all labs including one having minimum lab infrastructure and instruments. No wonder then that they may report on GE has been published immediately after the launching this GE tool. Thereafter many numbers of species for different interest have been modified by using CRISPR/Cas9 [3]. CRISPR/Cas9 motivated to develop and employ highly efficient plant transformation methods (Ran Y,). After delivery of CRISPR/Cas9 into plant system, it integrates into host genome. Such integration may become an obstacle in further excrements and in commercialization of developed mutant lines. To edit DNA inside cell the CRISPR/Cas9 is required for short period of time, by considering this a novel technique for non-DNA based system were developed. These new techniques comprise use of RNP, virus, and bacterial secretary system mediated transformation of Cas9/gRNA. Firstly, Je Wook Woo et al. successfully obtained DNA-free genome modification in plant [4], later multiple reports were published [5]. In this mini chapter we outlined the development and new trends in the field of DNA-free genome modification in plants.

11.2 CAS NUCLEASE VARIANT

11.2.1 THE NATIVE CAS9

The Cas9 is a 160 kilodalton big protein molecule that plays a significant role in the defense system of bacteria against invading viruses [6]. It is a part of CRISPR GE system that is being intensely used in the genetic engineering. Its main function is to cut DNA at a particular site which alters the cell's genome. Cas9, the nuclease, becomes active when it is linked with a guide RNA, a complex composed of two others naturally occurring RNA species, the trans-activating crRNA (tracrRNA) and CRISPR RNA (crRNA) [7]. The first 20 nucleotides on the targeted genome define the specificity of the nuclease activity, which occurs by complementary base pairing within the genomic DNA. Once the CRISPR-Cas9 system is activated inside the genome it will generate a two stranded break at the targeted site [8]. Cas9 uses two distinctive active sites, RuvC and HNH, generating site specific nicks on opposite DNA strands [9]. The double stranded breaks can lead to the gene activation or the introduction of heterologous genes through non-homologous end joining (NHEJ) and HR, respectively in different organisms [10]. Currently, the modifications in the Cas9 proteins are underway, in which the inactivation of nuclease activity of Cas9 in CRISPR system can be utilized in various other functions such as silencing of the genes, transcription deactivation of the genes, intense transcription of the genes, deacetylation of the targeted genes [11]. With the passage of time, the alterations in the CRISPR Cas system and changes in the Cas protein has enabled the humans to make changes in the DNA of the living organisms spanning over numerous generation which are free of transgenes [12].

11.3 GUIDE RNA VARIANTS

11.3.1 TRUNCATED GUIDE RNAS

CRISPR/Cas9 has been used in variety of organism due to its simplicity and targeted genome modification. But off targets effects may occur, sometime at higher frequency. Truncated gRNAs (trugRNAs) along with dimerization-dependent RNA-guided FokI-dCas9 nucleases (RFNs) known as tru-RFNs can be used to reduce off targets' effects. It can also maintain the full-on target activity of CRISPR/Cas9. About 17–20 nt targeted region can be modified complementarity at the 5′-end of a ~100-nt single guide RNA (gRNA) that made a complex with the protein Cas9 can be used to increase the specificity of CRISPR/Cas9 [13]. Tru-gRNA (double length target site) along with RFN (dimerization-dependent activity) can be designed to improve the sensitivity of off targets effects. RFN produced by dimerization of non-specific FokI cleavage domain with an amino terminus of Cas9 protein (dCas9) which helps in the DNA cleavage of spacer region efficiently resulting double the length target site [14]. Thus, undesirable mutagenic activity can be reduced by using RFNs.

11.4 RIBOZYME GRNA RIBOZYME

In 1988, Thomas R. Cech discovered the ribozymes that are catalytic in nature, recommended that ribozymes would be proved extremely valuable in biochemical and medical studies [15]. This approach was first developed in yeast to exploit its potential for multiplex genome editing (MGE). Ribozymes are RNA molecules that catalyzed reaction like that of enzyme. This system is based on the nuclease activity of modified ribozymes [16].

In 2014 an artificial known as ribozyme-gRNA-ribozyme gene (RGR) was developed by Gao & Zhao. In RGR, Hammerhead-type (HH) ribozyme is flanked with the gRNA. Initially, this system was developed to check the expression of single gRNA that after matured is self-cleaved by ribozymes. Researchers reported that by joining multiple cassettes of RGR can be used into a multiplexed system that can be expressed by single promoter and terminator. Recent studies have showed the successful processing of RGR system in zebrafish and in *Leishmania donovani* [17]. The RGR system has been also employed in human cell lines for targeting of two genes by using four gRNA, however the efficiency of gRNA processing was low. The studies demonstrated the limitation of this system with large number of SRNA [18]. But is expected that this system can be used for efficient expression of multiple gRNAs in plants.

11.5 POLYCISTRONIC TRNA-GRNA (PTG/CAS9)

Eukaryotic cells have variety of endogenous RNA-processing machineries. Through genetic engineering, few of them can be used to make multiple gRNAs from a single type of transcript. The polycistronic tRNA–gRNA (PTG)/Cas9 system was first developed for MGE in rice with highest efficiency [19] and also used in maize and kiwifruit with modification. This system has been developed for enhancing the capacity and efficiency of GE [20]. This system constructed by assembly of multiple gRNA into single expression cassette known as *PTG* gene, which can produce multiple gRNA by endogenous post-transcriptional RNA processing. It can produce large PCR-detectable deletions on target sites which can used for identification of mutation [16]. It can carry multiple spacer sequence, the host endogenous tRNA-processing system can excise tRNA fragments from PTG gene. And produce gRNA and guide the dCas9 for targeted gene editing. This system has advantage over other multiplex gene-editing systems as it is simpler in vector construction and transformation [20].

For CRISPR/Cas9 MGE, endogenous tRNA-processing machinery can be used to produce a large number of gRNAs [19]. This approach has been demonstrated in a variety of organism including microbes plants as well as in animals [21] because they all have enzymes essential for the tRNA-processing [22]. The polycistronic tRNA-gRNA (PTG) gene can be produce by accomplishment of these multiple pre-tRNA-gRNA which can cleave to co-expressed multiple gRNA at a time. This cleavage is caused by endogenous tRNA-processing enzymes known as RNases P and Z. Thus, a single PTG gene along with Cas9 can be used to for target modification at a time [19].

This approach has multiple advantages:

1. The tRNA can be used to enhance the expression of gRNA because the tRNA having cis elements can increase the activity of Pol III in process of transcription.
2. The PTG gene can be induce by inducible and tissue-specific promoters specifically Pol II promoters.
3. Moreover, the multiple gRNA can be efficiently packed with the help of small size tRNA.
4. For higher expression of gRNA and MGE. The tRNA can be used along with the CRISPR-Cpf1 system.

11.6 STRATEGIES TO GENERATE CROPS FREE FROM TRANSGENE DURING EDITING TARGET GENES

In agriculture, use of CRISPR is becoming a leading task for achieving transgene free plants and those plants have also been edited through favorable traits [16]. Every sustainable gene-edited crop must be free from transgenes due to various causes. Firstly, plants that comprise a construct based on CRISPR gene-editing are maybe risky towards the environment. Because when pollen or seeds having CRISPR will be released into the environment then neither it will gain acceptance from the public nor it, would gain endorsement from government regulatory organizations [23, 24]. Secondly, traits within the crops must be constant but the existence of the CRISPR constructs makes it challenging to evaluate the constancy as well as penetrance of the phenotypes [16]. As, gene-edited plants without the presence of any transgene is a requirement for the sake of commercialization of any plant having favorable traits based on CRISPR construct. The main objective is to produce gene edited crops free from transgenes with ideal costs of labor and time. Here, we will discuss the current novelties that enhanced the efficiency in separating plants based on CRISPR construct and free from transgenes:

1. Gene Editing Through Agrobacterium-Mediated Transient Expression of Transgenes: Agrobacterium-mediated plant transformation widely used for the addition of transgenes into plant genomes. However, it is also recognized that editing of genes by

mediating transient expression of transgenes can also be achieved through Agrobacterium [25–27]. As in this practice, Cas9 and gRNA can be produced transiently in plants without integrating the transgenes into the genome. So, transgene-free edited plants can be isolated through this procedure. This transient method offers one more advantage that there is no need of sexual segregation for removal of transgenes. Apart from these advantages, this practice is supposed to be time consuming for identification transgenes free desired plants. Proficiency of isolating of plants free from transgenes can be enhanced through more engineering the Agrobacteria, which lead towards reduction of transgene integration.

2. Transgene-Free Gene Editing Through Ribonucleoprotein (RNP): In purified recombinant Cas9 and *in vitro* produced gRNAs can be reconstructed into a Cas9-gRNA RNP complex. In plants, this RNP complex can be supplied through gene gun [4]. This RNP approaches offer various benefits as compared to plasmid DNA-based methods. Because in cellular machinery, various types of cells may not be capable to express sufficient quantity of CRISPR components. Furthermore, this RNP technique does not use any transgene. So, on that basis the edited plants can clearly be categorized as non-transgenic plants. Hence, The RNP compound is normally hosted into protoplasts, immature embryos, or calli via particle bombardment or transformation [28–30]. Conversion of protoplasts is not so challenging, however revival of a plant from a single protoplast is complex and very limited species of plants are responsive to this practice [31]. Plant cells having RNP and without RNP, will be able equally to regenerate without any pressure of selection. The fact is those plants are not transformed and edited which generated through bombardment of RNP into calli. Hence, it will make this practice very lengthy to separate edited plants. Therefore, RNP methods have not been commonly executed in biological laboratories.

3. Fluorescence Marker-Assisted Transgene Removal: For growing the plants, collection of leaf samples, prepare genomic DNA, and genotype through PCR are basic steps used for isolating plants free from transgenes. But this process is very laborious as well as time consuming. But any simple technique can minimize the amount of work for attaining plants free from transgenes by reducing the need

of growing the undesired progeny from T_0 plants. Gao et al. associate a fluorescence marker gene named as to CRISPR components in the similar plasmid [32]. Likewise, some natural color compounds can also be helpful in selection of plants free from transgenes [33]. Although this approach decreases amount of work at least 75%, but this technique still needs a lot of labor as well as time.

4. Gene Editing Through Nanotechnology: A latest exciting report point out that plasmid coated with carbon dots can also be delivered into the cells of plant through foliar application and then Cas9 or gRNAs formed through this technique can be effectively edit [34]. This novel technique possibly can be prolonged towards other plants, which will offer a simple and economical process for gene editing. The main requirement is active delivery of nucleases for CRISPR gene editing, but the problem is that, at present several plant species are not transformable. In plants various forms of nanomaterials have also been verified which includes magnetic nanoparticles, carbon nanotubes (CNTs), and carbon dots [34–37].
5. Self-Elimination of Transgenes: After the completion of gene editing tasks, a CRISPR construct would experience self-elimination process. This type of CRISPR design would significantly decrease the time span and labor which involved in separating plants free from transgenes. This Self-eliminating is named as TKC technology [38]. Total two functional units are planned for this TKC technology: one is a gene-editing unit and the other unit named as a suicide unit. The expression of these two units is temporally controlled. At first, activation of the gene-editing happened. After this editing, the suicide genes are activated for the death of cell. Subsequently, just those cells will stay alive which would be transgene free.

11.7 THE BENEFICIAL FEATURES OF CRISPR-CAS SYSTEM

In plant GE CRISPR/Cas systems provides an amazing change through wide genetic diversity. Some of the most beneficial tool for crop improvement are discussed here:

1. **Crop Trait Improvement Through Knockout:** Removal of negative elements is a favorable policy towards crop trait improvement. Hence, this knocking out of genes that converse unwanted traits, is

the easiest application of CRISPR/Cas9. A lot of traits have been modified by using this approach which includes yield, quality, biotic, and abiotic-stress resistance.

2. **Crop Trait Improvement Through Knock-In (KI) and Replacement:** KIs and replacements assist breeding through novel alleles lacking linkage drag [39]. This approach has unlimited benefits towards crop improvement as it modifies several traits through gene loading.
3. **Uses of Base Editors in Plants:** Base editing in plant breeding is relatively valuable as well as for crop improvement. In coding regions, the main use of base editing is to converse herbicide resistance.
4. **Gene Regulation Through Fine Tuning:** Modulating gene expression is a suitable method for investigating gene functions. At different stages expression of gene can affected. These stages include transcription, mRNA processing, and mRNA translation. All these procedures are under the control of a series of cis-regulatory elements. And then through GE these procedures can also be adjusted [40].
5. **Antiviral Plant Breeding Strategies:** It is thought that viruses are responsible for maximum diseases in plants, and as a result causes huge damages in agricultural production globally [41]. Since this CRISPR/Cas based system provides a good defense mechanism against these viruses.
6. **High-Throughput Plant Mutant Libraries:** The most valuable tools for functional genomics as well as for genetic Improvement are whole-genome-scale mutant libraries. By using different induction methods of mutations, traditional mutant libraries are also established. But, by using this approach many generations will be required to alleviate loss-of-function caused by mutations.

11.8 CURRENT LIMITATIONS OF CRISPR CAS SYSTEM

1. Because the size of the CRISPR/Cas9 system is comparatively vast, that's why, it is not appropriate for packing into viral vectors. The CRISPR system having small in size is obligatory for effective GE in crop species.

2. The need of a protospacer adjacent motif (PAM) location at the target sequence restrict the application of Cas9 because official spCas9 only identifies the NGG PAM location once every 8–16 bp. A composite PAM site selection system is necessary to rise target range.
3. Plant GE technology which are established on type II CRISPR/Cas system mainly depends upon Agrobacterium tumefaciens or direct gene transfer by using cultured plant tissues. However, in most plant species, genetic transformation and regeneration from tissue culture is no more effective because for selection it takes a lot of time and it can be the reason of production of somaclonal variation, which makes further mutations.
4. It is very essential to enhance the proficiency of KI HR. For this purpose, highly research is required for improving homology-directed repair (HDR) and viral vector effectiveness. Novel policies should be deliberated. Which includes inhibiting endogenous nonhomologous end-joining (NHEJ activity.
5. Escape of off-target effects through CRISPR/Cas9 is a need of time. To attain this, innovative and developed approaches are prerequisite.

11.9 THE BENEFICIAL FEATURES AND CURRENT LIMITATIONS OF CRISPR CAS SYSTEM

CRISPR is a useful tool that can alter the genetic constitution of the organisms. The most important advantage of this GE tool is its ease and proficiency. It can be used directly in the growing embryo that can alter the DNA in such a way that the mature organism contains more desirable characters [42]. CRISPR/Cas9 decreases the time which are required to modify targeted genes in comparison to other technologies which are also based on the use of embryonic stem cells. It can decrease the time of evolution from millions of years into a few months to years depending upon the kind of living organisms. With the advancements of bioinformatics tools, the highly appropriate sequences for designing guide RNAs as well as optimization of the experimental conditions has resulted in vigorous and reliable methodologies which ensure successful introduction of the desirable mutation in the targeted organisms [43].

CRISPR/Cas9 system has been thought-out the best gene editing tool in plants species till now, however there are also certain limitations that restricts its universal application. With the passage of time, many other better variants of CRISPR/Cas9 system have been introduced to plug the loopholes in the original gene editing tools such as spCas9-NG, base editing, xCas9, etc. [44]. However, a few problems that limits its widespread have been briefly discussed below [45]:

- The editing proficiency of CRISPR/Cas9 system is greatly compromised due to its large size as it is not fit for packing into viral vectors for delivery towards somatic tissues. A minor size CRISPR/Cas9 system is more desirable for the effective GE in plants.
- SpCas9 requires a 5′-NGG-3′ immediately adjacent to a 20-nt DNA target sequences where it only recognizes the NGG PAM site, and this can bound the efficacy as compared with new CRISPR/Cas9 variants.
- CRISPR/Cas9 can cause mutation at random places in the DNA, which is at non-specific loci, but at target sites it's not alike in homology.
- CRISPR/Cas9 can cause off target mutation in genome at multiple places that can complicate already present desirable characters. However, the new variants have greatly improved the off-target mutations.
- For creating a mutant plant. CRISPR/Cas9 require an Agrobacterium-mediated transformation system. But this system is lengthy as well as expensive. Yet, the use of tissue culture can provide efficiency.
- The commercialization of the crops that are edited with CRISPR/Cas9 in various countries is difficult. It is mainly due to the transgenic nature of these crops; therefore, it causes imposition of numerous constraints by regulatory systems for the field release.

11.10 CRISPR-CAS EDITED CROPS REGULATIONS AND SOCIAL ACCEPTANCE

The targeted mutagenesis in the plant will expose different vistas for the improvement of plant genetic potential. Erstwhile, the conventional way of mutagenesis was not able to access the complete genome due to nature of genes integration in the genome. CRISPR/Cas9 would be able to generate mutants from that part of genome which was previously inaccessible to the mutation. This GE tool can incur mutations at multiple loci and can cause large deletion which can pace up the plant breeding process without actually presenting the transgene. The genetically altered organisms could have been a plausible solution if the concerns about the health and environment have not been invoked by different organizations. The traditional approach of genetic engineering uses a DNA construct which is directly inserted in a random manner in one or multiple chromosomes for the manipulation of the genome into a more desirable output. Whereas the random insertions of the genes in the chromosomes can alter the already desirable genes and can leads to a more undesirable state. Moreover, insertion of large chunks of DNA into the chromosome, or triggering a whole new metabolic pathway is a very complicated and mostly unfavorable process in most of the living organisms [46].

The CRISPR/Cas system presents accurate changes towards plant genome which inherits firmly, and transgene section could be detached easily after the editing of target gene. The Cas9/sgRNA system is nowadays one of the innovative plant breeding techniques along with ZFNs (zinc finger nucleases) and TALENS (transcription activator like effector nucleases) [47]. Currently, the efficiency of NPBT is better and faster than the traditional tools available for the breeding methods for the introduction of different traits in the plants without insertion of transgenes. The plants that are developed by using plant breeding techniques are similar to the naturally propagated plants and ought to be estimated on the basis of their end products in spite of on the basis of process of production. Therefore, in this situation CRISPR edited plants might be out of GMO regulations. However, it still poses a serious challenge for its regulation and social acceptance for genetically edited crops. NPBT are being debate around the world in various organizations for its regulations in relation to its GMO legislation, though NPBT crops have their genome edited but they are free of any transgene so they can be publicized as non-GMO for social acceptance [48].

The extent of the potential of CRISPR technology in plant research can assist in widening the present gene pools available in the plants, moreover it will also be able to transfer desirable characters from the other crops into a desirable one without linkage drag. The pyramiding of the desirable characters in the plant can improve the yield and enhance the quality which can solve the problems of food security and malnutrition around the world [46].

KEYWORDS

- **CRISPR Cas9**
- **genome editing**
- **malnutrition**
- **mutants**
- **transcription activator like effector nucleases**
- **transgene free**
- **zinc finger nucleases**

REFERENCES

1. (a) Malzahn, A., Lowder, L., & Qi, Y., (2017). Plant genome editing with TALEN and CRISPR. *Cell & Bioscience, 7*(1), 1–18; (b) Mao, Y., Botella, J. R., Liu, Y., & Zhu, J. K., (2019). Gene editing in plants: Progress and challenges. *National Science Review, 6*(3), 421–437.
2. Wolter, F., & Puchta, H., (2017). Knocking out consumer concerns and regulator's rules: Efficient use of CRISPR/Cas ribonucleoprotein complexes for genome editing in cereals. *Genome Biol., 18*(1), 1–3.
3. Arora, L., & Narula, A., (2017). Gene editing and crop improvement using CRISPR-Cas9 system. *Frontiers in Plant Science, 8*, 1932.
4. Woo, J. W., Kim, J., Kwon, S. I., Corvalán, C., Cho, S. W., Kim, H., Kim, S. G., et al., (2015). DNA-free genome editing in plants with preassembled CRISPR-Cas9 ribonucleoproteins. *Nat. Biotechnol., 33*(11), 1162–1164.
5. Metje-Sprink, J., Menz, J., Modrzejewski, D., & Sprink, T., (2019). DNA-free genome editing: Past, present, and future. *Frontiers in Plant Science, 9*, 1957.
6. Sarma, A. P., Jain, C., Solanki, M., Ghangal, R., & Patnaik, S., (2020). Role of gene editing tool CRISPR-Cas in the management of antimicrobial resistance. In: *Sustainable Agriculture Reviews* (Vol. 46, pp. 129–146). Springer.
7. Barbour, A., Glogauer, J., Grinfeld, L., Ostadsharif, M. R., Fine, N., Tenenbaum, H., & Glogauer, M., (2021). The role of CRISPR-Cas in advancing precision periodontics. *Journal Periodontal Res., 56*(3), 454–461.
8. Chylinski, K., Le Rhun, A., & Charpentier, E., (2013). The tracrRNA and Cas9 families of type II CRISPR-Cas immunity systems. *RNA Biol., 10*(5), 726–737.
9. Jiang, F., & Doudna, J. A., (2017). CRISPR–Cas9 structures and mechanisms. *J. Annu. Rev. Biophys., 46*, 505–529.
10. Miao, J., Guo, D., Zhang, J., Huang, Q., Qin, G., Zhang, X., Wan, J., et al., (2013). Targeted mutagenesis in rice using CRISPR-Cas system. *Journal Cell Research, 23*(10), 1233–1236.

11. Wei, C., Liu, J., Yu, Z., Zhang, B., Gao, G., & Jiao, R. J. J. O. G., (2013). TALEN or Cas9–rapid, efficient, and specific choices for genome modifications. *J. Genet. Genomics, 40*(6), 281–289.
12. Hamilton, J. R., Tsuchida, C. A., Nguyen, D. N., Shy, B. R., McGarrigle, E. R., Espinoza, C. R. S., Carr, D., et al., (2021). Targeted delivery of CRISPR-Cas9 and transgenes enables complex immune cell engineering. *Journal Cell Research 35*(9), 109207.
13. Jinek, M., Chylinski, K., Fonfara, I., Hauer, M., Doudna, J. A., & Charpentier, E., (2012). A programmable dual-RNA–guided DNA endonuclease in adaptive bacterial immunity. *Science, 337*(6096), 816–821.
14. Guilinger, J. P., Thompson, D. B., & Liu, D. R., (2014). Fusion of catalytically inactive Cas9 to FokI nuclease improves the specificity of genome modification. *Nat. Biotechnol., 32*(6), 577–582.
15. Cech, T. R., (1988). Ribozymes and their medical implications. *JAMA, 260*(20), 3030–3034.
16. Gao, Y., & Zhao, Y. (2014). Self-processing of ribozyme-flanked RNAs into guide RNAs in vitro and in vivo for CRISPR-mediated genome editing. *J. Integr. Plant Biol., 56*(4), 343–349.
17. Lee, R. T. H., Ng, A. S. M., & Ingham, P. W., (2016). Ribozyme mediated gRNA generation for *in vitro* and *in vivo* CRISPR/Cas9 mutagenesis. *PLoS One, 11*(11), e0166020.
18. Xu, L., Zhao, L., Gao, Y., Xu, J., & Han, R., (2017). Empower multiplex cell and tissue-specific CRISPR-mediated gene manipulation with self-cleaving ribozymes and tRNA. *Nucleic Acids Res., 45*(5), e28–e28.
19. Xie, K., Minkenberg, B., & Yang, Y., (2015). Boosting CRISPR/Cas9 multiplex editing capability with the endogenous tRNA-processing system. *Proceedings of the National Academy of Sciences, 112*(11), 3570–3575.
20. Zhang, Z., Mao, Y., Ha, S., Liu, W., Botella, J. R., & Zhu, J. K., (2016). A multiplex CRISPR/Cas9 platform for fast and efficient editing of multiple genes in *Arabidopsis. Plant Cell Rep., 35*(7), 1519–1533.
21. Schwartz, C. M., Hussain, M. S., Blenner, M., & Wheeldon, I., (2016). Synthetic RNA polymerase III promoters facilitate high-efficiency CRISPR–Cas9-mediated genome editing in *Yarrowia lipolytica. ACS Synthetic Biology, 5*(4), 356–359.
22. Čermák, T., Curtin, S. J., Gil-Humanes, J., Čegan, R., Kono, T. J., Konečná, E., Belanto, J. J., et al., (2017). A multipurpose toolkit to enable advanced genome engineering in plants. *The Plant Cell, 29*(6), 1196–1217.
23. Callaway, E., (2018). CRISPR plants now subject to tough GM laws in European Union. *Nature, 560*(7716), 16–17.
24. Huang, S., Weigel, D., Beachy, R. N., & Li, J., (2016). A proposed regulatory framework for genome-edited crops. *Nat. Genet., 48*(2), 109–111.
25. Amoah, B., Wu, H., Sparks, C., & Jones, H., (2001). Factors influencing *Agrobacterium*-mediated transient expression of *uidA* in wheat inflorescence tissue. *J. Exp. Bot., 52*(358), 1135–1142.
26. Krenek, P., Samajova, O., Luptovciak, I., Doskocilova, A., Komis, G., & Samaj, J., (2015). Transient plant transformation mediated by *Agrobacterium tumefaciens*: Principles, methods and applications. *Biotechnol. Adv., 33*(6), 1024–1042.

27. Li, X. Q., Liu, C. N., Ritchie, S. W., Peng, J. Y., Gelvin, S. B., & Hodges, T. K., (1992). Factors influencing *Agrobacterium*-mediated transient expression of gusA in rice. *Plant Mol. Biol., 20*(6), 1037–1048.
28. Liang, Z., Chen, K., Li, T., Zhang, Y., Wang, Y., Zhao, Q., Liu, J., et al., (2017). Efficient DNA-free genome editing of bread wheat using CRISPR/Cas9 ribonucleoprotein complexes. *Nature Communications, 8*(1), 1–5.
29. Svitashev, S., Schwartz, C., Lenderts, B., Young, J. K., & Cigan, A. M., (2016). Genome editing in maize directed by CRISPR–Cas9 ribonucleoprotein complexes. *Nature Communications, 7*(1), 1–7.
30. Toda, E., Koiso, N., Takebayashi, A., Ichikawa, M., Kiba, T., Osakabe, K., Osakabe, Y., et al., (2019). An efficient DNA-and selectable-marker-free genome-editing system using zygotes in rice. *Nature Plants, 5*(4), 363–368.
31. Lin, C. S., Hsu, C. T., Yang, L. H., Lee, L. Y., Fu, J. Y., Cheng, Q. W., Wu, F. H., et al., (2018). Application of protoplast technology to CRISPR/Cas9 mutagenesis: From single-cell mutation detection to mutant plant regeneration. *Plant Biotechnol. J., 16*(7), 1295–1310.
32. Gao, X., Chen, J., Dai, X., Zhang, D., & Zhao, Y., (2016). An effective strategy for reliably isolating heritable and Cas9-free *Arabidopsis* mutants generated by CRISPR/Cas9-mediated genome editing. *Plant Physiol., 171*(3), 1794–1800.
33. Liu, Y., Zeng, J., Yuan, C., Guo, Y., Yu, H., Li, Y., & Huang, C., (2019). Cas9-PF, an early flowering and visual selection marker system, enhances the frequency of editing event occurrence and expedites the isolation of genome-edited and transgene-free plants. *Plant Biotechnol. J., 17*(7), 1191.
34. Doyle, C., Higginbottom, K., Swift, T. A., Winfield, M., Bellas, C., Benito-Alifonso, D., Fletcher, T., et al., (2019). A simple method for spray-on gene editing in planta. *BioRxiv,* 805036.
35. Bao, W., Wan, Y., & Baluška, F., (2017). Nanosheets for delivery of biomolecules into plant cells. *Trends Plant Sci., 22*(6), 445–447.
36. Demirer, G. S., Zhang, H., Goh, N. S., González-Grandío, E., & Landry, M. P., (2019). Carbon nanotube–mediated DNA delivery without transgene integration in intact plants. *Nat. Protoc., 14*(10), 2954–2971.
37. Wang, P., Lombi, E., Zhao, F. J., & Kopittke, P. M., (2016). Nanotechnology: A new opportunity in plant sciences. *Trends Plant Sci., 21*(8), 699–712.
38. He, Y., Zhu, M., Wang, L., Wu, J., Wang, Q., Wang, R., & Zhao, Y., (2018). Programmed self-elimination of the CRISPR/Cas9 construct greatly accelerates the isolation of edited and transgene-free rice plants. *Molecular Plant, 11*(9), 1210–1213.
39. Luo, M., Gilbert, B., & Ayliffe, M., (2016). Applications of CRISPR/Cas9 technology for targeted mutagenesis, gene replacement and stacking of genes in higher plants. *Plant Cell Rep., 35*(7), 1439–1450.
40. Chen, K., Wang, Y., Zhang, R., Zhang, H., & Gao, C., (2019). CRISPR/Cas genome editing and precision plant breeding in agriculture. *Annu. Rev. Plant Biol., 70*, 667–697.
41. Zaidi, S. S. E. A., Tashkandi, M., Mansoor, S., & Mahfouz, M. M., (2016). Engineering plant immunity: Using CRISPR/Cas9 to generate virus resistance. *Frontiers in Plant Science, 7*, 1673.
42. Young, S. A., Aitken, R. J., & Ikawa, M., (2015). Advantages of using the CRISPR/Cas9 system of genome editing to investigate male reproductive mechanisms using mouse models. *Asian Journal of Androl., 17*(4), 623.

43. Harrison, M. M., Jenkins, B. V., O'Connor-Giles, K. M., & Wildonger, J., (2014). A CRISPR view of development. *Genes Dev.*, *28*(17), 1859–1872.
44. Wang, H., La Russa, M., & Qi, L. S., (2016). CRISPR/Cas9 in genome editing and beyond. *Annu. Rev. Biochem.*, *85*, 227–264.
45. Manghwar, H., Lindsey, K., Zhang, X., & Jin, S., (2019). CRISPR/Cas system: Recent advances and future prospects for genome editing. *Trends in Plants Science, 24*(12), 1102–1125.
46. Khatodia, S., Bhatotia, K., Passricha, N., Khurana, S., & Tuteja, N., (2016). The CRISPR/Cas genome-editing tool: Application in improvement of crops. *Front. Plant Sci.*, *7*, 506.
47. Klann, T. S., Black, J. B., Chellappan, M., Safi, A., Song, L., Hilton, I. B., Crawford, G. E., et al., (2017). CRISPR–Cas9 epigenome editing enables high-throughput screening for functional regulatory elements in the human genome. *Nat. Biotechnol, 35*(6), 561.
48. Araki, M., & Ishii, T., (2015). Towards social acceptance of plant breeding by genome editing. *Trends Plant Sci. 20*(3), 145–149.

CHAPTER 12

HARNESSING THE POTENTIAL OF GENOME EDITING IN LARGE ANIMALS FOR FOOD SECURITY: PROSPECTS AND CHALLENGES

IBRAHIM BALA SALISU,[1,2] AMINA YAQOOB,[2] and SAHAR SADAQAT[2]

[1]*Department of Animal Science, Faculty of Agriculture, Federal University, Dutse, Nigeria*

[2]*Center of Excellence in Molecular Biology, University of the Punjab, Lahore, Pakistan*

ABSTRACT

To maximize production and satisfy increasing human demands for foods and other animal products, livestock farmers have been shaping the livestock genetically over a long period. Selection and conventional breeding were the only promising method for decades, which is an exasperating and gradual process of accruing incremental genetic gains. A more direct approach with the greater effects and possibility of creating genetic variations within a single generation is the advancement of technology known as transgenic livestock. Over the recent years, this technology faced a lot of methodological complications and limitations which have been greatly reduced by progressive developments. The dawn of state of art CRISPR/Cas genome engineering technology, in particular, has added a new level of efficacy and precision, and virtually allowed for making any changes in an organism's genome with extraordinary precision and specificity in addition

CRISPR-Based Gene Editing: A Solution for Global Food Security. Shakeel Ahmad, Asif Ali Khan, and Muhammad Arshad Javed (Eds.)

to being less laborious and more direct in comparison to conventional breeding strategies. Through genome editing (GE) technology, genetic improvement of large animals can be achieved easily by an understanding of the phenotypic effect of genetic sequence variants. In this chapter, the potential, as well as the challenges of GE in large animals for food security, will be discussed.

12.1 INTRODUCTION

The recent technological advances in agricultural practices including plant and animal breeding match well with the rising demand for food, feed, and fiber by the ever-increasing human population [1]. Feeding a balanced and nutritious diet to this burgeoning world population will be one of the human greatest challenges of this 21st century [2, 3]. According to the projection made by the Food and Agriculture Organization (FAO), by 2020, the world demand for animal-based eatables will rise by 70% [4]. By increasing the use of plant-based food and artificial meat may lead to the improved food quality but also be the reason nutritional scarcities and undernourishment in underdeveloped countries due to lack of basic proteins in human diet [2]. According to the FAO report on World Food Outlook, about 20 g of animal protein given to a person per day or 7.3 kg per year would effectively tackle malnutrition and undernourishment. This can be accomplished by eating 33 kg of lean meat or 45 kg of fish per year or 60 kg of eggs or 230 kg of milk per year. In the daily food intake, these sources are typically combined, but if these are not easily available, the intake of the other food stuff must be increased to overcome the deficiency [5]. As such, a significant improvement of livestock production is required to achieve this demand-driven livestock supply on a global scale [5]. Achieving such a requirement in a manner that has minimal environmental impact will require the employment of cutting-edge technologies. Thus, the introduction of advanced technology, such as genome editing (GE) would likely help in meeting this increasing demand for animal food products in a way having least effect on the surroundings to meet the standards of animal prosperity [2].

Over the past decades, genetic improvement in livestock is mainly achieved via the traditional breeding program in which genetic gain comes directly from the assortment of natural alterations existing within the breed. Typically, the livestock breeder strives to identify the animals which possess the superior genetic variations which are associated with the traits of financial implications [6]. Historically, this was done by cross-breeding accompanied by recurrent backcrossing to remove undesirable features that introgressed as selection strain from the other breed [7]. The main restriction in this method is the relation and the genetic variations within a breed apart from being exasperating in accruing incremental genetic gains [8]. These biological predicaments can be overcome easily through the application of GE technology, a new opportunity that allows for rapid attaining of genetic gain with higher precision [9]. Through GE, useful extant genetic variants can be accurately incorporated into organized livestock breeding programs. Haplotypes amongst breed and movement of beneficial alleles, complete knockout of undesirable genes and genetic variants can be fixed [10–12]. The earlier genetic engineering method was restricted to the transgene insertion into the genome in the form of recombinant DNA cassette, comprising of a promoter and protein-coding region along with the terminator region [7]. Due to the random gene integration as well as the variation of epigenetic environment among various regions of the genome, it will thus be difficult for a breeder to predict all the possible effects that the inserted gene would have on the animal. This also implies that each genetically modified (GM) founder animal had the gene incorporated into a numerous location within the genome. But with the recent GE methods, a breeder can precisely target the alteration of the existing genes and regulatory elements to delete detrimental mutations, like recessive lethal genetic variants, or introduce the desired traits of economic importance like disease resistance and heat tolerance without upsetting other existing genetic features of the animals [13–15]. The breeder can also predict the expected outcome in improved gain [16]. This chapter highlights the recent advances in GE technologies and their potential role in improving livestock production for food security.

12.2 LARGE ANIMALS' BREEDING METHODS

Conventional breeding and molecular breeding are two main methodologies being used in animal farming [17]. Soma clonal variations, selection, and mutagenic breeding come under the flag of conventional breeding. Whereas molecular breeding methods include genetic amendments in animals using gene integration, complete knockout of gene, producing mutation within

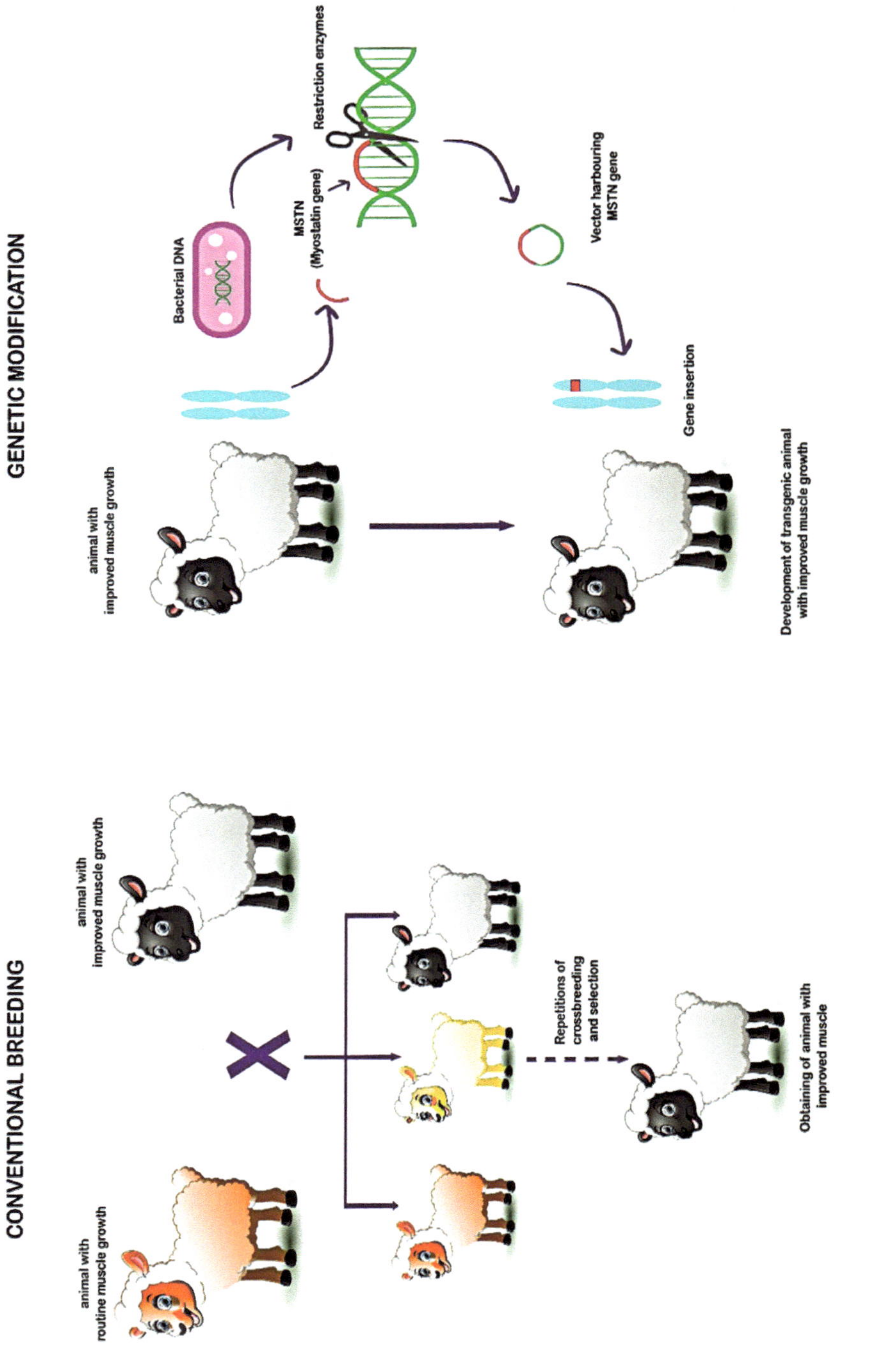
GENETIC MODIFICATION
animal with
improved muscle growth
Bacterial DNA
MSTN
(Myostatin gene)
Restriction enzymes
Vector harbouring
MSTN gene
Gene insertion
Development of transgenic animal
with improved muscle growth
CONVENTIONAL BREEDING
animal with
routine muscle growth
animal with
improved muscle growth
Repetitions of
crossbreeding
and selection
Obtaining of animal with
improved muscle

gene, interfering RNA, or silencing the gene. Among all the molecular methods, gene transformation approach is of great importance because of easiness, specificity, and being less time consuming [73]. Both marker-assisted breeding methods and advanced genetic approaches are significant; the primary goal is to recognize and isolate the relevant genes, which are involved in anticipated trait improvement and then sustainable propagation of those genes throughout the generations. We have already gotten a detailed concept of GE and its applications in agriculture, farming, and livestock. In this section, we are going to discuss and compare the conventional breeding with the modern GE methods in the context of livestock.

12.2.1 CONVENTIONAL BREEDING

Major types of conventional breeding are inbreeding, outbreeding, and assortative mating. The sexual meeting of two closely related organisms is called inbreeding or one can say that all the animals of a breed are related. While the outbreeding is the process of continuously mating different females of a herd to unrelated males of the same breed.

A livestock breeder generally practices the inbreeding among more closely related animals than the average of the breed [18]. As a result, the offspring is homozygous for a desired pair of gene, as he receives the identical genes from both parents, which lead to a superior trait performance. Contrarily, an inbred homozygous could potentially carry undesirable genes, which usually remain hidden in heterozygous and more likely to express producing undesirable traits, and the phenomenon is called inbreeding depression [19]. However, the inbreeding does not usually end up into such disasters due to the practices of line breeding on genetically superior individuals, at farms and breeding centers [20]. Line breeding is the most promising way to sustain a high relationship of inbred to some desirable ancestor. It involves the mating of more-distantly related animals. Line breeding has successfully attempted in cattle breeds. It not only maintains the genes from outstanding animals that are no longer capable of breeding, but also it helps in tagging and easy recognition of the ancestor at the inbreeding centers. On flip side, some drawbacks are associated with line breeding, such as the inescapable setup of inbreeding or the possibility of creating an inferior son of an outstanding bull [21]. When in the mated pairs, male and female phenotypes or genotypes are linked, assortative or nonrandom mating occurs [22]. There are some major evolutionary effects of assortative mating including positive assortment, the tendency of similar individuals to mate with each other, resulting in homozygosity within loci, encouraging relation disparity between loci, and thus expand the modification of quantitative characteristics. On the flip side, negative assortment increases heterozygosity, decreases distress of inbreeding, and may promote sexually antagonistic deviation maintenance. Different effects of assortative breeding, understanding of the selection power of natural population and its evolutionary origin is very important [23]. A weak characteristic positive assortative breeding is more common in animals than negative assortative breeding. A phylogenetic meta-analysis of assortative breeding for a varied variety of phenotypic and genotypic characteristics showed no relationship between the assortment intensity and species richness across and within the major taxonomic groups [24, 25]. It is evident from studies that animal's mate in a typical positive assortative manner. Although it evolves adaptively, but the disruptive or stabilizing selection solely couldn't drive the evolution of assortative breeding, except there is some exemption of indirect selection for adaptive assortative breeding [25]. Assortative mating is most potent in limiting blending when it is determined by a single locus or is essentially complete, or when there is a large mate exploration cost [23].

12.2.2 MOLECULAR BREEDING

Molecular breeding involves the use of genetic manipulation implemented at DNA level to enhance the characters in livestock animals. It includes the identification and characterization of: (i) metabolic genes; (ii) promoters for the expression of foreign genes; and (iii) desired expression of transgenes in organism. Several methods are used for this purpose, such as, transgenesis or gene insertion, mutagenesis, molecular marker-assisted selection, and GE [74].

12.2.2.1 TRANSGENESIS AND MUTAGENESIS

Transgenic animals have been developing for last few decades through the transformation or mutation of genes in the host animal. Transgenesis could be direct or indirect. There isn't any involvement of bacterial or peripheral assistance in direct method but transferring the DNA bit into the protoplast of plant by using micro-projectile bombardment system. Host cell wall is made competent enough to be integrated with transgene by using various chemicals or electrical impulses [17]. But protoplasts are sensitive towards chemicals and impulses resulting in no propagation. Gene guns have also been used for this purpose but resulted in low transformation efficiency, more time consuming, high cost of the apparatus and hindering the expression of transgene due to high copy number of gene. Many other direct approaches have been introduced afterward, i.e., silicon carbide fiber-mediated transformation, microinjections, embryo electrophoresis, infiltration, pollen tube transformation, and liposome-mediated transformation [26]. All these approaches were pretty successful but still have some limitations to either some specific tissues or under some circumstances along with high-cost and time consumption [17]. Furthermore, the ethical concern in consuming the transgenic animals is another burning topic of debate among people. Mutagenesis, on the other hand is claimed to be a safe alternative of transformation, as no foreign gene is inserted into the organism, instead the gene is knocked out or mutate for the desired trait expression. The aim of *in vivo* mutagenesis is to enhance either the intensity or the variety of mutations present within a population of animals. However, the group of transgenic animals is a burdensome process and remains problematic in the application of this technology [27].

12.2.2.2 GENOME-WIDE SELECTION (GWS)

Previously reported simulation studies have declared the genomic selection as the primary contributor to the increased breeding progress, especially by minimizing the generation intervals. Although the practical breeding certification for the use of genome-wide selection (GWS) is not available yet, yet a well-balanced selection practice is being used to increase the revenue of poultry and livestock production [28]. This advanced technology is helpful in eliminating the traits which are hard to deal through traditional breeding practices.

Simulation studies also calculated that genomic selection could increase the breeding yield by 20–40% if applied extensively. For instance, in poultry farms, conventional inbreeding allows the mating of one pure line male with >10 females and eggs are kept for seven days; the progenies are hatched on weekly basis. In the absence of a DNA marker, a random male sample from several hatches is used to be grown as "selection candidate." All the females are tested to carry out the required data input for breeding assessment. Sex ratio from hatch to hatch produces significantly different number of progenies. Hence, only a limited number of sons/dams are attained. By using

DNA markers sons having higher genetic potential from each family can be moved to breeding farm and used as candidates for selection to produce more generations [28]. However, the theoretical advantages of genomic selection are currently being challenged with substantial expenditure. Besides a significant launching cost for the establishment of the method in each line/ gene pool, there are also costs for on-going genotyping of candidates in each generation of selection candidates.

12.2.2.3 GENOME EDITING (GE)

Genome editing (GE) efficiently alters the genome of an organism to get the targeted integration of exogenous genes or the desired modification in endogenous genes. This advanced molecular tool has quickly swept through the entire livestock breeding. Animals created through GE provide the promise of precision and accuracy [29]. In comparison to the conventional GM, approaches which are primarily based on gene transfer between organisms [30], modern precision engineering of the genome provides a chance to make targeted changes in the existing genes and regulatory elements, without actually implanting the foreign DNA [31]. Traditional breeding methods have several limitations like time consuming mating cycle and non-employment of genetic resources. The cost and time efficient genetic editing offers resolutions to these issues. Early hope for the prospects of GM animals for instance [32], faded when it was proved to be more difficult than it seemed at first to produce commercial GM animals. Over the past few years, several improvements in molecular biology have been made, including genetic modification, approaches of making genetic alterations into organisms, and better comprehension of the underlying genetics [16, 33, 34]. In contrast to classical breeding, this genome engineering comprises of the molecular scissors, which have great potential application in large animal's improvement for food security. For instance, through the use of GE tools, genetic disorders and other diseases can be cured by modifying the genetic makeup of the diseased phenotype. A less wanted allele of a gene can also be modified to a more desirable one without any necessary introgression or by breeding the animal with the mate having desirable traits. With the use of this technology, a breeder can turn-off a gene efficiently via non-homologous end-joining breaks (NHEJ) or introduce unique allelic variants [35, 36], and conceptually whole genes or transgenes [37–39] as determined by the nucleic acid sequence template (i.e., HR template), which they want in their target population employing site-directed nuclease (SDN) which eliminates the use of 'artificial' selection, cross-mating, and/or gene editing using random insertion/or gene integration. Gene-edited livestock, such as, *CD163* gene-edited pigs resistant to porcine reproductive respiratory syndrome and SP110 gene knocked cow resistant to tuberculosis, have been produced at feasible and valuable standards. By providing the high precision of genetic engineering and manipulating less cost, CRISPR/Cas9 technology will be of enormous help in future for producing livestock having more desirable traits (Table 12.1).

TABLE 12.1 Comparison between Conventional and Molecular Breeding Methods Used for Large Animals Improvement

Sl. No.	Conventional Breeding	Transgenesis	Random Mutagenesis	Genome-Wide Selection	Genome Editing
Precision	Intermediate	High	None	High	High
Time to achieve.	Years	Months to a few years.	Months with existing backcrossing years.	Months to a few years.	Months
Requires genetic transformation.	No	Yes	No	No	Sometimes
Requires genetic and molecular understanding of trait.	No	Yes	No	Yes	Yes
Examples	Cattle and poultry breeding.	Transformation of insulin expressing genes in Cow's milk.	Knock out disease related genes.	Cattle and poultry breeding.	Disease resistant cow.

12.3 METHODS OF GENOME EDITING (GE)

As it is most commonly practiced, different components are being used for the recognition and accurate cutting of the genetic material in the cells [40].

These components are commonly called SDNs and are also named as sequence-specific nucleases (SSNs). SDNs normally create a double stranded break (DSB) specifically at a given locus within the organism's genome. This DSB break can be repaired in two different ways by the cell.

12.3.1 HOMOLOGOUS RECOMBINATION (HR)

This can be done by using the DNA template which comprises of the sequences matching the flanking site of either side of the DSB break to repair the damage. By providing the appropriate DNA template, HR can be used to precisely add or delete or replace sequence at the site of break.

12.3.2 NONHOMOLOGOUS END JOINING (NHEJ)

This repair mechanisms are applied when there is no available DNA template that is homologous to either side of the breaks. Thus, the cell uses NHEJ to repair the break. In this method, the results of the repair processes may lead to an accurate gene edits or random mutations, respectively. Even though the site of the break is clearly defined, however, the exact change that is taking place is actually random so when the DNA repair itself a number of indels can be seen in the outcomes [41]. The methods which are devised to achieve DNA modifications are briefly discussed in this section.

12.3.3 ZINC FINGER NUCLEASES (ZFNS)

Genome editing (GE) technology was initially started with the introduction of zinc finger nucleases (ZFNs). ZFNs are artificially engineered restriction enzymes that will lead to site-specific editing and can produce DSBs at specific sites [42].

ZFNs have a DNA-binding domain (DBD) which is made of a series of zinc finger domains, each of which can identifies 3–4 base pairs of DNA. Zinc fingers having different DNA recognition site can be threaded to recognize new DNA targets. ZFNs are hybrid heterodimeric proteins, consists of zinc finger and Fok1 endonuclease function as a dimer so that two ZFNs will bind to DNA target closely. This hybrid then creates DSBs by binding with target DNA. The specificity depends mainly on the configuration of the neighboring zinc fingers and DNA target and not all DNA triplets are likely to possess the corresponding zinc finger domains available [43].

ZFNs can be designed easily by using already existing zinc fingers having known recognition sequences. This designing method, however, is not suitable for routine practice. Context-dependent specificity is one of the major disadvantages of the ZFNs. Thus, such specificities can depend on the framework in the head-to-head zinc fingers and DNA targets. In other words, the head-to-head sequences and itself the target sequence determines the specificity. Non-specific breaks lead to the disintegration of the genome and make it volatile [44].

12.3.4 TRANSCRIPTION ACTIVATOR-LIKE EFFECTOR NUCLEASES (TALENS)

Transcription activator-like effectors (TALEs) are Xanthomonas proteins which are obtained naturally (i.e., a genus of Proteobacteria and a plant pathogen). TALENs are comparable to ZFNs in their principle that every targeted DNA sequence necessitates the re-engineering of the proteins [45, 46]. They possess a DNA-binding sites which consist a highly conserved repeats of 33- to 35-aa-long units each of which defines a particular base pair centered on two variable residues located near the unit's center [47].

To produce TALE nucleases, the TALEs have been fused to the FokI DNA-cleavage domain and are used in pairs to identify and cut target DNA just as in the case of ZFNs. The major advantages of TALEs over ZFNs are that a single nucleotide is just recognize by one module as compared to three base pairs recognized by the first zinc finger domain in its DBD [48].

Thus, even when several modules are joined, interference of the recognition sequence will not occur. Since, TALENs cleavage target sequence is more precise than ZFN, any DNA sequence in any organism's genome can potentially be targeted. Such variation allowed for TALENs systems that recognize more target sequences are being created. In terms of efficiency, TALEN appears to be more efficient than ZFNs at producing DSBs in both somatic and pluripotent stem cells [49].

Furthermore, TALENs tend to be less toxic in human cell lines due to off-target breaks in the genome that causes unwanted modifications and toxicity. In comparison to ZFN, TALENs have a higher success rate of GE in livestock embryos through cytoplasmic injection of TALEN mRNA [48]. Simplicity in design [49] and cost effectiveness [50] are additional advantages of TALEN system over ZFNs. Despite these qualities, TALEN system seems to be a bit complex particularly for those that are not conversant with recent molecular biology techniques. Besides, the system also is faced with another limitation of size (obstructing delivery) as compared to ZFN system [48, 51].

12.3.5 CLUSTERED REGULARLY INTERSPACED SHORT PALINDROMIC REPEATS (CRISPR)/CRISPR-ASSOCIATED PROTEIN 9 (CAS9)

CRISPR-Cas a GE technique derived from an antiviral defense mechanism found in a variety of bacteria and archaeal organisms, is the most recent addition to the GE toolkit [52].

This adaptive immunity mechanism targets a Cas nuclease to a specific DNA sequence using a short non-coding RNA. Presently, CRISPR/Cas9 is most commonly used system inspired by the Streptococcus scheme, which employs a small guide RNA complexed with Cas9 nuclease. The first 20 nucleotides at the 5′ end of the small guide RNA that anneal with the corresponding target genomic DNA, which must also contain a protospacer adjacent motif (i.e., PAM: 5′-NGG-3′) immediately following the small guide RNA homology region, direct Cas9 to cleave the target DNA. To create the double stranded break (DSB), target DNA is cut three nucleotides upstream of PAM by Cas9 [53].

Unlike ZFN and TALEN, where the identification of the target DNA site is based on the sequence recognition by artificial proteins which also involves a protein-DNA interaction, the CRISPR/Cas system relies on RNA-DNA interactions which has many advantages over ZFNs and TALENs. Easy design for any genomic targets, simple prediction regarding off-target sites, and the possibility of multiplexing are just a few of the benefits (i.e., altering a number of genomic sites concurrently). Moreover, CRISPR/Cas system is more efficient as compared to ZFNs and TALENs, since modifications can be introduced directly into developing embryos by injecting RNAs encoding the Cas protein and small guide RNA. Despites these qualities, the system also has certain limitations. Despite the fact that the CRISPR/Cas9 system is much easier to execute and build TALEN, the off-target effect in CRISPR/Cas9 is higher than in TALEN. The off-target effect is considered to be low because the DSB occur only after proper binding of a pair of TALEN to the target sequence (Table 12.2).

12.4 THE POTENTIAL OF GENOME EDITING (GE) IN LARGE ANIMALS TO COMBAT THE FOOD INSECURITY

Some of the most recent application of GE in large animals for improved food security are discussed in subsections.

TABLE 12.2 Comparison Various System of Genome Editing

Feature of Comparison	Common Genome Editing Tools		
	ZFNs	**TALENs**	**CRISPR/Cas9**
Origin	Eukaryotes	Bacteria	Bacteria/archaea
Structure	Dimer	Dimer	Monomer
Simplicity in design	Moderate	Somewhat complex	Easier
Target specificity	Low	Higher	Higher
Mode of action	Induce double-strand breaks (DSBs) in target DNA.	Induce double-strand breaks (DSBs) in target DNA.	Induce DSBs in target DNA by wtCas9 or single-strand nicks by Cas9 nickase.
Cleavage efficacy	Efficient	Efficient	Very efficient
Multiplexing	Difficult	Difficult	Simpe
Success rate	Low (~ 24%)	Very high (> 99%)	High (> 90%)
Cytotoxicity	Variable to high	Low	Low
Cost	Low	High	Low

12.4.1 PRODUCTION OF ALLERGEN-FREE MILK

Improving the milk quality is one of the most important milestones achieved through GE, i.e., reducing the level of β-lactoglobulin (BLG) which is the major whey protein and dominant allergen in cow, sheep, and goats milk causing allergy symptoms that vary from mild to life-threatening. Two widely used methods to minimize the allergenic potential of BLG are heat processing and enzymatic hydrolysis, however, these biochemical methods are not only expensive but affect the nutritional value of milk. A more direct approach to lowering the levels of BLG in ruminant milk has been successfully achieved through the application GE tools (i.e., ZFNs and TALENs) by mutating of BLG in cattle [54, 55]. Such cattle having mutated BLG were found to be free from any matured BLG. Similarly, TALENT system was used to knock-out the BLG in goats as mammary gland bioreactors for large-scale production of human lactoferrin [56]. A similar strategy was also used by Luo and his co-workers to produce high expression levels of human serum albumin in cow's milk [57].

12.4.2 PRODUCTION OF MUSCULAR ANIMALS WITH IMPROVED MEAT QUALITY

Production of more muscular animals by creating a positive mutation in the myostatin (MSTN) gene in a number of livestock species, including cattle

[58, 59], sheep [59, 60], goats [61] and pigs [17, 62] has also been achieved magnificently through use of various GE tools.

12.4.3 PRODUCTION OF DISEASE-RESISTANT ANIMALS

Improved disease resistance is another significant field where GE has been used with perceived animal welfare benefits. For instance, production of pigs' tolerance to deadly pig infection African Swine Fever [63] as well as pig resistance to a severe respiratory and reproductive disease, porcine reproductive and respiratory syndrome (PRRS) [64]. A Zinc-finger nickase-mediated insertion was also used to produce a resistant cow against a potentially fatal mammary gland infection (i.e., mastitis) either by inserting human lysozyme gene [38] or lysostaphin gene [37] at the beta-casein locus (an SDN-3 edit) in the dairy cattle. GE produced resistant cattle against infectious bovine tuberculosis by inserting the mouse SP110 nuclear body protein gene into the cattle's genome via SDN-3 system [39].

12.4.4 PRODUCTION OF ANIMALS WITH IMPROVED THERMOREGULATION

Improvement of livestock thermoregulation has also been possible through GE technology. In a cold environment, brown adipose tissue remains as the most essential tissue for the survival of mammal [65]. Uncoupling protein 1 (UCP1) is a mitochondrial inner membrane protein that is involved in brown adipose tissue-mediated thermogenesis, as well as cold tolerance and homeostasis energy regulation. Pigs, on the other hand, lack a functional UCP1, rendering them more resistant to cold and fat deposition resulting in neonatal mortality and low production effectiveness. The advent of GE technology allowed successful insertion/knock-in (KI) of mouse adiponectin-UCP1 into the pig endogenous UCP1 locus via CRISPR/Cas9-mediated HR and the resultant UCP1-KI pigs displayed a significant improvement in body temperature maintenance, improved carcass lean percentage and reduced accumulation of fat [66]. Which, in turn, makes the animals more valuable in the livestock industry.

12.4.5 PRODUCTION OF POLLED COW

Producing hornless cows is one of the early applications of GE [35, 67]. Most of the dairy cows are naturally born to possess huge horns which can easily harm not only their handler but any other farmworker that may come near them. Thus, the dehorning and/or disbudding (i.e., removal of horn-producing cells by chemical means) in calves less than two months of age become a common practice among the livestock owner for safety and economic purposes [68]. However, this chemical disbudding is a painful and unpleasant process that needs to stop for animal welfare [69]. Though naturally hornless (polled) cows do exist, but the majority appears to be beef cattle with much worse milking capacity. Increasing the incidence of the polled gene variant (allele) in the dairy herd via selective breeding will take a very long process especially if the milk yield has to be preserved at the same time. This gene mutation, however, has been specifically introduced into dairy cows through GE [35]. The production polled cows ensured not only the safety of the farmers/farm workers but also allowed for easy handing and milking (Figure 12.1).

FIGURE 12.1 Various application of gene editing in large animals for improve food security.

12.5 GENOME EDITING (GE) PROSPECTS AND CHALLENGES

Genome editing (GE) is a recent and emerging technology that allowed for unprecedented control of the genetic material, providing an opportunity to make rapid developments in basic and applied biology. This technology offers revolutionary approaches to modify, control, determine, and imagine genomes in large animals, possibly providing new applications in agriculture biomedicine. It is anticipated that in the near future, more numbers of applications such as genome-edited NHPs combined with SCNT, livestock organ xenotransplantation used in clinical trials, and genome-edited livestock-derived meat making its way to the food table will be materialize. However, integrating GE into agriculture especially its livestock aspect and biomedicine is faced with various challenges. It is evident that the efficiency and specificity of the current available tools used in GE requires some kind of upgrade but concerns regarding ethical and safety of using GM tissues, organs, and animals remain the major point of discussion. The approach of producing large founder animals with a desired allele in one stage, without a long period of breeding, is in high demand in large animal improvements. However, mosaic mutations, which are commonly found in zygote injection-based GE, are another possible stumbling block in large animal GE. These problems could possibly be resolved by tagging Cas9 with ubiquitin proteasomal degradation signals [70] and introducing editing components in a suitable format for instance;(introducing a Cas9 protein/small guide RNA complex into very early-stage zygotes [71, 72].

It also appears to be difficult to perform GE in cells with low transfection efficiency or in certain cultured cells such as primary cultured cells. Genotoxicity is a characteristic problem of enzymes that operate on nucleic acids, despite the fact that highly specific endonucleases are supposed to minimize or even eradicate this problem.

Other issues that could potentially affect the progress of this novel technology as it relates large animal improvement include how GE compares to other approaches of genetic manipulation, environmental, and animal welfare impacts of specific applications, values of producers and consumers, and economic impacts, among others. Much more needs to be learned concerning the variety of sociocultural factors that affect risk perception and technological acceptance of this new innovative technology.

KEYWORDS

- **CRISPR/Cas**
- **food security**
- **genome editing**
- **homologous recombination**
- **large animals**
- **transcription activator-like effector nucleases**
- **zinc finger nucleases**

REFERENCES

1. Bogdanove, A., Donovan, D., Elorriaga, E., Kuzma, J., Pauwels, K., Strauus, S., & Voytas, D., (2018). *Genome Editing in Agriculture: Methods, Applications, and Governance* (Vol. 60, pp. 25–42). CAST Issue Paper.
2. McFarlane, G. R., Salvesen, H. A., Sternberg, A., & Lillico, S. G., (2019). On-farm livestock genome editing using cutting edge reproductive technologies. *Front. Sustain. Food Syst. 3*, 106. doi: 10.3389/fsufs.
3. Giller, K. E., (2020). The food security conundrum of Sub-Saharan Africa. *Global Food Security, 26*, 100431.
4. Alexandratos, N., & Bruinsma, J., (2017). *The 2012 Revision World Agriculture Towards 2030/2050: The 2012 Revision.*
5. FAO, (2014). *World Food Outlook.*
5. Michalk, D. L., Kemp, D. R., Badgery, W. B., Wu, J., Zhang, Y., & Thomassin, P. J., (2019). Sustainability and future food security—A global perspective for livestock production. *Land Degradation & Development, 30*, 561–573.
6. Zhao, J., Lai, L., Ji, W., & Zhou, Q., (2019). Genome editing in large animals: Current status and future prospects. *National Science Review, 6*, 402–420.
7. Bishop, T. F., & Van, E. A. L., (2020). Genome editing approaches to augment livestock breeding programs. *Journal of Experimental Biology, 223.*
8. Lee, K., Uh, K., & Farrell, K., (2020). Current progress of genome editing in livestock. *Theriogenology.*
9. Ruan, J., Xu, J., Chen-Tsai, R. Y., & Li, K., (2017). Genome editing in livestock: Are we ready for a revolution in animal breeding industry? *Transgenic Research, 26*, 715–726.
10. Rexroad, C., Vallet, J., Matukumalli, L. K., Reecy, J., Bickhart, D., Blackburn, H., Boggess, M., et al., (2019). Genome to phenome: Improving animal health, production, and well-being–a new USDA blueprint for animal genome research 2018–2027. *Frontiers in Genetics, 10*, 327.
11. Raguž, N., & Lukić, B., (2020). *Potential Gain of Genome Editing for Improved Animal Breeding* (p. 464). sa55.

12. Van, E. A. L., Silva, F. D. F., Trott, J. F., & Zilberman, D., (2020). Genetic engineering of livestock: The opportunity cost of regulatory delay. *Annual Review of Animal Biosciences, 9.*
13. Schlegel, G., Ringseis, R., Windisch, W., Schwarz, F., & Eder, K., (2012). Effects of a rumen-protected mixture of conjugated linoleic acids on hepatic expression of genes involved in lipid metabolism in dairy cows. *Journal of Dairy Science, 95*, 3905–3918.
14. Davis, S., Spelman, R., & Littlejohn, M., (2017). Breeding and genetics symposium: Breeding heat tolerant dairy cattle: The case for introgression of the "slick" prolactin receptor variant into *Bos taurus* dairy breeds. *Journal of Animal Science, 95*, 1788–1800.
15. Proudfoot, C., Lillico, S., & Tait-Burkard, C., (2019). Genome editing for disease resistance in pigs and chickens. *Animal Frontiers, 9*, 6–12.
16. Laible, G., Wei, J., & Wagner, S., (2015). Improving livestock for agriculture–technological progress from random transgenesis to precision genome editing heralds a new era. *Biotechnology Journal, 10*, 109–120.
17. Rao, S., Fujimura, T., Matsunari, H., Sakuma, T., Nakano, K., Watanabe, M., Asano, Y., et al., (2016). Efficient modification of the myostatin gene in porcine somatic cells and generation of knockout piglets. *Molecular Reproduction and Development, 83*, 61–70.
18. Szulkin, M., Stopher, K. V., Pemberton, J. M., & Reid, J. M., (2013). Inbreeding avoidance, tolerance, or preference in animals? *Trends in Ecology & Evolution, 28*, 205–211.
19. Doekes, H. P., Veerkamp, R. F., Bijma, P., De Jong, G., Hiemstra, S. J., & Windig, J. J., (2019). Inbreeding depression due to recent and ancient inbreeding in Dutch Holstein–Friesian dairy cattle. *Genetics Selection Evolution, 51*, 54.
20. Northcutt, S. L., Buchanan, D. S., & Clutter, A. C. (2004). *Inbreeding in Cattle.* Oklahoma Cooperative Extension Service.
21. Lush, J. L., (2017). Linebreeding. *Bulletin, 25*, 1.
22. Kopp, M., Servedio, M. R., Mendelson, T. C., Safran, R. J., Rodríguez, R. L., Hauber, M. E., Scordato, E. C., et al., (2018). Mechanisms of assortative mating in speciation with gene flow: Connecting theory and empirical research. *The American Naturalist, 191*, 1–20.
23. Irwin, D. E., (2020). Assortative mating in hybrid zones is remarkably ineffective in promoting speciation. *The American Naturalist, 195*, E150-E167.
24. Alexander, R. M., Jayes, A., Maloiy, G., & Wathuta, E., (1979). Allometry of the limb bones of mammals from shrews (*Sorex*) to elephant (*Loxodonta*). *Journal of Zoology, 189*, 305–314.
25. Janicke, T., Marie-Orleach, L., Aubier, T. G., Perrier, C., & Morrow, E. H., (2019). Assortative mating in animals and its role for speciation. *The American Naturalist, 194*, 865–875.
26. Saeed, T., & Shahzad, A., (2016). Basic principles behind genetic transformation in plants. *Biotechnological Strategies for the Conservation of Medicinal and Ornamental Climbers* (pp. 327–350). Springer.
27. Dyck, M. K., Lacroix, D., Pothier, F., & Sirard, M. A., (2003). Making recombinant proteins in animals–different systems, different applications. *Trends in Biotechnology, 21*, 394–399.
28. Preisinger, R., (2012). Genome-wide selection in poultry. *Animal Production Science, 52*, 121–125.

29. Bruce, A., (2017). Genome edited animals: Learning from GM crops? *Transgenic Research, 26*, 385–398.
30. Clark, J., & Whitelaw, B., (2003). A future for transgenic livestock. *Nature Reviews Genetics, 4*, 825–833.
31. Fahrenkrug, S., Blake, A., Carlson, D., Doran, T., Van, E. A., Faber, D., Galli, C., Gao, Q., Hackett, P., & Li, N., (2010). Precision genetics for complex objectives in animal agriculture. *Journal of Animal Science, 88*, 2530–2539.
32. Bulfield, G., (1990). Genetic manipulation of farm and laboratory animals. *The Bio-revolution: Cornucopia or Pandora's Box.*
33. Bruce, A., Castle, D., Gibbs, C., Tait, J., & Whitelaw, C. B. A., (2013). Novel GM animal technologies and their governance. *Transgenic Research, 22*, 681–695.
34. Tan, W., Proudfoot, C., Lillico, S. G., & Whitelaw, C. B. A., (2016). Gene targeting, genome editing: From Dolly to editors. *Transgenic Research, 25*, 273–287.
35. Carlson, D. F., Lancto, C. A., Zang, B., Kim, E. S., Walton, M., Oldeschulte, D., Seabury, C., et al., (2016). Production of hornless dairy cattle from genome-edited cell lines. *Nature Biotechnology, 34*, 479–481.
36. Lillico, S. G., Proudfoot, C., King, T. J., Tan, W., Zhang, L., Mardjuki, R., Paschon, D. E., et al., (2016). Mammalian interspecies substitution of immune modulatory alleles by genome editing. *Scientific Reports, 6*, 21645.
37. Liu, X., Wang, Y., Guo, W., Chang, B., Liu, J., Guo, Z., Quan, F., & Zhang, Y., (2013). Zinc-finger nickase-mediated insertion of the lysostaphin gene into the beta-casein locus in cloned cows. *Nature Communications, 4*, 1–11.
38. Liu, X., Wang, Y., Tian, Y., Yu, Y., Gao, M., Hu, G., Su, F., et al., (2014). Generation of mastitis resistance in cows by targeting human lysozyme gene to β-casein locus using zinc-finger nucleases. *Proceedings of the Royal Society B: Biological Sciences, 281*, 20133368.
39. Wu, H., Wang, Y., Zhang, Y., Yang, M., Lv, J., Liu, J., & Zhang, Y., (2015). TALE nickase-mediated SP110 knockin endows cattle with increased resistance to tuberculosis. *Proceedings of the National Academy of Sciences, 112*, E1530–E1539.
40. Voytas, D. F., (2013). Plant genome engineering with sequence-specific nucleases. *Annual Review of Plant Biology, 64*, 327–350.
41. Nemudryi, A., Valetdinova, K., Medvedev, S., & Zakian, S., (2014). TALEN and CRISPR/Cas genome editing systems: Tools of discovery. *Acta Naturae (англоязычная версия), 6.*
42. Khalil, A. M., (2020). The genome editing revolution. *Journal of Genetic Engineering and Biotechnology, 18*, 1–16.
43. Paschon, D. E., Lussier, S., Wangzor, T., Xia, D. F., Li, P. W., Hinkley, S. J., Scarlott, N. A., Lam, S. C., Waite, A. J., & Truong, L. N., (2019). Diversifying the structure of zinc finger nucleases for high-precision genome editing. *Nature Communications, 10*, 1–12.
44. Gupta, A., Hall, V. L., Kok, F. O., Shin, M., McNulty, J. C., Lawson, N. D., & Wolfe, S. A., (2013). Targeted chromosomal deletions and inversions in zebrafish. *Genome Research, 23*, 1008–1017.
45. Bogdanove, A. J., & Voytas, D. F., (2011). TAL effectors: Customizable proteins for DNA targeting. *Science, 333*, 1843–1846.
46. Christian, M., Qi, Y., Zhang, Y., & Voytas, D., (2013). Targeted mutagenesis of *Arabidopsis thaliana* using engineered TAL effector nucleases. *G3 3*, 1697–1705.

47. Kim, M. S., & Kini, A. G., (2017). Engineering and application of zinc finger proteins and TALEs for biomedical research. *Molecules and Cells, 40*, 533.
48. Khan, S. H., (2019). Genome-editing technologies: Concept, pros, and cons of various genome-editing techniques and bioethical concerns for clinical application. *Molecular Therapy-Nucleic Acids, 16*, 326–334.
49. Huo, Z., Tu, J., Xu, A., Li, Y., Wang, D., Liu, M., Zhou, R., et al., (2019). Generation of a heterozygous p53 R249S mutant human embryonic stem cell line by TALEN-mediated genome editing. *Stem Cell Research, 34*, 101360.
50. East-Seletsky, A., O'Connell, M. R., Burstein, D., Knott, G. J., & Doudna, J. A., (2017). RNA targeting by functionally orthogonal type VI-A CRISPR-Cas enzymes. *Molecular Cell, 66*, 373–383. e373.
51. Gaj, T., Sirk, S. J., Shui, S. L., & Liu, J., (2016). Genome-editing technologies: Principles and applications. *Cold Spring Harbor Perspectives in Biology, 8*, a023754.
52. Wright, A. V., Nuñez, J. K., & Doudna, J. A., (2016). Biology and applications of CRISPR systems: Harnessing nature's toolbox for genome engineering. *Cell, 164*, 29–44.
53. Ran, F. A., Hsu, P. D., Wright, J., Agarwala, V., Scott, D. A., & Zhang, F., (2013). Genome engineering using the CRISPR-Cas9 system. *Nature Protocols, 8*, 2281–2308.
54. Yu, S., Luo, J., Song, Z., Ding, F., Dai, Y., & Li, N., (2011). Highly efficient modification of beta-lactoglobulin (BLG) gene via zinc-finger nucleases in cattle. *Cell Research, 21*, 1638–1640.
55. Wei, J., Wagner, S., Maclean, P., Brophy, B., Cole, S., Smolenski, G., Carlson, D. F., et al., (2018). Cattle with a precise, zygote-mediated deletion safely eliminate the major milk allergen beta-lactoglobulin. *Scientific Reports, 8*, 1–13.
56. Cui, C., Song, Y., Liu, J., Ge, H., Li, Q., Huang, H., Hu, L., et al., (2015). Gene targeting by TALEN-induced homologous recombination in goats directs production of β-lactoglobulin-free, high-human lactoferrin milk. *Scientific Reports, 5*, 10482.
57. Luo, Y., Wang, Y., Liu, J., Cui, C., Wu, Y., Lan, H., Chen, Q., et al., (2016). Generation of TALE nickase-mediated gene-targeted cows expressing human serum albumin in mammary glands. *Scientific Reports, 6*, 1–11.
58. Luo, J., Song, Z., Yu, S., Cui, D., Wang, B., Ding, F., Li, S., et al., (2014). Efficient generation of myostatin (MSTN) biallelic mutations in cattle using zinc finger nucleases. *PloS One, 9*, e95225.
59. Proudfoot, C., Carlson, D. F., Huddart, R., Long, C. R., Pryor, J. H., King, T. J., Lillico, S. G., et al., (2015). Genome edited sheep and cattle. *Transgenic Research, 24*, 147–153.
60. Crispo, M., Mulet, A., Tesson, L., Barrera, N., Cuadro, F., Santos-Neto, P. D., Nguyen, T., et al., (2015). Efficient generation of myostatin knock-out sheep using CRISPR/Cas9 technology and microinjection into zygotes. *PloS One, 10*, e0136690.
61. Wang, X., Niu, Y., Zhou, J., Zhu, H., Ma, B., Yu, H., Yan, H., et al., (2018). CRISPR/Cas9-mediated MSTN disruption and heritable mutagenesis in goats causes increased body mass. *Animal Genetics, 49*, 43–51.
62. Cyranoski, D., (2015). Super-muscly pigs created by small genetic tweak: Researchers hope the genetically engineered animals will speed past regulators. *Nature, 523*, 13–15.
63. Lillico, S. G., Proudfoot, C., Carlson, D. F., Stverakova, D., Neil, C., Blain, C., King, T. J., et al., (2013). Live pigs produced from genome edited zygotes. *Scientific Reports, 3*, 2847.

64. Whitworth, K. M., Rowland, R. R., Ewen, C. L., Trible, B. R., Kerrigan, M. A., Cino-Ozuna, A. G., Samuel, M. S., et al., (2015). Gene-edited pigs are protected from porcine reproductive and respiratory syndrome virus. *Nature Biotechnology, 34*, 20.
65. Hou, L., Shi, J., Cao, L., Xu, G., Hu, C., & Wang, C., (2017). Pig has no uncoupling protein 1. *Biochemical and Biophysical Research Communications, 487*, 795–800.
66. Zheng, Q., Lin, J., Huang, J., Zhang, H., Zhang, R., Zhang, X., Cao, C., et al., (2017). Reconstitution of UCP1 using CRISPR/Cas9 in the white adipose tissue of pigs decreases fat deposition and improves thermogenic capacity. *Proceedings of the National Academy of Sciences, 114*, E9474–E9482.
67. Tan, W., Carlson, D. F., Lancto, C. A., Garbe, J. R., Webster, D. A., Hackett, P. B., & Fahrenkrug, S. C., (2013). Efficient nonmeiotic allele introgression in livestock using custom endonucleases. *Proceedings of the National Academy of Sciences, 110*, 16526–16531.
68. Mueller, M., Cole, J., Sonstegard, T., & Van, E. A., (2019). Comparison of gene editing versus conventional breeding to introgress the POLLED allele into the US dairy cattle population. *Journal of Dairy Science, 102*, 4215–4226.
69. Thompson, N. M., Widmar, N. O., Schutz, M. M., Cole, J. B., & Wolf, C. A., (2017). Economic considerations of breeding for polled dairy cows versus dehorning in the United States. *Journal of Dairy Science, 100*, 4941–4952.
70. Tu, Z., Yang, W., Yan, S., Yin, A., Gao, J., Liu, X., Zheng, Y., et al., (2017). Promoting Cas9 degradation reduces mosaic mutations in non-human primate embryos. *Scientific Reports, 7*, 1–11.
71. Hashimoto, M., Yamashita, Y., & Takemoto, T., (2016). Electroporation of Cas9 protein/sgRNA into early pronuclear zygotes generates non-mosaic mutants in the mouse. *Developmental Biology, 418*, 1–9.
72. Park, K. E., Powell, A., Sandmaier, S. E., Kim, C. M., Mileham, A., Donovan, D. M., & Telugu, B. P., (2017). Targeted gene knock-in by CRISPR/Cas ribonucleoproteins in porcine zygotes. *Scientific Reports, 7*, 1–10.
73. Keshavareddy, G., Kumar, A. R. V., & Ramu, V. S. (2018). Methods of plant transformation-a review. *Int. J. Curr. Microbiol. Appl. Sci, 7*(7), 2656–2668.
74. Ribaut, J. M., De Vicente, M. C., & Delannay, X. (2010). Molecular breeding in developing countries: challenges and perspectives. *Current Opinion in Plant Biology, 13*(2), 213–218.

CHAPTER 13

REGULATORY ASPECTS, SAFETY, AND ETHICAL ISSUES OF GENOME EDITING IN ANIMALS

MUAWUZ IJAZ,[1] MUBARIK MAHMOOD,[1] MUHAMMAD KASHIF YAR,[1,2] and IFTIKHAR HUSSAIN BADAR[2]

[1]*Department of Animal Sciences, College of Veterinary and Animal Sciences, Jhang, University of Veterinary and Animal Sciences, Lahore, Pakistan*

[2]*Department of Meat Science and Technology, Faculty of Animal Production and Technology, University of Veterinary and Animal Sciences, Lahore, Pakistan*

ABSTRACT

Genome editing (GE) is an emerging field of biotechnology involving a change in the nucleotide sequence of the living cells. Now a day, GE is considered one of the rapid solutions to the problems related to animal breeding and genetics. However, the use of this technology has also brought some issues related to ethics and animal welfare. Further, we should also be aware of the difficulties related to its application on large scale. One of the important welfare problems of GM animals is poor efficiency and other associated undesirable effects. These issues need to be addressed by all stakeholders so that relevant policies can be developed regarding the creation and usage of GM animals. Veterinarians may play a vital role in developing the policies and monitoring the research setting especially when new GM animal strains are being developed. Therefore, in this chapter, we will address all the regulatory, safety and ethical aspects of GE in animals.

CRISPR-Based Gene Editing: A Solution for Global Food Security. Shakeel Ahmad, Asif Ali Khan, and Muhammad Arshad Javed (Eds.)

13.1 INTRODUCTION

Globally, products of animal such as meat, milk, and eggs provide approximately 28% of the proteins and 13% of the energy, however, in developed countries these numbers rise to 48% and 20%, respectively [1]. Meat and milk from cow and buffalo contribute about 45% of the animal protein supply, followed by chicken with 31% and pigs with 20% in whole world [2]. Despite the innovation in animal production practices over the past few decades, predictions propose that demand of beef and pork could increase to 66% and 43% to feed the estimated world's population of 9 billion in 2050. The biggest rise in demand is expected for poultry products, with demand for eggs increasing by 65% and poultry meat by 121% [3]. In order to meet such big demands, it is need of the hour to enhance the speed of genetic gain for dairy and beef animals. Therefore, scientists are integrating the genome editing (GE) with the genome selection program to manipulate and alter the required traits such as generation interval, growth rate and other genetic variation related with the animals' performance in order to increase the speed of genetic gain.

Since many years, scientists are looking for using the cascades of cellular reparation to manipulate and change the DNA of the organisms by GE for correcting the mutation or introducing a new function. The First brick of this technique was established in 1953 when Watson and Cricks gave double helix structure of DNA. This was the big breakthrough in the history of genetics. The Idea of the Double helix came into existence after Rosalind Franklin's work, who got X-ray diffraction images of DNA proteins. In the 1970s, Genetic engineering took a high flight when researchers successfully created chimeric recombinant DNA. Cloning SV40 molecule into plasmid DNA. After that, john Hopkins successfully purified Restriction Enzyme type 2, which can cut DNA at specific sites. One discovery leads to another, and in the 1980s, the first transgenic animal was created. The development of new approaches of GE such as transcription activator-like effector nucleases (TALENs), meganucleases (MNs), and zinc finger nucleases (ZFNs) has made it more efficient, flexible, and less expensive. Recently, CRISPR/Cas9 has developed as latest preferred method for GE due to its high reliability, rate of diffusion, simple manufacturing, and low cost. All these GE technologies have been used successfully and extensively in the field of crop sciences. However, some of them like CRISPR/Cas9 is predominantly used in livestock GE for curing the disease, correcting the mutation and for breeding purposes [4–6].

The scientific methods have not extensively been applied to animal research as it demands an eagerness to integrate and correct previous knowledge, based on measurable observations, and subjected to laws of justification and reasoning. However, use of animals in the field of science started decreasing and their use was dropped to 50% in the mid of 1970–1980s [7]. The trend was reversed with the emergence of genetically modified (GM) animals of which genetic material has been changed deliberately by substitution, insertion, or deletion of the DNA. At that time, although, the use of non-GM animals continued to decline, however, due to increase in use of GM animals, overall number of using the animals in field of science have been rising [8]. However, it is difficult to determine the trends of using the GM animals due to different reporting requirements and it is estimated that annually approximately 13 million animals used in research in the European Union and 115 million used globally [9, 10]. After approving the GM tomato variety in 1994 [58], Food and Drug Administration (FDA) has recently approved the GM salmon for food consumption in 2015 [11]. No citizens and environmentalist loudly oppose to decision of FDA or raised the issue of safety. Nevertheless, psychological investigations have been shown that GM animals are less acceptable by the society as compared to the GM plants.

13.2 GENOME EDITING (GE) IN ANIMALS

Animals benefit human beings in numerous ways, for instance, they provide companionship, clothing, entertainment, and sports, transportation, and rescue services in addition to providing valuable food items like milk, meat, and eggs. According to FAO estimates, the demand of animal food products is expected to grow by 70% until 2050 [12], which concomitantly requires innovations in improving the health status and productivity of the animals. Selective breeding processes considerably uplifted the production of animals during 1960 to 2005, as observed by 67% more milk production, 80% higher egg production on consuming unit volume of diet, and attainment of 2 kg mass in broiler in 40 days instead of 100 days [13]. However, these processes have been constrained by longer time required and loss of species variety by standing genetic variation in addition to other factors. In this context, GE which is a technique based on inducing precise changes to the animal genome by engineered nucleases, can be beneficial in terms of providing specific and precise gene alterations to enhance productivity and disease resistance [14]. In addition to being faster than the normal selection procedures, GE is highly efficient in knocking in, knocking out or replacing the specific part of genetic material [15].

The animal models are integral part of the biomedical research due to sharing a number of physiological similarities with humans, but still differences exist and there is need to develop optimized animal models which can be closer to humans in terms of their physiology, anatomy, and metabolism.

Although, the history of GE is not too old in animals, and the commercial application is under development, nevertheless, it has been expanding at faster rate due to high profile benefits and availability of modern GE techniques like clustered regularly interspaced short palindromic repeat-associated nuclease Cas9 (CRISPER/Cas9) technology. The other available techniques are TALENs, ZFNs and mega nucleases (MGs). The progress is so fast that a number of animal genetics laboratories have been equipped with CRISPR technology which was initiated just seven years ago [16]. Moreover, more than 500 publications of GE using this technique have been published in top quality journals. Therefore, genome edited livestock and pet animals have been gaining much popularity. As GE is a tool to make targeted genome changes at site-specific locations, it can be used as a tool to intensify the traits of interest, eliminate the unwanted traits and to conserve the genes prone to extinction. It has been successfully done in a number of animals focusing few of the many relevant traits (Table 13.1).

TABLE 13.1 Animal Traits Focused Through Genome Editing

Animals	Focused Traits	Techniques	References
Goat	• Removal of the allergic component from milk; • Enhance meat production; • Enhance meat quality; • Increase in litter size.	CRISPR/Cas9	[17–20]
Sheep	• Enhance meat production; • Increase in wool length.	TALEN CRISPR/Cas9	[21, 22]
Cattle	• Removal of the allergic component from milk; • Polled character development; • High milk production.	TALEN, ZFN TALEN ZFN	[23–25]
Poultry	• Carcass and meat quality; • Increased growth; • Hypoallergenic eggs; • Disease resistance.	CRISPR/Cas9	[12, 26, 27]
Cats	• Elimination of cat allergen.	CRISPR/Cas9	[28]
Pigs	• Development of micro-pigs; • Increase in muscle mass; • Improved feed efficiency and reduced environmental impact; • Thermoregulation.	TALEN TALEN SCNT CRISPR/Cas9	[29–32]
Dogs	• Muscularity development.	CRISPR/Cas9	[33]

Abbreviations: CRISPR/Cas9: Clustered regularly interspaced short palindromic repeats; TALEN: transcription activator-like effector nucleases; ZFN: zinc-finger nucleases; and SCNT: somatic cell nuclear transfer.

13.2.1 FUTURE PROSPECTS AND SCOPE

It is safe to say that GE has tremendous scope in animal kingdom, and it is expected to have far-reaching impacts on human welfare and well-being. Currently, the commercial application of GE in animals is in infancy due to a short but intense history, however, one can foresee the multifold benefits in the near future. On one hand, GE in animals can present the optimized animal models which would be beneficial in better understanding of pathogenesis of human diseases and their treatments through development of precise drugs [34]. On the other hand, this area of gene therapy can innovate agriculture and play role in eradicating hunger and malnutrition challenges by considerably improving genetic value of livestock for production parameters [13].

To meet the rising demands of animal food products, GE can be a breakthrough in upgrading the production of milk, meat, and eggs from the animals. It can be achieved by targeting more and more genes of interest and correspondingly their functions and associated phenotypes in different animals [33]. For example, genome edited livestock animals can better tolerate heat stress and can be resistant to a number of potential diseases. Additionally, the genes controlling wool characteristics can be focused on small ruminants. Likewise in pet animals, the target genes may be those controlling the aggression, hunting behavior, eye colors, and emotional attachment with the owners. The racing stamina and sport performance of the animals can be targeted, for instance, by modifying the "gait keeper gene" and "speed gene" in horses. Home aquariums can be beautified by introducing the desired colors and patterns of the fish body. Application of genome technology have full potential to conserve a number of animal breeds which are in danger of extinction in addition to contribution to the longer life span of the animals which is one of the prime goals in case of almost all animals. Last but not the least, GE can be a way of inter-species organ transplantation called xenotransplantation.

13.2.2 CHALLENGES AND LIMITATIONS

Although area of GE has potential to provide a lot of associated benefits, there are still certain safety and ethical related limitations which must be considered in parallel. The technology can raise mosaic mutations in the founder animals which may provide false-positive genotyping results [35]. A lot of off target mutations have been observed in a number of studies of gene edited animals [36], therefore, it can introduce off targeting events leading to the development of unintended mutations. As off targeting is associated with cell death, higher responsibility is required for its execution raising the safety issues. The process of xenotransplantation has low acceptability among majority of the population and raises ethical concerns [33], which mainly include animal rights, resources distribution, and public health concerns.

13.3 REGULATORY ASPECTS OF GENOME EDITING (GE) IN ANIMALS

Regulation is the rules and norms that apply on social, institutional, or economic levels. These regulations are vital for the development process of different countries. In the production sectors such as agriculture and livestock, regulatory tools could be used to ensure the welfare of society. It is the responsibility of policymakers to formulate the rules and regulations which ensure welfare and development on different economic levels.

Recently much research is in progress regarding the introgression of valuable alleles through GE in the livestock sector. However, until now, genome-editing has not produced any animal item for general community. Furthermore, conventional genetic engineering has received only one consent: the fast-growing Aquadvantage salmon. The slow progress is due to this sector's technical and regulatory obstacles [37]. Although technical obstacles have been reduced with time, the regulatory aspects are still complicated [38].

13.3.1 REGULATORY ASPECTS OF GENOME EDITING (GE)

The regulatory characteristics of genome-editing in livestock are unclear. However, attempts to minimize or avoid the regulatory aspects of genome-editing in livestock could mislead the general perception of avoiding the public discussion related to this sector. By stowing away the type of genome-altering advancements, individuals can be avoided. Indeed, if monitoring aspects of genetically modified organisms (GMOs) are judged over the top for genome-edited creatures, a few kinds of opportunities for open talk about and arrangement of strategies for recognizing goods from genome-edited organisms appears compulsory. For case, the guidelines for genome-edited organisms, may illustrate the regulatory aspects of the medical-related instruments. Assumed that creating genome-edited organisms will need a genuine venture. Inventers will likely look to recover costs by offering the sale of resultant breeding animals at best. It will not hide that the organism may be a due to genome-editing. In any case, such open discernments can endure.

13.3.2 REGULATORY ASPECTS OF GENE EDITING IN ARGENTINA

Argentina remained the primary nation to distribute its projected regulatory aspects for genome-editing and other modern breeding strategies [39]. Argentine approach is if there is no new genetic blend, and the ultimate item is free of transgenes that item would not be subjected to regulatory aspects like the genetically altered animals. This framework creates no difference between GM animals and plants. Argentina's GM tilapia fish that posed no foreign DNA was exempted from regulatory aspects in 2018.

13.3.3 REGULATORY ASPECTS OF GENE EDITING IN USA

In 2018, USDA declared that they do not control or have any strategies to control plants that seem to be produced using the traditional breeding technologies. The US, FDA has engaged a diverse approach to GM creatures. They announced that all intended animal GE alterations would be directed as novel animal drugs [40]. This direction explains that each change ought to follow a required premarket multigenerational security review, independent of that either the change currently be present within the species or might have been accomplished utilizing routine breeding. Probably this will harm GE studies related to food animals. The pay inferred from excess animals and animal products delivered by the institutional research is necessarily counterbalanced to the ample costs of research. Classifying the genome-edited products, independent of an oddity, as drugs, subsequently blocking marketable items from GM animals, will considerably increase the costs related to this study. This will drastically increment the advancement costs related to commercializing genome-edited animals.

13.3.4 REGULATORY ASPECTS OF GENE EDITING IN EUROPEAN UNION

Court of Justice of the European Union (CJEU) in 2018 judged that genome-edited harvests would be regulated similar to conventional GMOs [41]. Probably this will ruin the utilization of genome altering by animal and plant investigators within the European Union and the selection of this innovation in European Union agriculture. Numerous researchers described the effect these anticipated regulatory tactics will have on the capacity of public division researchers and little corporations, specifically, to utilize genome-altering in food animals. Hundreds of researchers have signed petitions in European Union and USA concerning the scientific-based and product risked focused GE in food animals [42, 43].

13.3.5 REGULATORY ASPECTS OF GENE EDITING IN CANADA

Genetically edited animals are regulated under the Canadian Environmental Protection Act in Canada. However, there is no specific GE policy, and the Canadian Government is effectively locked in with interested partners. In Canada, the regulatory approach to genome-editing is outcome-based. It focuses on the final product rather than how it is made. By centering on the result, product developers are given the flexibility to use the most suitable techniques to attain their goals [44]. To cope up with the potential risks of environment, livestock, and human consumption regarding genome-edited products, the Canadian regulatory authorities require the pre-market assessment of the novel products. However, Canada does not require the pre-market assessment of products that are not novel. In order to better assess the novelty of the product, the Canadian regulators discuss all the necessary aspects to address the pre-market assessment of a peculiar product with the product supporter.

13.3.6 RELATIONSHIP BETWEEN COMPANY POLICIES AND REGULATION

Regulations do not only impact the safety and efficacy-related matters of innovation and affect the company policies [45]. Research articles announcing the accomplishment of many genome-edited creatures are critical because of logical progressions. To create a commercially practical item, the number of animals ought to be expanded, and the animal products should be acknowledged by different partners as well as effectively manage any controlling boundaries. A conceivable course to advertise is required. Within the genome-edited animals, recent improvements show up to incorporate those where a conceivable course of advertising is developing. For example, one of leading worldwide animal breeding companies, Genus plc, has declared an elite permit to utilize genome-editing to yield PRRS-resistant pigs commercially [46].

The worldwide animal breeding segment is well organized while having a few heterogeneities. Several worldwide breeding firms overwhelm the division. Expert breeding firms embrace genetic selection to yield better livestock. Dissimilar to crops, where the complete populace may be quickly supplanted, presenting genome-editing into an animal population may be a complex procedure.

In spite of the fact that the section of one chief breeding firm into genome-editing is popular, the breeding firm concept is not the as it were one present within breeding stock supply, as state mediation in animal breeding is common in numerous EU. There is proof the breeding segment is formed by directions. For example, in France, state intercession, and after that, the removal of the government from the breeding division has a significant effect on the viable partners [47]. Hannachi and Tichit contend the cow breeding in France has shifted from a joint workout practiced exclusively and characterized by competition [48].

The type of genome-editing procedures may additionally have an influence on the development direction and its effect on diverse segments of the livestock population framework. Cattle that do not create horns are still fair cattle, and the livestock business is not disturbed by not having horns. The hornlessness specifies better animal wellbeing, labor savings in maintaining a strategic distance from an unsavory assignment, and financial decrease in materials and labor.

The capability to evade the necessity for neutering animals may have better animal welfare paybacks, along with extra benefits such as reduced labor, and better execution from animals that do not endure a development check from the method. The method may be troublesome if it permits novel meat foods to be made as pigs might be butchered at higher weights if the males taste perception is missing. An application that has the capacity of, as it was, creating commercially valuable animals of the desired sex has the possibility to be troublesome in ways that are difficult to foresee, as the reproducing procedure can be reduced at certain farms. There could have to be novel institutional regulations that permit support of the parent animals of both genders, resulting in production costs, dispersion of control and possession of breeding stock.

Diverse partners will likely be influenced completely differently inside the animal production frameworks. End users are not offered genome-edited animals, rather ranchers. For agriculturists, genome-edited animals are just like incremental advancements whose administration is path-specific. After all, the product is either cattle or pig, like every other. There may be extra recording and recognizable proof prerequisites on the off chance that tagging and co-existence necessities are comparative to those for hereditarily altered crops within the European Union. Though, the requests for monitoring individual creatures and their movement are already considered in European Union. The product is basically a bovine or pig, like every other. There could be extra monitoring and recognizable proof prerequisites on the off chance that naming and co-existence necessities are comparative to those for hereditarily altered crops within the EU. Though, the requests for monitoring personal creatures and creature developments within the EU are now, significant.

For breeders deprived of research facility science ability, creating genome-edited creatures is just like path-breaking, requiring securing modern abilities. Be that as it may, bigger companies in this division have a higher absorptive capacity. In any case, assembling the necessities of intellectual property partnered with genome-edited animals may be troublesome. Up to this point, the intellectual property in animal breeding has been fundamentally ensured by the privacy and utilization of crossbreeding. The veterinary industry is one that could be significantly impacted and for which this could be a ground-breaking innovation [46].

13.3.7 VETERINARY SECTOR

Disease management and control is a beginning attention for a few genome-editing ventures; the example of PRRS was given above. These infections have few or no elective medicines conjointly posture a noteworthy financial danger to pig production. Hereditary disease resistance is planning to be utilized in disease inhibition. On the basis of verifiable examination of UK veterinarians, Woods [48] contends that veterinarian have utilized rhetoric of disease avoidance but observed it troublesome to provide in practice for various reasons. In its place, they have sustained focus on treatment approaches. Woods recommends that for compelling anticipation, a suitable combination of veterinary, disease, and cultivation, all these endorse disease prevention. A vital footstep is to involve the veterinarian and other suitable partners to distinguish the drivers required for genome-edited animals to be saddled for animal wellbeing enhancement.

Genome-edited creatures are implausible to be embraced by all agriculturists. Thoughtful epidemiological suggestions of presenting a new resistance characteristic to a few creatures could be the questions that should be addressed properly. Veterinarians can execute infection control measures, especially where there are particular lawful necessities. This could be the reason of possible rise of African Swine Fever within the UK. Genome-editing hence might have much more extensive applications since the doubt that a notifiable illness of worldwide noteworthiness is present can be reason for a major disturbance to universal trade. Some stakeholders might subsequently investigate the possible suggestions of genome addition for infection resistance within the occasion of upcoming outbreaks of these types of diseases.

13.4 SAFETY AND ETHICAL ISSUES OF GENOME EDITING (GE) IN ANIMALS

There has been a significant advancement in gene editing, which also brought safety and ethical issues that must be considered.

13.4.1 TERRORIST ISSUES

Genome editing (GE) of the non-human being is one of the most significant safety issues, and it can lead to the bio-war in the world. GE of non-human organisms through CRISPR/Cas9 is a potent tool to make animals used pharmaceuticals, biochemical, biofuel, reduction in pollution, disease models, and treatment regimes. However, there is a safety risk in this technology. If this technology goes in the wrong hand, it will be destructive for human existence. Modified viral genome or bacterial genome can result in pathogen which can create havoc in human life. Similarly, this pathogen, if attacked by animals and plants, will have food safety and security issues in this world. CRISPR is an easy and efficient tool for gene editing. The efficiency and easiness of this technique raise a big problem that with this technology setup, anyone can create harmful species of pathogens. These harmful species can result in bio-war on earth.

13.4.2 BIOETHICAL ISSUES

Gene editing aims to remove disease genes, clean the environment, and save human life by creating disease models of animals, but all these benefits put us under a moral value to respect these animals. The unnecessary pain these animals suffer during experimentation is unethical. This makes humans the only moral living organism with the power to manipulate nature for its benefits. Due to human technological intervention, natural things are turned into unnatural, and most of them harms a human being [49]. Genomic editing affects the natural behavior of animals, including the feeding behavior and the interaction of animals with other animals. Technological interventions by humans are modifying the natural behavior of animals. As a result of this, humans are intervening in the hand of nature and people's religious beliefs.

13.4.3 ISSUES RELATED TO ANIMAL WELFARE

We humans as sentient beings should consider animal welfare seriously. In gene, editing animals suffer confinements, surgeries, and injectable. These animals are subject to death if something goes wrong, like non-targeted mutations, loss of function of organs, or fatal abnormalities due to the adverse reaction of gene editing. The public should be aware of the fact that this technology must be regulated under strict limitations. The benefit of research should be justified about the research value for reducing animal and human diseases or the learning and understanding the basic process of life. Research should focus on live experimental alternatives and clearly state the meaningfulness of experiments. These experimental animals should not suffer only for the personal use of some riches, but they can be used as a model for generalized conditions in human beings. In order to maintain the welfare of experimental animals following points must be kept under considerations:

- There must be adequate space and food for the animals;
- The animal should not be kept isolated from a member of its species or even other animals;
- The animal must have the ability to interact with another member of its species after experimentation;
- Modification of genes must be limited;
- Duration of confinement must be within a proper limit after which an animal can adopt its natural behavior.

13.4.4 OFF TARGET MUTATIONS

Although the CRISPR/Cas9 system is very efficient, it may create unwanted or off-target mutations in the genome [50]. These Mutations happen when the CRISPR/Cas9 system target genes at different stages of embryonic development [51]. When CRISPR/Cas9 system breaks DNA within the genome sequence, off-target mutations occur at that time. These off-target mutation results are deleteriously ended in transformation or cell death [52].

13.4.5 LACK OF REGULATIONS OF GENE EDITED PRODUCTS

Several agencies like the FDA and the Nation Health Institute (NIH) regulate GMOs in the USA, but these agencies have no check and control capacity over GMOs. The national health institute has some guidelines, but these guidelines are only implemented over the NIH research. It also lacks regulatory authority. Regulatory agencies must be required from researchers and scientists a complete control mechanism to use the CRISPR system for gene editing in organisms. Now a day's some biotechnological companies have regulatory control over genetically edited organisms. Internationally, it is the subject matter of interest of the United Nations Convention on Biological Diversity. There must be regulations about genetically edited organisms by the UN.

13.4.6 ISSUES OF PATENTING THE GENOME EDITED ORGANISMS

Once a genome-edited organism came into existence, patenting it is a significant issue. There may be a dispute between the parties over new research. There was a significant dispute between intellectual parties over discovering the CRISPR-Cas9 system [53]. Regulations regarding the patent of genome-edited organisms are still yet to document.

13.4.7 IDENTIFICATION OF NORMAL OR GENOME EDITED ORGANISMS

When an edited genomic product is in the market, it is challenging to segregate it from normal genomic animals. There is a safety issue of using genetically edited organisms, and regulation/criteria should be devised to check the safety level of modified animals. Criteria should be devised to check the genome-edited organisms are safe for human consumption. For instance, the genetically improved weight of meat animals, improved nutrition context of products, and improved production and nutritional contents of crops must be assessed before entering to market [54]. Experiments that involve genomic modification may have serious bio-safety issues for researchers themselves [55]. GMOs are only approved to enter the market after solid testing and people consultation. There must be rules and regulations that estimate the harmful effect of genomic edits on other organisms, humans, and the environment when these genomic edits proved lethal afterward.

13.4.8 TRANSPLANTATION OF NON-HUMAN ORGANS IN HUMAN (XENOTRANSPLANTATION)

Transplantation of non-human organs into the human body is Xenotransplantation. Some scientists believe that animal organs can be transplanted, but there is a safety issue that animal microorganisms can cause infection in the human body. There is another risk of rejection of the organ by the human body's immune system towards the transplanted organs. This risk is evident when their species of donor and recipient are different. There is a significant risk of zoonosis, i.e., transferring disease from animals to humans and vice versa. Xenotransplantation is helpful if it is used in a controlled and mechanized way. Scientists are making it possible to transplant animal organs to humans with CRISPR by reducing the risk of infection transfer and immune response backlash [56].

13.4.9 DISEQUILIBRIUM OF ECOSYSTEM

As we know, there is a significant risk of unintended mutation during genomic editing. These off-target mutations amplify in every upcoming generation, and there is also a possibility of that gene modification transfer in other members of species. It will result in the change of natural behavior of organisms to a typical ecosystem. In the end, the equilibrium of the ecosystem flopped out. Therefore, it is necessary to control the off-target mutation in order to sustain Ecosystem equilibrium. Otherwise, it will be tough to control the disequilibrium of the ecosystem [55]. If the genetically modified organism is released accidentally, it will create a drastic change in the ecosystem. It may cause the extinction of the entire population from the ecosystem [57].

13.5 CONCLUSION

Although edited genomic animals will be helpful to humans in many aspects of life, there are many hurdles ahead in the practicality of this technique. Ethical issues and safety issues related to edited genomic organisms are the primary concern of people. Regulatory authorities can control and monitor the whole technique and make regulatory guidelines for the researchers and scientists. Genome-editing guarantees the plausibility of creating commercial animals carrying exact hereditary changes, which are troublesome or outlandish to realize with specific breeding. Eventually, long-term of these improvements will depend on political choices, administrative necessities, public worthiness, and innovative capabilities. The more extensive situation of livestock production is additionally imperative to ponder. The burden on the livestock sector is increasing to lessen its influence on the environment whereas keeping up efficiency.

KEYWORDS

- **animal welfare**
- **disequilibrium**
- **ethics**
- **genome edited organisms**
- **genome editing**
- **public worthiness**
- **regulatory aspects**
- **safety issues**

REFERENCES

1. FAO (Food and Agriculture Organization), (2009). *Awakening Africa's Sleeping Giant: Prospects for Commercial Agriculture in the Guinea Savannah Zone and Beyond.* World Bank, Washington, DC; Food and Agriculture Organization, Rome.
2. Mottet, A., De Haan, C., Falcucci, A., Tempio, G., Opio, C., & Gerber, P., (2017). Livestock: On our plates or eating at our table? A new analysis of the feed/food debate. *Global Food Secur., 14*, 1–8.

3. Mottet, A., & Tempio, G., (2017). Global poultry production: Current state and future outlook and challenges. *World's Poult. Sci. J., 73*(2), 245–256.
4. Wang, Y. Y., Du, B., Shen, X., Zhou, J., Li, Y., Liu, J., Wang, J., et al., (2015). Efficient generation of gene-modified pigs via injection of zygote with Cas9/sgRNA. *Sci. Rep., 5*, 8256.
5. Proudfoot, C., Carlson, R., Huddart, C. R., Long, J. H., Pryor, T. J., King, S. G., Lillico, A. J., et al., (2015). Genome edited sheep and cattle. *Transgenic Res., 24*, 147–153.
6. Whitworth, K. M., Rowland, C. L., Ewen, B. R., Trible, M. A., Kerrigan, A. G., CinoOzuna, M. S., Samuel, J. E., et al., (2016). Gene-edited pigs are protected from porcine reproductive and respiratory syndrome virus. *Nat. Biotechnol., 34*, 20–22.
7. UK Home Office, (2016). *Statistics of Scientific Procedures on Living Animals: Great Britain 2015*. [Online]. Available at: http://www.gov.uk/government/statistics/statistics-of-scientific-procedures-on-living-animals-great-britain-2015 (accessed on 16 December 2023).
8. Ormandy, E. H., & Schuppli, D. M., (2009). Worldwide trends in the use of animals in research: The contribution of genetically-modified animal models. *Alternatives to Laboratory Animals., 37*(1), 63–68.
9. Taylor, K., & Rego, L., (2016). EU statistics on animal experiments for 2014. *Alternatives to Animal Experiments, 33*(4), 465–468.
10. Taylor, K., Gordon, N., Langley, G., & Higgins, W. (2008). Estimates for worldwide laboratory animal use in 2005. *Alternatives to Laboratory Animals, 36*(3), 327–342.
11. FDA, (2015). *FDA has Determined that the AquAdvantage Salmon is as Safe to Eat as nonGE Salmon.* https://goldrushcam.com/sierrasuntimes/index.php/news/local-news/5329-fda-has-determined-that-the-aquadvantage-salmon-is-as-safe-to-eat-as-non-ge-salmon (accessed on 16 December 2023).
12. Park, T. S., Park, J., Lee, J. H., Park, J. W., & Park, B. C., (2019). Disruption of G0/G1 switch gene 2 (G0S2) reduced abdominal fat deposition and altered fatty acid composition in chicken. *The FASEB Journal, 33*, 1188–1198.
13. Tait-Burkard, C., Doeschl-Wilson, A., McGrew, M. J., Archibald, A. L., Sang, H. M., Houston, R. D., Whitelaw, C. B., & Watson, M., (2018). Livestock 2.0–genome editing for fitter, healthier, and more productive farmed animals. *Genome Biology., 1*, 1–11.
14. McFarlane, G. R., Salvesen, H. A., Sternberg, A., & Lillico, S. G., (2019). On-farm livestock genome editing using cutting edge reproductive technologies. *Frontiers in Sustainable Food Systems, 3*, 106.
15. Li, Q., Qin, Z., Wang, Q., Xu, T., Yang, Y., & He, Z., (2019). Applications of genome editing technology in animal disease modeling and gene therapy. *Computational and Structural Biotechnology Journal, 17*, 689–698.
16. Cong, L., Ran, F. A., Cox, D., Lin, S., Barretto, R., Habib, N., Hsu, P. D., et al., (2013). Multiplex genome engineering using CRISPR/Cas systems. *Science, 339*, 819–823.
17. Zhou, W., Wan, Y., Guo, R., Deng, M., Deng, K., Wang, Z., Zhang, Y., & Wang, F., (2017). Generation of beta-lactoglobulin knock-out goats using CRISPR/Cas9. *PloS One, 12*, 0186056.
18. Ni, W., Qiao, J., Hu, S., Zhao, X., Regouski, M., Yang, M., Polejaeva, I. A., & Chen, C., (2014). Efficient gene knockout in goats using CRISPR/Cas9 system. *PloS One, 9*, 106718.
19. Zhang, J., Cui, M. L., Nie, Y. W., Dai, B., Li, F. R., Liu, D. J., Liang, H., & Cang, M., (2018). CRISPR/Cas9-mediated specific integration of fat-1 at the goat MSTN locus. *The FEBS Journal, 285*, 2828–2839.

20. Niu, Y., Zhao, X., Zhou, J., Li, Y., Huang, Y., Cai, B., Liu, Y., et al., (2018). Efficient generation of goats with defined point mutation (I397V) in GDF9 through CRISPR/Cas9. *Reproduction, Fertility and Development, 2*, 307–312.
21. Li, H., Wang, G., Hao, Z., Zhang, G., Qing, Y., Liu, S., Qing, L., et al., (2016). Generation of biallelic knock-out sheep via gene-editing and somatic cell nuclear transfer. *Scientific Reports*, (1), 1–12.
22. Hu, R., Fan, Z. Y., Wang, B. Y., Deng, S. L., Zhang, X. S., Zhang, J. L., Han, H. B., & Lian, Z. X., (2017). Rapid communication: Generation of FGF5 knockout sheep via the CRISPR/Cas9 system. *Journal of Animal Science, 95*, 2019–2024.
23. Yu, S., Luo, J., Song, Z., Ding, F., Dai, Y., & Li, N., (2011). Highly efficient modification of beta-lactoglobulin (BLG) gene via zinc-finger nucleases in cattle. *Cell Research, 21*, 1638–1640.
24. Carlson, D. F., Lancto, C. A., Zang, B., Kim, E. S., Walton, M., Oldeschulte, D., Seabury, C., et al., (2016). Production of hornless dairy cattle from genome-edited cell lines. *Nature Biotechnology, 34*, 479–481.
25. Liu, X., Wang, Y., Guo, W., Chang, B., Liu, J., Guo, Z., Quan, F., & Zhang, Y., (2013). Zinc-finger nickase-mediated insertion of the lysostaphin gene into the beta-casein locus in cloned cows. *Nature Communications, 4*, 1–11.
26. Kim, G. D., Lee, J. H., Song, S., Kim, S. W., Han, J. S., Shin, S. P., Park, B. C., & Park, T. S., (2020). Generation of myostatin-knockout chickens mediated by D10A-Cas9 nickase. *The FASEB Journal, 34*, 5688–5696.
27. Oishi, I., Yoshii, K., Miyahara, D., Kagami, H., & Tagami, T., (2016). Targeted mutagenesis in chicken using CRISPR/Cas9 system. *Scientific Reports, 6*, 1–10.
28. Brackett, N., Riedy, J., Adli, M., Pomes, A., & Chapman, M., (2020). Gene editing the major cat allergen, Fel d 1, using CRISPR-Cas9. *Journal of Allergy and Clinical Immunology, 145*, AB156.
29. Cyranoski, D., (2015). Gene-edited' micropigs' to be sold as pets at Chinese institute. *Nature News, 526*, 18.
30. Kang, J. D., Kim, S., Zhu, H. Y., Jin, L., Guo, Q., Li, X. C., Zhang, Y. C., et al., (2017). Generation of cloned adult muscular pigs with myostatin gene mutation by genetic engineering. *RSC Advances, 21*, 12541–12549.
31. Zhang, X., Li, Z., Yang, H., Liu, D., Cai, G., Li, G., Mo, J., et al., (2018). Novel transgenic pigs with enhanced growth and reduced environmental impact. *Elife, 7*, 34286.
32. Zheng, Q., Lin, J., Huang, J., Zhang, H., Zhang, R., Zhang, X., Cao, C., et al., (2017). Reconstitution of UCP1 using CRISPR/Cas9 in the white adipose tissue of pigs decreases fat deposition and improves thermogenic capacity. *Proceedings of the National Academy of Sciences, 114*, 9474–9482.
33. Zhao, J., Lai, L., Ji, W., & Zhou, Q., (2019). Genome editing in large animals: Current status and future prospects. *National Science Review, 6*, 402–420.
34. Lee, H., Yoon, D. E., & Kim, K., (2020). Genome editing methods in animal models. *Animal Cells and Systems, 24*, 8–16.
35. Mehravar, M., Shirazi, A., Nazari, M., & Banan, M., (2019). Mosaicism in CRISPR/Cas9-mediated genome editing. *Developmental Biology, 445*, 156–162.
36. Perisse, I. V., Fan, Z., Singina, G. N., White, K. L., & Polejaeva, I. A., (2020). Improvements in gene editing technology boost its applications in livestock. *Frontiers in Genetics, 11*, 1713.

37. Van, E. A. L., & Muir, W. M., (2011). Transgenic salmon: A final leap to the grocery shelf? *Nature Biotechnology, 29*, 706–710.
38. Van, E. A. L., (2018). The importance of a novel product risk-based trigger for gene-editing regulation in food animal species. *The CRISPR Journal, 1*, 101–106.
39. Whelan, A. I., & Lema, M. A., (2015). Regulatory framework for gene editing and other new breeding techniques (NBTs) in Argentina. *GM Crops & Food, 6*, 253–265.
40. Maxmen, A., (2015). Gene-edited animals face US regulatory crackdown. *Nature News*.
41. Callaway, E., (2018). CRISPR plants now subject to tough GM laws in European Union. *Nature, 560*, 16, 17.
42. Wight, A. J., (2018). Strict EU ruling on gene-edited crops squeezes science. *Nature, 563*, 15, 16.
43. Van, E. A. L., Wells, K. D., & Murray, J. D., (2019). Proposed US regulation of gene-edited food animals is not fit for purpose. *NPJ Science of Food, 3*, 1–7.
44. Ellens, K. W., Levac, D., Pearson, C., Savoie, A., Strand, N., Louter, J., & Tibelius, C., (2019). Canadian regulatory aspects of gene editing technologies. *Transgenic Research, 28*, 165–168.
45. Mittra, J., Tait, J., & Wield, D., (2011). From maturity to value-added innovation: Lessons from the pharmaceutical and agro-biotechnology industries. *Trends in Biotechnology, 29*, 105–109.
46. Bruce, A., (2017). Genome edited animals: Learning from GM crops? *Transgenic Research, 26*, 385–398.
47. Hannachi, M., & Tichit, M., (2016). Does biotechnological innovation require organizational innovation? Learning from the cattle breeding industry in France. *Animal Frontiers, 6*, 80–85.
48. Woods, A., (2013). Is prevention better than cure? The rise and fall of veterinary preventive medicine, C.1950–1980. *Social History of Medicine, 26*, 113–131.
49. Singer, S. S., Singer, B. D., & Vaughan, D. E., (2017). Molecular and physiological manifestations and measurement of aging in humans, *Aging Cell, 16*, 624–633.
50. Bortesi, L., Zhu, C., Zischewski, J., Perez, L., Bassié, L., Nadi, R., Forni, G., et al., (2016). Patterns of CRISPR/Cas9 activity in plants, animals, and microbes. *Plant Biotechnol. J.,* 142203–142216.
51. Xin, L. U., Min, L. I., & Bing, S. U., (2016). Application of the genome editing tool CRISPR/Cas9 in non-human primates. *Zool. Res., 37*, 241.
52. Zhang, F., Wen, Y., & Guo, X., (2014). CRISPR/Cas9 for genome editing: Progress, implications, and challenges. *Hum. Mol. Genet., 23*, 40–46.
53. Smith-Willis, H., & San, M. B., (2015). Revolutionizing genome editing with CRISPR/Cas9: Patent battles and human embryosle. *Cell Gene Ther Insight, 1*, 253–262.
54. Ledford, H., (2015). CRISPR, the disruptor. *Nature News, 522*, 20.
55. Esvelt, K. M., Smidler, A. L., Catteruccia, F., & Church, G. M., (2014). Emerging technology: Concerning RNA-guided gene drives for the alteration of wild populations. *Elife, 3*, 3401.
56. Fung, R. K. F., & Kerridge, I. H., (2016). Gene editing advance re-ignites debate on the merits and risks of animal to human transplantation. *Intern. Med. J., 46*, 1017–1022.
57. Oye, K. A., Esvelt, K., Appleton, E., Catteruccia, F., Church, G., Kuiken, T., Lightfoot, S. B. Y., et al., (2014). Regulating gene drives. *Science, 345*, 626–628.
58. Bruening, G., & Lyons, J. (2000). The case of the FLAVR SAVR tomato. *California Agriculture, 54*(4), 6–7.

INDEX

D

H

T

U

V